D0667172

# Space 2000

Meeting the Challenge of a New Era

# Space 2000

## Meeting the Challenge of a New Era

Harry L. Shipman

Plenum Press • New York and London

Library of Congress Cataloging in Publication Data

Shipman, Harry L.
  Space 2000.

  Includes bibliographical references and index.
  1. Outer space—Exploration. I. Title. II. Title: Space two thousand.
TL790.S49   1987                      919.9′04                      87-2401
ISBN 0-306-42534-3

© 1987 Harry L. Shipman
Plenum Press is a Division of
Plenum Publishing Corporation
233 Spring Street, New York, N.Y. 10013

Printed in the United States of America

# Preface

Space is the next frontier. It was not until the 1950s that the exploration of space began with the dramatic launch of *Sputnik 1*. Since then, the space program has taken people to the moon, shown us the exotic, fascinating beauties of the near and distant universe, and created a global village by weaving a communications net over the earth. Human activities in space range from planetary exploration to star wars, from mapping to spying, from astronomy to space-based materials-processing.

What we are doing, and will do, in space will determine where we go next. The American space station, initially inhabitable in the mid-1990s, will open the door to a wide range of applications. At first, those pioneers who have already reached toward the heavens, who have begun to use the space environment for various purposes, will remain as the major players in the space game. Others will soon follow. Their wants and needs will determine the next major goal beyond the space station.

The explosion of the space shuttle *Challenger* in 1986 sent resounding shock waves through the entire space program, and the effects of this tragedy will be felt for the better part of a decade. In the short run, NASA and the space community will have their hands full rebuilding the shuttle fleet, constructing the space station, and maintaining a balanced program of space science and applications so that America and its partners can do something with the space station once it is built. The next ten years will be difficult, but we can look beyond them to a frontier of opportunity.

The focus of this book is on the enormous variety of past and future human activities in space. Before America made the initial symbolic step of landing humans on the moon in 1969, the space program was a single race toward a single goal. In the 1970s and 1980s, the space programs of the United States, the Soviet Union, a consortium of European nations, Japan, and other countries then began to simultaneously compete and cooperate with each other as they sought to make use of the space environment in a much broader way. Private companies have become heavily involved in space activity. The military presence in space has grown considerably, especially in the wake of Star Wars, the popular name for the Strategic Defense Initiative (SDI). As a result, what we do in space is more significant than where we go.

I've been involved with the space program in one capacity or another since 1974, and my purpose in writing this book is to give the general reader a perspective on the breadth of our activities in space. I have been a guest investigator on many different satellites operated by NASA and by the European Space Agency, and I have served on various review panels as a NASA consultant (including a stint on a team negotiating with the European Space Agency on a future satellite mission). However, I am only a NASA grantee and consultant and have never been an employee of the agency. The views I present here are my own, based on my experience and my reading of the open literature. The discussions of military space programs presented in this book are all based on unclassified material; I have no security clearance.

Many people have helped me as I have worked on this book. Andrea Dupree, Nancy Evans, Paul Evenson, Don Kent, Vic Klemas, Norman Ness, John Nousek, Ron Pitts, Martin Pomerantz, and Richard Wasatonic read preliminary drafts of particular chapters. Cheryl Thompson of the University of Delaware library was always able to chase down the most obscure government publications; Henry Fuhrmann of JPL, Paul Maley, and Jeff Rosendhal of NASA helped provide photographs and other material. I've received grants from NASA, from the National Science Foundation, from the John Simon Guggenheim Memorial Foundation, and from the Research Corporation over the years; those grants have helped keep my research program going and have given me the perspective that I'm sharing with you, the reader, in this book. Linda Regan, my editor at Plenum, has the patience and the persistence necessary to keep an author not too far behind schedule. My children, Alice and Tom, and my wife, Wendy, have had to put up with my spending many weekend and evening hours on this book; it is only now that Tom has an answer to his question: "How many more pages, Daddy?"

# Contents

# PART I

# Voyaging into Orbit and Beyond

"Five . . . four . . . three . . . two . . . one . . . ignition!" The space vehicle rises on a pillar of fire and brimstone. The ascent into space begins, first slowly, and then rapidly as the tremendous force of the rocket engines pushes the payload faster and faster, ever closer to the speed needed to hurl it away from the powerful grasp of the earth's gravity. In a minute or so, the powerful first-stage rockets drop away, having done their job, and the smaller second-stage rockets take over. These, too, burn out, and in a few minutes the powerful roar of rocket engines is only a memory, to be replaced by the silence of outer space.

During most of the thirty-year history of the space program, getting into orbit or some other particular destination was the primary focus of effort. Although more and more of our effort in space involves a special activity—be it building space stations, growing crystals, or using telescopes—the technology that is available to launch astronauts or equipment is still a major factor in determining what kinds of space ventures are feasible. The development of the space shuttle was a major technological achievement of the 1970s, a dramatic increase in our capability to take heavy payloads into orbit and, what is more important, back down to earth again. The tragic explosion of the space shuttle *Challenger* and its aftermath demonstrated that the shuttle was not the space truck that NASA had claimed it to be, however.

The effects of *Challenger* and the central role of space transportation technology suggest that a book about the present and future state of the space program should begin with a discussion of rockets, space shuttles, and other space vehicles of the present and future. The *Chal-*

*lenger* tragedy and the ensuing investigations will have a major effect on the space program for the next ten years. But we must look beyond *Challenger*. The current state of aerospace technology may allow us to create a vehicle that will be the low-cost highway to orbit that the space shuttle is not. If we look still further forward to the twenty-first century, some clever technologies can develop bridges between worlds that may permit the settlement of the inner solar system.

Space travel requires institutions as well as rockets. What we can and will do in space will depend on NASA's approach to space, not just on rocket technology. How has NASA become the agency that it is, and what will it be like in the future? How does the political climate affect what NASA can and cannot do? The international situation, too, will affect the future. Twenty years ago, America and the Soviet Union were embroiled in cutthroat competition, with the rest of the world largely standing on the sidelines. But in the mid-1980s, the race was over, and a number of other players have joined in. The members of the European Space Agency, the Japanese, the Chinese, and the Indians all participate in space to a greater or lesser degree. Hence, American predominance is no longer ensured. The decisions we make in the next few years may well determine whether our leadership position in space will be maintained or whether America will become a second-class citizen in the international community of the twenty-first century.

# CHAPTER 1

# *Challenger*

"You won't believe this, Harry. The space shuttle just exploded!"

On 28 January 1986, I telephoned Pam Hawkins of the American Astronomical Society about some work I might do for the society's Committee on Manpower and Employment. We were suddenly and tragically interrupted. Neither Pam nor I could say anything. Whatever business we had was quickly forgotten. Once I collected my thoughts, I immediately shared news of this tragedy with my colleagues down the hall. We speculated then, as many of us still wonder, what will happen to our space program.

I suspect that many people, certainly those of us who have been associated with the space program in one way or another, will remember exactly where we were and what we were doing when the space shuttle exploded. Most people who are old enough to remember can recite exactly what they were doing when they heard that John Kennedy was shot, in much the same way. Seven brave people, including schoolteacher Christa McAuliffe, perished in a ball of fire when the shuttle's external fuel tank exploded with the force of a small nuclear bomb. Many people in and out of the space program had come to think of the space shuttle as a "truck," something that could routinely and safely carry things into space. The TV networks had begun to view shuttle launches as so routine that they didn't deserve live coverage. But this tragedy brought us back to reality, reminding us that a space shuttle hurled into the hostile vacuum of space atop a pillar of flame is a lot more dangerous method of transportation than a Mack truck barreling down an interstate highway.

In the days and weeks after the accident, the producers and viewers of network news programs, people who had all but forgotten about NASA, were treated to more and more instant replays of the tragedy. First, we were shown what seemed to be endless repetitions of films of the shuttle exploding, accompanied by a voice from the Kennedy Space Center inexplicably describing this tragedy as a "major malfunction." A few days later, we saw that one of these launch pictures contained what could be a key clue: a foul plume of black smoke spitting out of the side of one of the solid rocket boosters. These oversized Roman candles provide much of the power to the space shuttle, provided that they burn in a well-controlled way. And then we heard about giant rubbery rings called *O-rings*, and how a "no" can turn into a "yes" when it moves up the chain of command.

The immediate response to the *Challenger* tragedy showed that many Americans were still intrigued by the space program. Of course, this launch was a special one because it had the first "civilian" passenger, but polls taken later on confirmed the continued surge of public interest. I was a guest on a radio call-in show a couple of days after the accident, and the calls never stopped. Although many people were and are questioning just how the space program was and should be conducted, no one said we should stop. A poll taken in August 1986 confirmed what I had felt in January, with 89% of the American public wanting to resume shuttle flights despite the risks.[1] This deadly accident will set us back in our quest to expand humanity's grasp to include outer space, but it won't bring it to a halt.

Space is still a frontier, and a dangerous frontier just like the American West a hundred years ago or the oceans of our entire planet in the seventeenth century. It's easy to forget that explorers who play a major role in shaping the future of humanity often pay with their lives for their place in history. Something like half the leaders of major expeditions in the early stage of discovering America were killed in action,[2] and the names of Dick Scobee, Michael Smith, Judith Resnik, Ellison Onizuka, Ronald McNair, Gregory Jarvis, and Christa McAuliffe will be honored in memory in the same way as the names of others. Ferdinand Magellan was killed in the Philippines; Henry Hudson was marooned by a rebellious crew in the bay that bears his name; John Cabot was lost at sea. We're far from the stage where buying a ticket on the space shuttle will be as routine as getting on a jet airplane. This is the major lesson of *Challenger:* we have not yet built a highway to heaven.

## WHAT HAPPENED?

Flight 51-L of the space shuttle *Challenger* was originally scheduled for July 1985. It was the twenty-fifth flight in the space shuttle program, and the ninth scheduled for the fiscal year 1985. (The numbering system NASA uses is rather complex; the first digit designates the fiscal year of the scheduled launch, the second the launch site [1 = Kennedy Space Center, Florida, and 2 = Vandenberg Air Force Base], and the letter is an alphabetical sequence for the fiscal year.) The launch of this particular mission was rescheduled six times because NASA and its customers kept changing the payload that was to fly in the shuttle's cargo bay. By June 1985, mission plans seemed to be fairly well set for a launch in January 1986.

The NASA astronauts assigned to this mission were selected in January 1985. Commander Francis R. (Dick) Scobee had piloted one previous shuttle mission, in 1984. This was to have been pilot Michael J. Smith's first shuttle flight. The three mission specialists, astronauts whose job is to keep the shuttle working well, were Ellison Onizuka, Judith Resnik, and Ronald McNair, all of whom had flown on the shuttle before.

The crew of each shuttle flight often includes a number of "payload specialists." These people are not astronauts, though they do go through NASA training. Their job is usually to operate the scientific instruments that are on board. Gregory Jarvis was assigned to flight 51-L only a few months before launch, in October 1985, as a representative of the Hughes Aircraft Company, a company that designs and builds satellites. Jarvis was to do a number of experiments in fluid dynamics that would be used by the Hughes satellite design program. These payload specialists are generally nominated by companies or are individuals with an interest in a shuttle mission; they must all be approved by NASA.

Payload specialist Christa McAuliffe made this *Challenger* flight special for many Americans. In an effort to bring the space program closer to the American people, President Reagan and NASA decided to fly a teacher in space on this twenty-fifth shuttle mission. After a nationwide search, NASA selected McAuliffe to represent the country's hard-working teachers. She left her junior high school classrooms in New Hampshire, underwent astronaut training, and planned to conduct a series of classroom lessons from orbit. NASA also chose 100 teachers, 2 from each state, to support NASA's Teacher in Space project, visiting schools in their home states and teaching classes about space.

The major payload or cargo on Flight 51-L was a Tracking and Data Relay Satellite (TDRS). This communications satellite was to be a key part of a new NASA scheme for communicating with its rapidly growing collection of orbiting payloads. Messages would be relayed from individual satellites to the TDRS and then to ground antennas, eliminating the need for a vast, complicated, and expensive global tracking network. *Challenger* carried in the cargo bay a telescope that was to look at Halley's comet. The astronauts would toss this telescope out of the cargo bay and retrieve it after it looked at the comet and other targets for a couple of days.

NASA postponed the launch several times. Some of the delays resulted from the heavy work load at the Kennedy space center; the final postponement was due to bad weather. On 27 January, the crew were strapped into their seats at 7:56 A.M., but the countdown was halted at 9:10 when there was a problem with an exterior hatch handle. Although this problem was solved, the winds at Cape Kennedy freshened a bit, and the launch had to be called off. If problems develop during the shuttle's ascent, one abort option is to return to the launch site, and so space shuttles are not launched if heavy crosswinds would prevent a shuttle's landing on the Kennedy Space Center's runway. Countdown was rescheduled with a launch the next day.

The night of 27 January was cold and wintry on the east coast of Florida. NASA had not apparently identified any serious problems that the cold weather would cause, even though the weather forecast predicted temperatures in the high twenties or low thirties, well below the temperatures of other shuttle launches. The coldest it had been on previous shuttle launches was 53°, on 24 January 1985, when the shuttle *Discovery* carried a U.S. Defense Department payload into orbit. Later investigations revealed that the warnings of engineers working for Morton Thiokol, the manufacturer of the solid rocket boosters, were not communicated further up the chain of command. Icicles covered many of the structures on the launch pad, and the launch was delayed two hours beyond the scheduled lift-off time of 9:38 A.M. to allow the ice to melt. The last ice inspection was completed at 11:15 A.M., and *Challenger* began its tragic last voyage at 11:38 A.M. The air temperature at this time was 36°.

At the time of the accident, no one knew that anything was wrong until the shuttle exploded seventy-three seconds after launch. But subsequent analyses of photographs explained what happened. Six seconds before lift-off, the shuttle main engines were ignited while the craft was still bolted to the launch pad. (The illustration shows a schematic diagram of the space shuttle and its parts.) The strains on the shuttle assembly are released in the first few seconds after launch, putting the maximum strain on the joints on the solid rocket boosters. Only 0.678 seconds after lift-off, a strong puff of gray smoke spurted from the aft (or bottom) field joint on the right solid rocket booster. (In the perspective shown in the illustration, this joint is behind the orbiter's wing.) The smoke came from the side of the booster facing the giant external fuel tank, an orange cylinder that contains the hydrogen and oxygen to power the shuttle's main engines. Eight more puffs of blacker and blacker smoke were recorded in the next two seconds.

We now know that those ominous puffs of black smoke came from disintegrating O-rings, key parts of the joints between different parts of the solid rocket boosters. On the chilly winter morning of 28 January, the O-rings were not as flexible as they should be, and they failed to seal properly. The hot gases inside the solid rocket boosters penetrated through the seals and burned them.

For the next minute, the flight of *Challenger* was not very different from the twenty-four previous space shuttle flights. Four seconds after launch, the shuttle's main engines were throttled up to 104% of their rated thrust level. Surpassing 100% seems illogical, but the "rated" thrust was set fairly early in the shuttle program, and modifications in the main engines allow them to exceed the rated thrust. Between 7 and 21 seconds after launch, *Challenger* executed a roll

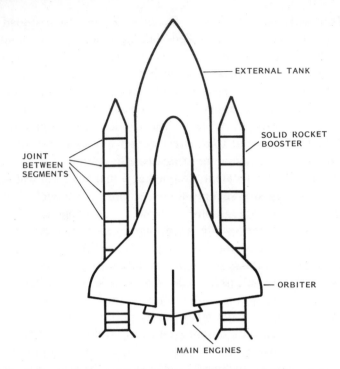

Schematic (not exactly to scale) illustration of the space shuttle and its parts. The orbiter, the external tank, and the two solid rocket boosters are all launched as a unit. The boosters and the external tank separate from the orbiter during the launch, and only the orbiter returns to earth. The term *space shuttle* is used to refer to both the assembly at launch and the orbiter, the only fully reusable part of the system.

maneuver to put the astronaut's heads toward the ground, as is standard practice on a shuttle launch. At about 37 seconds, *Challenger* encountered high-altitude crosswinds, and the automatic steering system changed the direction that the main engines were pushing in to counteract the winds. At 45 seconds, pictures of the launch show a few bright flashes of light, which have been seen on other flights; perhaps these are simply the effects of the sun reflecting off various parts of the spacecraft.

The main engines were still at 104% of their rated thrust, and the solid rocket boosters were increasing their thrust, when a small flame showed up on a computer-enhanced image, which, it was later determined, was taken 58.788 seconds after launch. In the next frame, the flame becomes larger. At 60 seconds, the shuttle's sensors detected a difference in pressure inside the right- and left-hand rocket boosters. The crew did not, however, know of this problem; it was discovered only later when data telemetered to the ground were analyzed. Pressure in the interior of the right booster dropped because the exhaust gases had another place to leak from. *Challenger*'s steering system valiantly kept the craft on course, so that neither the astronauts nor the flight controllers suspected that there was anything wrong. Then, 64.660 seconds after launch, the flame became much larger as it ate through the wall of the external tank, burning the hydrogen fuel. And 8 seconds later, the flames cut through the strut linking the right booster to the external tank; the right booster began to rotate around the upper attachment strut. At 73.137 seconds, the wildly spinning solid rocket booster banged into the external tank, spilling hydrogen and liquid oxygen into the air. This fuel soon burned almost explosively, igniting the propellants carried on board the orbiter.[3]

The orbiter broke apart, and the screens at mission control that had been displaying housekeeping data from the shuttle went blank. Pictures of the ascent, analyzed later, show the main engine section with the tail, one wing, and the crew compartment emerging from the fireball. In what must be a historic misstatement, what happened was described initially as a "major malfunction." About half a minute later, the Air Force Range safety officer gave a command to blow up the solid rocket boosters so that they would not fly out of control and land on a Florida city.

Did the astronauts survive the explosion? The last words spoken by the crew to the ground were Dick Scobee's confirmation, "Go at throttle up." But words spoken on the spacecraft intercom were not received by the ground. What little we know of the final moments of *Challenger* came from tapes and other debris painstakingly recovered from the shuttle wreckage. Full analysis of these tapes took months, and the results were not released until the end of July 1986.

The tapes recorded the last words of Michael Smith, the shuttle pilot: "Uh-oh." Smith, speaking seventy-three seconds after launch, might have known that there was a major fire, or he could have seen something on his control panel. Four emergency air packs were recovered from the wreckage, and three (including Smith's) had been turned on. Scobee's had not been activated; it's not known which astronauts wore the other two. What's troubling is that the gauges on two of the packs (including Smith's) showed that three-quarters to seven-eighths of the air had been used up through normal breathing.

We don't know and never will know how long individual astronauts lived and how long they remained conscious. For six months after the tragedy, NASA officials, along with all the rest of us, believed that it was impossible that the crew had survived the fiery explosion that marked the end of *Challenger*. But the tapes and the air packs suggest a troubling possibility, that at least some of the astronauts may have been aware of what was happening when the crew cabin plunged eleven miles into the sea. Physician-astronaut Joseph Kerwin suggests that they might have had six to fifteen seconds of "useful consciousness" after a sudden decompression of the cabin at an altitude of 65,000 feet.[4]

The possible survival of the crew has prompted many people to wonder whether the shuttle should contain some sort of ejection mechanism to allow the crew to escape. On the first four test shuttle flights, the orbiter *Columbia* was equipped with conventional ejection seats. However, these seats are intended for use during controlled gliding flight, when the shuttle is coming down for a landing. The difficulty with providing a way to survive something like the *Challenger* accident is that the solid rocket boosters can't be turned off once they start burning, in the way that a liquid-fueled rocket engine can be shut down by cutting off the fuel supply. If the astronauts had been able to eject from the *Challenger* explosion, they would almost certainly have been fried by the hot exhaust from the solid rocket boosters. It is hard to imagine how a parachute mechanism, attached to the entire crew compartment, could still have worked after the explosion broke up the *Challenger*. Because of the limited utility of ejection seats, NASA removed them from the orbiter after the first four flights. Astronaut Robert Crippen expressed a universally accepted viewpoint in testimony before the presidential commission investigating the space shuttle accident: "I don't know of an escape system that would have saved the crew from the particular incident that we just went through [referring to the *Challenger* accident]."[5]

But this doesn't mean that all discussions of crew escape have stopped. There are a number of other possible situations in which crew ejection might help them survive. After two minutes of ascent, the solid rocket boosters drop off the shuttle, eliminating the primary obstacle to some kind of successful recovery from engine failure. It is possible that an abort at this time would send the shuttle gliding down to earth toward an unsuitable landing site, either an unprepared airport or the ocean. An escape mechanism might make sense for this situation, though crew escape would lead to loss of the orbiter. In response to recommendations by the president's commission, NASA is studying whether some kind of crew escape mechanism should be installed in the shuttle. But NASA needs to decide whether the weight penalties of including an ejection mechanism could be justified by the likelihood that they will be useful. No human pursuit is absolutely safe, and space flight is no exception. If the shuttle contained every conceivable safety device to protect the crew from any possible accident, it would be far too heavy to fly.

## THE DIRECT CAUSE OF THE ACCIDENT: O-RINGS

After the accident, the investigations began. Less than a week had elapsed when Ronald Reagan issued Executive Order 12546 establishing the Presidential Commission on the Space Shuttle Challenger Accident. Former Secretary of State William Rogers headed the commission. Rogers and the commission members quickly realized that they had to establish their independence of the NASA managers whose decisions had preceded the accident. The Rogers Commission successfully established itself as *the* group of people who could independently investigate what had happened, forestalling separate major investigations by other groups like Congress. Working quickly and efficiently, they presented their report to the President in June 1986. (This report has been published and is the basis for much of the material in this chapter.)

The right solid rocket booster was quickly identified as the cause of the disaster. The shuttle is lifted into space by two of these giant Roman candles, 149-foot-long and 12-foot-diameter steel cylinders containing aluminum powder and ammonium perchlorate fuel. Once the fuel starts burning, it can't stop, because there are no pumps to feed the fuel to the motor. Solid rockets are really quite simple in construction, being basically long, solid tubes, filled with rocket fuel, with a nozzle at the end. A liquid-fueled rocket, in contrast, has to have lots of plumbing to feed the fuel and oxidizer to the rocket engine in carefully measured doses. The simplicity of solid rockets makes them inexpensive, and before the *Challenger* accident, they had proved to be quite reliable. The U.S. Air Force's Titan III rocket had been launched thousands of times without failure, although as it turns out that there are some significant differences between the design of the solid rockets used in Titan and those used in the shuttle, especially in the joints between rocket segments.[6]

The manufacturer, Morton Thiokol, ships these rockets to Cape Canaveral in a number of pieces. At the space center, these separate pieces are assembled, and the joints between the segments are sealed by zinc chromate putty, O-rings, and pins to hold them together. The diagram illustrates the old design of the joints, along with a modified design that NASA adopted later. These joints, which are sealed at the Kennedy Space Center, are called *field joints* to distinguish them from the joints that are sealed by Thiokol at its plant in Utah before the rockets are shipped to the Cape. The O-rings are rubbery washers that are supposed to flex under pressure, being squashed into the gaps between one solid rocket segment and another. They function in somewhat the same way as the washers in ordinary household faucets. Each joint of the solid rocket boosters contains two O-rings, a primary and a secondary. The secondary is supposed to serve as a backup. But because it sits in a groove that is parallel to the primary O-ring, it is subject to much the same stresses; if something keeps the primary O-ring from sealing, it's quite possible that the second O-ring won't seal either.

On the morning of 28 January, the two O-rings in the aft field joint did not seal the joints in the way that they should have. Flames from burning rocket fuel burned through the O-rings, and perhaps it was the burning rubber that generated the puffs of ugly black smoke less than a second after launch. It's not clear why the joint didn't fail immediately, exploding *Challenger* on the launch pad and producing

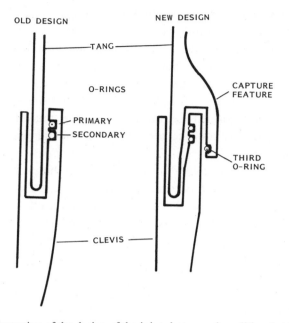

OLD DESIGN       NEW DESIGN

TANG

O-RINGS

CAPTURE
FEATURE

PRIMARY
SECONDARY

THIRD
O-RING

CLEVIS

Schematic illustration of the design of the joints between the solid rocket boosters used in *Challenger* and previous space shuttles (left) and the new design that NASA has decided to use in future space shuttles, subject to rigorous testing (right). Only the outer casings of the rockets are shown; there are also some significant differences in how the insulation in the two segments fits together. In the old design, any bending of the joint that tends to open up the slot that the primary O-ring sits in also opens the slot of the secondary O-ring. The new design adds a third O-ring, which will be compressed if the slots holding the first two O-rings open up. (Redrawn from an illustration provided by NASA.)

an even worse disaster. Perhaps the leak at lift-off was only a small one, and it took some time for the flames to burn completely through. Another possibility is that combustion debris coming from the inside of the rocket resealed the joint, and that upper altitude winds pushed the solid rocket booster around, opening the joint up again.

The Rogers Commission identified a number of factors that could have contributed to the failure of the O-rings. The low temperature on 28 January is a prime suspect. In a rather theatrical performance, Commissioner Richard Feynman dunked an O-ring into some ice water during a break in testimony to demonstrate experimentally that O-rings are not as flexible when they are chilled. Another factor was that the rocket segment was not perfectly round and had expanded as a result of prior use. (During a normal shuttle flight, the solid rocket boosters are parachuted into the ocean, taken apart, refueled, and reassembled for use in a future shuttle flight.) Measurements of the shape of the particular segments that formed the faulty joint show that the O-ring was squashed very tightly into the joint at the particular place where the joint failed.

Either of these causes, or both working together, could have been the direct cause of the *Challenger* disaster. However, the roots of the problem lie deeper. A joint design that fails under circumstances like these has to be regarded as faulty in engineering terms. The limitations of the design were not fully understood before *Challenger*.

When you think about how an O-ring works, the whole picture makes sense. What's supposed to happen is that immediate, gradual pressure from the burning rocket pushes the O-ring into the proper place so it will seal the joint. The O-ring sits in a little groove that is somewhat bigger than it is. When the rocket is turned on, the combustion gases create moderate pressures on the inside of the rocket about 0.1 seconds later. These gases push on the putty that separates the fuel from the O-rings, and the putty exerts pressure on the air that surrounds the O-ring in its little groove. This pressure forces the O-ring into the gap between the piece of the top rocket booster (called the *tang*) and the piece of the bottom rocket booster (which, in the joint, is called the *clevis*).

What actually happened is not known in detail. Something like the following sequence of events would account for the failure: The low temperature made the O-ring much less flexible than usual, so that it would not move into the gap as easily as it should have. The tightness of the joint at this particular point further compressed the O-ring, limiting its flexibility. Thus, either the O-ring was not pressurized as soon as the rocket was turned on, or its response to pressure was insufficient to seal the joint. The mounting forces from the burning rocket fuel kept pushing on the joint, forcing the joint open before the O-ring was able to seal it. (For reasons unclear to me, this phenomenon is called *joint rotation*.) Because the joint was open, gases blew by the first O-ring and escaped from the rocket booster. The same forces acted on the second O-ring, and the burning gases blew by that one, too, eventually eating a hole through the side of the joint.

A possible factor contributing to the accident, discussed at some length in the Rogers Commission report, is the behavior of the putty. This putty lies between the joint and the burning rocket fuel and is supposed to keep the hot rocket gases away from the O-rings. There was some concern that either the effects of the burning rocket fuel or the testing procedures opened up some holes in the putty that combustion gases could flow through. The joints are tested by pumping compressed air into the gap between the two O-rings and seeing that the air pressure remains high for a period of time. This testing pressure might conceivably create weak spots in the putty.

So what can be done to prevent such an accident from happening again? This particular joint design is, from an engineering standpoint, flawed in that it is too sensitive to various types of problems. In the joints used in the Air Force's Titan missile program, the insulation of one segment fits tightly against the insulation of the next one, not relying on putty to make the design gas-tight. The secondary O-ring was basically useless, as the forces that keep the primary O-ring from sealing act in virtually the same way on the secondary O-ring.

As this is written, NASA has selected a new joint design that will be much safer but that will still allow the agency to use the existing segments for the solid rocket boosters that provide much of the force needed to hurl the shuttle into orbit. But this design must be tested and must still pass some detailed reviews both inside and outside NASA, before it is deemed safe enough. The previous illustration shows a picture of the new design as well as the old. A third O-ring will be added to the system, and forces that tend to open the joints that the two existing O-rings sit in will tend to shut the joint containing the third O-ring. A new configuration of insulating materials and adhesives, not shown in the picture for reasons of clarity, will provide more shielding between the O-rings and the combustion gases. The new design calls for a different, more resilient material to be used in the O-rings themselves; O-rings are available that work better at low temperatures.[7] The redesigned joint will no doubt be built and tested extensively in real engines before it is used on a shuttle.

## A LONG HISTORY OF O-RING PROBLEMS

If it were just a question of fixing a faulty joint, the effects of the *Challenger* tragedy on the space program would be bad enough, because of mission delays. In such a situation, NASA's reputation as a uniquely competent government agency would remain. But the most significant finding of the Rogers Commission was that NASA needed fixing, too. The agency's files contained a sorry history of documents relating to O-ring problems. The mere presence of such documents does not necessarily mean that NASA made a mistake. Whenever you build anything, you accept a certain element of risk. Hammer a nail into a board and there is always the chance that you will hammer your finger instead. Still, we haven't outlawed hammers (not to mention power tools, which are far more dangerous). But even granting the tendency of 20-20 hindsight to find fault with what are basically reasonable management decisions, NASA blew it this time. The Rogers Commission certainly said so, and I tend to agree.

There were some early warnings of potential problems with the O-rings. As part of its safety procedures, NASA compiles a "critical items list" of parts of the shuttle design that are most critical to the success of the mission. Initially, they classified the solid rocket joints as "1R," second behind a classification of "1." Criticality "1" items are those that could result in the "loss of life or vehicle" if they fail. The "R," meaning "redundant," was there because of the second O-ring, which supposedly could provide a margin of safety if the first O-ring failed. Earlier warnings from Leon Ray, an engineer with NASA's Marshall Space Flight Center, generated some strong language in the 1980 critical items list: "It is not known if the secondary O-ring would successfully reseal if the primary O-ring should fail after motor case pressure reaches or exceeds 40 percent of maximum expected operating pressure."[8] Ray's arguments were finally accepted, and the status was changed to criticality "1," supposedly meaning that problems with this joint would receive more attention. However, there are a host of criticality "1" items, and it would not be possible to fly a shuttle that had none of them. The purpose of the list of such items is to highlight those parts of the space shuttle that must be inspected most thoroughly before launch.

In 1984 and 1985, O-ring performance on various shuttle flights began to show signs of weakness in the joint design. The simple erosion of the O-rings themselves is not necessarily a sign of real trouble, for the rings can still be sealing very adequately even though hot gases contact them momentarily and erode them a bit. As long as the O-rings are only partially eroded, they can still work. Such erosion is cause for concern because the putty is supposed to keep the gas away from the O-rings. On shuttle Flight STS 41-B (the tenth one), engineers disassembling the solid rocket motors noticed a tiny charred area, 1 inch long, 0.1 inches wide, and 0.03–0.05 inches deep. (The fact that such a tiny erosion area was noticed shows how careful these workers are.) However, lab tests had shown that an O-ring with a simulated erosion depth of 0.095 inches would still hold, and the shuttle flights went on. The erosion was only one-third of the highest acceptable value.

More serious problems arose on Flight 51-C, which was launched in a temperature of fifty-three degrees in January 1985. There was soot in the gap between the O-rings, indicating that combustion gases had blown by the first one, which evidently didn't seal. The secondary O-ring was slightly eroded, too. On the next mission, STS 51-B, a primary O-ring in the left nozzle joint had eroded a full 0.17 inches and did not seal; the secondary one had eroded 0.03 inches and did seal.

A group of about ten engineers at Morton Thiokol became increasingly concerned about these events. The tone used in various memos escalates in pitch. These memos are reprinted in the Rogers Commission report. A 31 July memo from Thiokol's Robert M. Boisjoly begins, "This letter is written to insure that management is fully aware of the seriousness of the current O-ring erosion problem in the SRM joints from an engineering standpoint." Three weeks later, A. R. Thompson was more forceful: "The O-ring seal problem has lately become acute." And on October 1, Thiokol's R. V. Ebeling begins a memo with "HELP!" and finishes the same memo with "This is a red flag."[9] On 19 August 1985, senior NASA and Thiokol officials briefed NASA headquarters on the seal problem, concluding that the "lack of a good secondary seal in the field joint is most critical" but judging anyway that it was still safe to fly.[10]

Particularly troubling in retrospect are the meetings held at Thiokol's plant in Utah the day before the launch of the *Challenger*. When it became clear that *Challenger* was to be launched in cold weather, Robert Ebeling, Roger Boisjoly, and other Thiokol engineers met for an hour and concluded that because the shuttle had never been launched in such a cold temperature, a launch delay was called for. They did not at the time really know why it was unsafe to launch in the cold; they based their arguments primarily on the history of problematical shuttle launches. Two successive teleconferences resulted in similar recommendations from the engineers who had conducted the O-ring tests. However, possibly responding to the urging of NASA managers at the Marshall Space Flight Center, Thiokol management reversed its initial position, overruled the engineers, and approved the launch of the *Challenger*.

The Rogers Commission feels that the response of senior management at both NASA and Thiokol to all these warnings was inadequate. It's difficult to determine from the commission report just who should have called a halt to shuttle flights until the O-ring problem was fixed, and at what point. Such an exercise in finger pointing does not help to show what's wrong with the organization that let these warnings go unheeded. What seemed to be happening was a subtle but significant change in the way that some managers approached safety questions. Previously, it had been up to the people who built the hardware to convince program managers that it was safe to fly. But now, as Roger Boisjoly stated:

> One of my colleagues that was in the meeting [where Thiokol management overruled their engineers and recommended launching the *Challenger*] summed it up best. This was a meeting where the determination was to launch, and it was up to us to prove beyond the shadow of a doubt that it was not safe to do so. This is in total reverse to what the position usually is in a preflight conversation or a flight readiness review. It is usually exactly opposite that.[11]

Richard Feynman, a member of the Rogers Commission, offered a similar perspective, adopted by the commission as a whole in their report:

> [The decision making was] a kind of Russian roulette. . . . [The shuttle] flies [with O-ring erosion] and nothing happens. Then it is suggested, therefore, that the risk is no longer so high for the next flights. We can lower our standards a little bit because we got away with it last time.[12]

Any agency or company in charge of a technological venture needs to devise an organizational system that is responsive to appropriate early warnings but that, at the same time, is not paralyzed by every undotted *i* and uncrossed *t,* or there would be no program. Perhaps responding to pressures on the shuttle's launch schedule, NASA had moved away from its earlier attitude of being hyperconscious about safety to take a different position where anyone who wanted to stop a shuttle launch had to prove that it would be unsafe. That's what most of us do with our cars: go ahead and drive them unless some nasty noises from under the hood or some warning lights on the dash indicate trouble. But cars and trucks are simpler than space shuttles, and we know more about their potential failures.

Another potential problem is the absence of a regular maintenance schedule or adequate supply of spare parts for the shuttle. Conscientious automobile owners, airline companies, and trucking companies regularly maintain their vehicles. And automobiles are sufficiently forgiving technologically so that those of us who don't follow regular maintenance schedules to the letter don't end up trashing our cars. NASA had declared the space shuttle "operational"—whatever that means—after only four flights. I heard people at NASA use the word *truck* to describe the shuttle as early as 1977. But the shuttle is not a truck and can't be treated like one when it comes to scheduling, operation, or maintenance.

## NASA'S RESPONSE TO THE SHUTTLE TRAGEDY

NASA has responded quite positively to the Rogers Commission's criticisms of the way it operates. Safety offices have been created, the joints are being redesigned, and management problems are being addressed. There is every indication that NASA will go back to the old way of doing business where safety is paramount but the job still gets done.

With all the stories about O-ring failures and so forth, it's very easy to forget just how far this remarkable government agency has come in the last thirty years. Other government agencies give us four-hundred-dollar hammers or mountains of red tape; NASA has given us a number of tremendous achievements that are a source of national pride.

Consider what NASA has done. In 1958, a small group of engineers who were to become the nucleus of NASA reacted to the *Sputnik* challenge and put up an orbiting satellite two months after they had been given the go-ahead to try. President Kennedy said in 1961 that we should land on the moon in the next ten years, and NASA did it. For all of its faults, the space shuttle does allow us to gain access to orbit in a new and valuable way, taking heavy payloads up and bringing them back again. In time, this capability will make it much easier to do experiments of all types in space. The venerable, durable *Voyager* spacecraft has lasted for ten years in its remarkable tour through the outer solar system, past the strange satellites and rings of Jupiter, Saturn, and Uranus.

Some of the specific actions NASA took in the first few months after the release of the Rogers report are indications of the seriousness of their efforts at reform. The Rogers Commission recommended a separate safety organization that would carry some clout within the agency and with its contractors. Many of the people who work on the space program are not NASA civil servants but are employees of companies that work for the agency under contract and are thus called *contractors*. On July 8, 1986, NASA administrator Fletcher created a new top-level position: Associate Administrator for Safety, Reliability, and Quality Assurance. Astronauts have been brought into management-level positions within the agency, as recommended. Fletcher has assigned General Sam Phillips, general manager of the Apollo program, the task of doing a complete review of agency management. Other detailed recommendations are being addressed in a similar way.[13]

Another reason for my optimism is the comprehensive nature of NASA's response. A common reaction to any disaster is to triple-lock one particular barn door, after the horse has got out, and to forget about all the others. I'm not worried about the solid rocket booster joints on future flights because I know that those will receive the attention that they deserve. It's the other things that might have slipped by that can cause problems. NASA has established a new landing-safety team, responding to the concerns raised by the Rogers Commission (and others) about the brakes, tires, and nose wheel steering. On 13 March 1986, well before the Rogers Commission had completed its work, NASA started a full-scale review of all of the criticality "1" items and all shuttle failure modes—all of the ways in which the shuttle can crash. If these reviews are done in the way that they should be, the shuttle should be as safe as it can be and still get off the ground.

In the decade ahead, there will be some frustrating times for those of us who participate in the space program. NASA was hoping to fly fifteen shuttle flights in 1986 and flew none. Still more flights were slated for 1987, and these have also been scrubbed. Once the shuttle does fly, there will be fewer flights, partly because only three orbiters are available, and partly because the pressures of a high flight rate were held responsible for the *Challenger* accident. The backlog of missions waiting for a lift-off into orbit is growing longer and longer. Some can be sent up on expendable rockets, but the supply of those isn't very big either. Despite the existence of a space shuttle flight list that extends through 1994, no one really knows exactly how long it will take to clear the backlog. I suspect that the tight schedule will last for ten years after *Challenger,* to the mid-1990s.

It's perhaps too easy to think of ten years as being forever and to become depressed about the future of the space program. People who started with the space program in the late 1950s have reason to be melancholy, for they are facing retirement in the mid-1990s and can look forward only to a decade of launch constraints and, perhaps, budgetary uncertainties. But I take the long view, and I hope other Americans will, too. The tragedy of seven astronauts will remain forever, but ten years from now, the pace of the space program will have recovered, provided that we as a nation have the guts to stay the course.

\*     \*     \*

The *Challenger* accident was the biggest disaster NASA has had to confront. The O-rings, other parts of the shuttle, and the organization of NASA itself needed some redesigning and rethinking. This work is now beginning, and in a couple of years, the space shuttle will be flying again.

But NASA exists in a world rather different from the world it faced in 1967, when a fire in the *Apollo* spacecraft claimed the lives of three astronauts. The past successes of the space program have molded NASA's traditions and, what is more important, the public expectations of the space program. The international arena is much more complex as well: the Europeans, the Japanese, the Indians, the Chinese, and others have their own national space programs.

# CHAPTER 2

# NASA and Its Foreign Competitors in the 1980s

The American space program stands at a critical point, in part because of *Challenger* and in part because of a number of trends that have been building since the late 1970s. Space activity is no longer a two-horse race with a finish line on the moon. The American preeminence, which had been taken for granted ever since NASA won the moon race, is fading. A complex web of relationships between various countries that are simultaneously competing and cooperating has replaced the American-Soviet race to the moon.

Yet, despite these complexities, many national attitudes about space are rooted in the *Apollo* days. NASA remains, in large part, an agency that was put together to send astronauts to the moon. Because our space activities are so multidimensional, we Americans cannot expect to have a preeminent space program, like the one we had when we won the moon race. The attitudes of the *Apollo* days may well be mismatched to the needs of the 1980s. Although the focus of this book is on the future rather than on the past, a little bit of history can illuminate how NASA came to be the type of agency that it is.

## *SPUTNIK* AND THE RACE TO THE MOON

The roots of the American and Soviet space programs lie in the 1950s. During the Cold War, all three branches of the U.S. military had their own rocket programs. They launched scientific instruments as well as warheads into the upper atmosphere on sounding rockets. More ambitious visions of orbiting satellites were kindled in both the United States and the Soviet Union, partly because scientists lobbied for their use for peaceful purposes, and also because the military recognized their use in reconnaissance. By the mid 1950s, Americans had learned that the Soviets were working on an ICBM program. We then felt the need for a spy plane or spy satellite in order to monitor Soviet progress.

In 1955, President Eisenhower announced that the United States would celebrate its participation in the International Geophysical Year by launching a satellite, *Vanguard 1*. This particular commitment seemed rooted in pure science, but in reality, it was a military, not scientific, decision. One reason for beginning with a scientific satellite was to establish the legitimacy of flying satellites over other countries' territories.[1]

A parallel but much more clandestine effort was under way in the Soviet Union. A then politically insecure Nikita Khrushchev finally succumbed to the entreaties of Sergey Korolev and gave the go-ahead for *Sputnik*. Korolev had been developing a powerful rocket booster, ostensibly for a Soviet missile program, but one that could also be used to launch payloads into earth orbit and beyond.

Few in the West realized that the Soviet Union was planning a satellite launch, and *Sputnik*'s appearance was a real shock. The Soviets successfully launched the first earth satellite, *Sputnik 1*, on 4 October 1957. The little "beep-beep-beep" from this, the first human artifact to be launched into orbit, could be heard on radio sets around the world and served as an audible reminder of the technological body blow that Khrushchev and Korolev had administered to the West. American journalists and politicians panicked.[2]

The exploration of space often serves as a symbol of many things: human enterprise and daring, technological prowess, national prestige, and the courage to reach for the unknown. Although there's a lot more to space exploration than that, the effect of *Sputnik* on American minds stemmed from its symbolism of how good a particular nation's technology really was. Then, as now, most Americans knew little of Russian culture or history. Many Americans, perhaps, saw the Russians (incorrectly) as a nation of illiterate peasants. After all, if they couldn't write their $R$'s in the right direction, how could they possibly launch a satellite before we did?

*Sputnik* and its aftermath illustrate a major theme that has pervaded American space policy from the beginning. The space program is an instrument of global politics. The degree and type of involvement have varied over time, but this theme recurs again and again in the story of human activities in space. In the early days, the focus was on international competition, primarily with the Soviet Union. Now we're competing with some countries and cooperating with others, sometimes competing and cooperating with the same country in different parts of the space program. Many American space scientists feel uncomfortable with the idea that the space program is part of our posture in the global playground. We scientists are trained from the beginning to disregard international borders as much as possible, and to look on foreign colleagues as friends and collaborators who just may not happen to speak the same language. Other participants in the space program may similarly resent their role as pawns in a global chess game. But anyone who wants to go up into space, or to use orbiting equipment, these days has to be involved with NASA or a foreign governmental agency, which means that global politics enter in no matter what.

The effect of *Sputnik* on America was probably the exact opposite of what Khrushchev expected. Rather than shocking the West into submission, it galvanized our leaders, who suddenly started listening to the science teachers who had been telling them for years that our science education was backward. New curriculums sprouted like mushrooms, and the federal dump truck would pour money into just about any respectable university research program. The American effort to launch a scientific satellite suddenly became quite urgent indeed. The world awaited the launch of *Vanguard 1* in December 1957.

The United States had confidence; we televised the countdown. American technology always worked, and there was no reason to wait until afterward to let the media in. The familiar chant of "... seven ... six ... five ... four ... three ... two ... one ... ignition" was not followed by the expected lift-off. One second after ignition, when the rocket had risen a mighty yard and a third, the bottom half of the rocket exploded. A cloud of ugly black dust and smoke immediately enveloped the launching pad. The tip of the rocket, which should have climbed upward toward the sky, plopped downward instead, hiding shamefully in the foul fumes. The satellite survived the explosion and beeped happily away, resting on the ground. Kaputnik. Flopnik. Humiliating.

Soon after the *Vanguard* failure, Wernher Von Braun's U.S. Army rocket team was given the go-ahead to proceed with its own satellite project. This group of German rocketeers, transplanted from Peenemunde, Germany, to Huntsville, Alabama, had deeply resented the decision to give the *Vanguard* project to the U.S. Navy, which may have had more scientists on board but which had little rocket experience. Von Braun's team worked rapidly, refitting their Army missile with a satellite, and did launch the *Explorer 1* satellite, the American response to *Sputnik 1,* in January 1958. At last, America was in orbit. The *Vanguard* project kept going and finally made it into space in March 1958.

Still, the pace in America seemed slow, and the failures too numerous. In the following years, commentators kept score of every time Soviet Union beat America in those things that could be easily counted: total satellite weight, first dog in space, first man in space, first woman in space, and so on. National prestige was on the line each time a new mission went up on either side. While the Soviet Union was taming the skies, we were left apparently stumbling in the dust.

The space crisis deepened in the next few years because of military developments. The *Sputnik* success increased American respect for Soviet prowess in rocketry. The U.S. military had suspected for some time that the Soviet Union had a well-developed ICBM program and now felt an urgent need for more information about it. For reconnaissance, the United States had relied on a high-flying aircraft, called the U-2, which flew over the Soviet Union supposedly beyond the reach of antiaircraft weapons. The U-2 photographed Soviet territory, concentrating on missile sites. On 1 May 1960, the apparent invulnerability of these high-flying aircraft was shattered by a Soviet surface-to-air missile, which shot down pilot Francis Gary Powers.

What was perceived to be American incompetence embarrassed the country once again. In this case, it was a Soviet missile rather than a satellite that foiled American desires, but the political system did not draw a distinction between the two. John Kennedy exploited the so-called missile gap in his 1960 presidential campaign against Richard Nixon. Kennedy argued that the nation that could orbit the heaviest satellite could also do a better job in hurling bombs around the world on nuclear-tipped missiles. This belief is not entirely wrong, but in fact, the American and Soviet missile programs were more-or-less equivalent at the time of the 1960 election. The missile gap was a phantom.

Thus, a second theme for the space program emerged even before NASA got off the ground. Despite the organizational separation between NASA and the military space program, the two have been intertwined from the beginning. There have been times when the military role in space was easy to forget. In the *Apollo* decade of the 1960s, the military uses of space seemed to be a long way from NASA; the astronauts were identified as human heroes, not the military officers that most of them were. But the military role emerged again with the U.S. Air Force's role in the selection of a compromise design for the space shuttle in 1972. Even when NASA's astronaut program moved away from picking only military test pilots, the Department of Defense was still represented in the astronaut selection process.[3] Many space shuttle missions involve military payloads. And, of course, there was and is Star Wars, the Strategic Defense Initiative (SDI), which would make space into the battleground for Armageddon.

The relationship between the military and NASA has been an uneasy one on both sides, and it's not just a question of a pure and good NASA versus a "tainted" military. Much of what the military does in space is not directly related to war. In many ways, one of the most important military space programs involves the reconnaissance satellites, which have made arms control agreements possible without on-site inspection. The primary conflict between the military and civilian space programs involves who has access to, and who should pay for, technology and facilities that both use, like space shuttles.

## THE RACE TO THE MOON

Four years after *Sputnik,* the charismatic young President John Kennedy sought to rally American spirits by challenging NASA to put a human astronaut on the lunar surface before the end of the decade. This challenge committed America to a space race with the Soviet Union. It also demonstrated that politics drove the American space program. The participants' reasons for going into space were really unimportant, be they seizing the military "high ground," doing scientific research, forwarding economic exploitation of space, or pursuing the romance of exploring the last frontier. Soon after Kennedy became President in January 1961, he decided to reject Eisenhower's approach of slow progress on many fronts and to commit the nation to a lunar landing. In May, he challenged the country: "Now it is time to take longer strides—time for a great new American enterprise—time for this Nation to take a clearly leading role in space achievement which in many ways may hold the key to our future on earth."[4]

This decision to initiate the *Apollo* program was by far the most important influence on the character of the American space program. Like *Sputnik,* it permanently established the role of the space program as part of the field of international politics. Kennedy's advisers were quite explicit: "The non-military, non-commercial, non-scientific but 'civilian' projects such as lunar and planetary exploration are, in this sense, part of the battle along the fluid front of the cold war."[5]

*Apollo* also established the dominant role of highly visible megaprojects in the minds of many citizens and decision makers. This consideration places a premium on projects that are technologically challenging and imaginative, and that appear to the average person to represent extreme challenges to the human spirit. There is a widespread perception both inside and outside the agency that a space program, to be successful, must include at least some kind of *Apollo*-like major project and should perhaps focus on such a venture to the exclusion of all other space activities.

But space has changed since *Apollo,* and a narrow focus on a single objective is incompatible with what we are actually doing now. Space is being used for a wide variety of purposes, including materials processing, pure scientific research, spying, communications, mapping, and weather forecasting. A number of policy statements have referred to the need for a wide-ranging space program rather than one that is narrowly focused on an objective, like a lunar landing. President Reagan's statement on a national space policy in 1982 said that

"United States Government programs shall continue a balanced strategy of research, development, operations, and exploration for science, applications, and technology."[6] I wonder how many people, both inside and outside the space community, appreciate the significance of this shift from a single-minded to a balanced program.

There is an advantage to a single, inspiring mission of overriding priority, visibility, and technological challenge. A mission like the moon race attracts public attention and, in the case of *Apollo,* generous funding. Congress and the American people responded to Kennedy's challenge and gave NASA the money it needed to go to the moon in ten years. NASA's budget exploded from about $0.5 billion in 1960 to $5 billion in 1965, a level that is (in real terms) about three times the current level, as is shown in the graph.[7]

But a focus on a single goal can create some problems, because an agency that focuses on a single, short-term goal can find itself at a loss when that goal is achieved. The *Apollo* program, in the way that it was conducted, did not lay the groundwork for subsequent activities in space.

Once we had landed on the moon, the question was what to do next? Because the focus of leadership in NASA had been on meeting the Kennedy timetable for a lunar landing, post-*Apollo* planning was not, apparently, seriously debated at high levels until just before the lunar landing in 1969. The focus of the space program during the *Apollo* era was on a single goal, with little consideration of the long term. A commission headed by Vice President Spiro Agnew tried to sell the nation a bold space program, which included the space shuttle, a space station, and a trip to Mars. But because, as Senator Matsunaga of Hawaii puts it, Vietnam was "devouring the federal budget and the national will,"[8] only the shuttle survived as NASA's budget plummeted.

Another *Apollo* legacy is best expressed as an often asked question: Should our space activities be manned or unmanned? (The sexist language is as traditional as it is unfortunate.) The objective of the race to the moon was to put an astronaut, not a machine, on the lunar surface. The widespread perception persists that one needs to involve an astronaut in order to sell an expensive space program to Congress. "No Buck Rogers, no bucks" is a widely quoted aphorism.

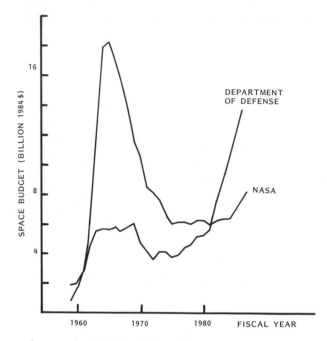

The amount of money that NASA and the U.S. Department of Defense spent on space programs in various years shows the rising and falling fortunes of each as well as the relative emphasis on civilian and military space activities. In the 1960s, NASA's budget was at an all-time high because of the *Apollo* program. Defense expenditures increased slowly through the late 1970s and more rapidly in the 1980s, partly because of the Strategic Defense Initiative, but also because of increased activity in antisatellite weapons and other areas. (Source: Paul B. Stares, *The Militarization of Outer Space,* and others.)

But this often-asked question really misses the point. There is no such thing as an "unmanned" space program. The absence of astronauts is not the same thing as the absence of human beings. An automated spacecraft is controlled virtually nonstop by human controllers, who command its movements minute by minute. The excitement of spaceflight is there to a great extent, even if one is just sitting on the ground, controlling a robot eye in the sky. Planetary probes don't aim their cameras by themselves; mission controllers have to decide such questions as where to orbit, where to land, and where to point the cameras. To be sure, there are some aspects of space travel that a spacecraft controller cannot enjoy: the giddiness of weightlessness and the view of the earth below outside the window.

The astronauts play a very special role in the public perception of the space program, as we were reminded by the outpouring of national grief following the *Challenger* accident. Astronauts do become folk heroes, as I have seen personally. Jim Lovell piloted the crippled *Apollo* 13 spacecraft back from the moon, freezing in the dark because the shortage of electrical power compelled the crew to turn off everything that they didn't need to use. Lovell visited Bamberger's department store in Delaware as part of a Halley's comet promotion in October 1985. I stood in astonishment when I saw him mobbed by hundreds of people, seeking his autograph on a photograph of him wearing a flight suit. Now president of a telecommunications company, he had not been in space for fifteen years, but the magic luster of having actually been in space still remains.

## FROM *APOLLO* TO *CHALLENGER:* A MULTIFACETED SPACE PROGRAM

The space program of the 1960s included far more than the highly visible *Apollo* program. The communications satellite industry, the American and Soviet planetary exploration programs, sophisticated spy satellites, space astronomy, weather satellites, and the use of space photography to study the Earth's environment all had their beginnings during this period. The nature of NASA and the political environment that nurtured it were such that this wide variety of activities was all but invisible, eclipsed in the shadow of *Apollo*.

During the early 1970s, many of these other ways of using the space environment were explored in a surprising amount of depth. There is a widespread perception that the rejection of Mars as a long-term goal left NASA completely adrift in space, but this perception is incorrect. In a wide variety of disciplines, space scientists and engineers investigated what possibilities awaited them in earth orbit. None of these missions was particularly expensive, and for this and other reasons, none of the missions was anywhere near as visible as the *Apollo* missions. Many of these projects were started in the looser fiscal environment of the 1960s and came to fruition through the 1970s.

The variety of space activities that have been carried out since the 1970s generates many different perceptions of what the space program actually is, or should be. Broadly speaking, there are three major categories of space activities: practical applications, space science, and major technological feats like the lunar landing. Within these three broad divisions, there is a much larger number of viewpoints shared by various groups of individual scientists, engineers, and policy analysts within the space community. Any individual involved with the space program will come into direct contact with only one, or at most a few, aspects of what humans are accomplishing in space. I sometimes find myself wondering whether I am dealing with the same space program when I hear others discuss it. It reminds me of the parable of the six blind men and the elephant, each of whom visualized a very different beast from his limited contacts with it.

Allow me to give you a tour of the space program as seen through the eyes of a number of different observers:

Space scientists see one of the major opportunities of the space program as being the exploration of the near and the distant universe. The orbital vantage point offers us a unique view of the earth as a planet, producing revolutions in some areas of geology, meteorology, and oceanography. The ability to send probes to other planets has unveiled some beautiful, exotic new landscapes and has provided some remarkable insights into how our own earth works. Telescopes from above the earth's atmosphere can use new windows on the universe to unveil unexpectedly violent cosmic phenomena. This space science work is more than pure research; some work in the earth sciences in particular has had practical applications in areas like weather forecasting.

Many of the people who work for NASA are engineers, who have a perspective on the space program that is similar to, yet curiously different from, that of the space scientists. They deal with a hostile yet challenging environment in which to display their technological expertise. First of all, the remoteness of spacecraft from the earth requires that equipment launched on spacecraft missions must have a degree of reliability very rarely found in other areas of engineering. The engineers must design equipment that will work in the absence of air and the absence of gravity. They must consider very extreme conditions of heat and cold; a spacecraft is fried on the side pointing toward the sun and freezes on the shady side. There is radiation. There are the tremendous stresses created by the launch. All these considerations and more create a difficult and exciting challenge to the spacecraft designers.

Hard technical challenges are part of what the engineer and writer Samuel Florman calls "the existential pleasures of engineering."[9] Someone who builds a gadget that does something outrageously difficult and makes it work feels really good inside, proud of his or her role in this tremendous achievement. Florman finds that the joy of building, of fabricating, of creating useful objects is rooted in sources as fundamental to our culture as the Old Testament. The engineering perspective on space values the challenges that space presents and the technically sweet means of meeting those challenges. The scientist is concerned with what can be discovered in space and the engineer enjoys surmounting the challenges of building the device to make the discovery.

To some politicians and policy analysts, the preeminent justification for the *Apollo* lunar landing program—enhancing national pride and power—remains an important if not the most important justification for the space program.[10] The changing status of various national space programs as well as changing perspectives within the United States have placed a premium on international cooperation as well as on international competition. The competitive aspects are well illustrated by the race to the moon discussed earlier. Cooperative projects apparently have a value of their own, apart from any scientific findings. The hope is that two countries that work together on something like a space project will develop a basis of mutual trust that will lead to cooperation in other areas.[11] Congress's Office of Technology Assessment cites other benefits of international cooperation, such as expanding research opportunities for American scientists, promoting greater access to information, increasing American prestige by making American science and technology visible worldwide, giving the United States political access to other countries, developing global markets for American aerospace products, and sharing the costs of expensive programs.[12] The unique aspect of this political perspective on the space program is the high value attached to international cooperation for its own sake, irrespective of any particular gains that a cooperative arrangement might produce.

The military, too, has its own perspective on space. In the more extreme forms of this perspective, space militarists view space as the "new high ground" or the "battlefield of the future" and conjure up scenarios that sound like something straight out of the movie *Star Wars*. Although some of these visions are undoubtedly unrealistic, it would be foolish to dismiss the military usefulness of space. We would probably have no arms control agreements at all were it not for reconnaissance satellites. The key role of space in civilian and military communications forces us to examine the possibility that an antisatellite attack could quickly bring America to its knees. So, despite the organizational separation between NASA and the U.S. Defense Department's space programs, the military application of technology to space and the common use of space technology and facilities by the two space programs have to be considered.

Another perspective on space comes from American industry. In some cases, this perspective simply consists of seeing space as just one more arena containing business opportunities. A number of visionaries see space achievement as possibly determining the dominant economic nation of the twenty-first century. This vision is at least partly drawn from the history of European exploration of the earth. Those nations that did the best job of exploring, exploiting, and colonizing the continents they discovered in the sixteenth and seventeenth centuries, most notably Britain and France, were dominant world powers in the eighteenth and nineteenth centuries. It is true that colonialism exploited people as well as natural resources and is thus morally tainted. However, space industrialization will not involve exploitation of the local inhabitants (as there are none) and therefore does not carry the moral connotations that colonialism did.

Peter Glaser of Arthur D. Little, an observer of the space program for many years and a promoter of the idea of solar power satellites, ably described this view of the space program at the 1984 meeting of the American Association for the Advancement of Science:

> The space station opens the door to American industry, and in that sense
> it far surpasses the importance of the space shuttle. It will be crucial to

future U.S. businesses, because the successful development of commercial products and services is essential to exploiting, as the president said in his State of the Union message, the enormous potential of space commerce. I believe that space commerce will be as significant in determining political and commercial relationships in the twenty-first century as developments in aviation, electronics, and computers in the twentieth century, and will determine economic growth, industrial expansion, and international influence.[13]

Glaser's vision assumes that the great potential for space industrialization will be realized in the relatively near future. However, the extent to which this potential is realized will depend on the discovery of sustainable markets for space products and a significant reduction in the cost of doing things in orbit, not just on who builds a space station first.

The most commonly publicized perspective on the space program is that of the "space enthusiasts" who focus primarily, if not exclusively, on astronaut adventures. Space enthusiasts take the analogy between the space program and the European exploration of the earth quite seriously, if not literally. I think they sometimes go overboard in overlooking the important differences between space exploration and global exploration, especially the much higher costs of mounting an expedition in space and the availability of automated equipment that can scout the territory. The analogy between the two programs nevertheless illustrates a number of very valid points. The astronauts appeal to our sense of heroism, our sense of adventure, our sense of approaching and conquering the unknown. James and Alcestis Oberg ably represented this viewpoint in introducing their recent book: "The space experience is an essentially human one, deeply rooted in our history; the technological trappings are merely secondary, however much easier they are to display to the public. Space is being pioneered by our souls as well as our machines."[14]

I've often wondered why it is that the American space program has emphasized manned space flight to such an extreme degree in comparison with, say, the European space program. Ever since the *Apollo* program, putting people in space, on the moon, or on Mars is an end in itself. I wonder if part of the American preoccupation with astronauts is related to the peculiar geography and history of the United States. A widely quoted theme in American history is the importance of the frontier. Most of the time that America has existed, people have been able to move on, to explore new lands, to create new towns and villages, to explore the unknown. The existence of this frontier has powerfully shaped American attitudes and institutions. Natural barriers have been overcome again and again and are perhaps even now seen by Americans as challenges that can be met once more rather than as permanent obstacles to progress. This theme is often identified as the "Turner thesis" because Frederick Jackson Turner proposed it in 1893 in opposition to the prevailing view that European institutions were the germ of everything American.[15] And so perhaps it is our heritage that has given us a generation of heroic astronauts who are the modern counterparts of swashbuckling pioneers such as Davy Crockett and Daniel Boone.

Thus, the space program has more facets than one might assume. Anyone who writes on it brings his or her own set of perspectives and prejudices. I should pause to set forth my position among the various observers and participants in this national venture. Trained as an astronomer, I first became involved in the space program in the early 1970s, as a guest investigator with the *Copernicus* astronomical satellite. As a space scientist, I'm convinced that the challenge of and the prospects for space science are great. Science is an important human endeavor, and there's a lot of science to be done in space.

However, I believe that scientists must not forget that space exploration is a multifaceted, pluralistic effort. One reason I wrote this book is that I kept hearing of these other aspects of the space program in random conversations with NASA managers and wanted to gain a more comprehensive perspective on what we're doing in space. I don't claim that this book is perfectly balanced; a comprehensive, balanced treatment of the entire space program would take thousands of pages. But I seek to give you at least a taste of most important space activities.

It is in recognizing the many faces of the space program that I part company with space scientists like James Van Allen of the University of Iowa, who argued, even before *Challenger*, that we should abandon plans for a space station and concentrate on unmanned activities.[16] I accept the multidimensional aspects of the U.S. space program and recognize that the manned space program, space stations, and other major projects are an important part of the American effort. Many space scientists whom I talk to share my recognition that we are part of a major national effort extending well beyond the piece of the space program that we see in our professional lives.[17] Its direction is governed by many considerations that transcend the narrow boundaries of any of the perspectives I describe above. For example, if the national vision is that NASA should build a space station, I can support such a project even though the science, by itself, cannot justify the cost of building a space station. I share Van Allen's concern that a massive program like the space station might, if it were to become the exclusive focus of NASA's activities, leave everything else that NASA is or should be doing in the dust. But does this concern mean that we shouldn't build a space station, particularly as the concept of a "balanced program" has been endorsed in many high-level policy statements? Surely not, in my view.

The wide variety of NASA activities in space has clouded the meaning of the concept of American preeminence. Because space exploration involves so many areas of human endeavor, it is difficult for any one nation to be preeminent in all areas. The word *preeminent* first appeared in a 1962 speech by John Kennedy: "Whether it [space science] will be a force of good or ill depends on man, and only if the United States occupies a position of pre-eminence can we help decide whether this new ocean will be a sea of peace or a new, terrifying theater of war."[18] This concept has haunted the space program ever since. In the 1960s, it was easy for one nation to be preeminent in space: all(!) we had to do was to be first on the moon.

In the 1980s, it is much harder to decide which country is "ahead" in space. People still try. Soviet astronauts have spent much more time in space than American ones, so in some views, the Soviets have a lead.[19] Space science, the ability to bring large cargoes back from orbit via the shuttle, assembling structures in space, the use of cryogenic engines, and the ability to build reliable, long-lived spacecraft are all areas in which America is first. With space covering such a wide range of activities, how does one decide who is ahead or behind? It's hard to tell. There are other participants as well, each of which has begun to establish its own niche.

## THE GROWING INTERNATIONAL COMMUNITY

Through the *Apollo* years, there were only two major players in the space game: the United States and the Soviet Union. Both of these countries developed a sophisticated capability of launching heavy rockets into earth orbit and beyond, of flying astronauts in space, and of landing spacecraft on the moon. (The Soviet Union did place a number of remotely operated landers on the lunar surface and used several of them to return samples to earth.) Virtually unnoticed were the pioneering efforts of several other countries in this same period. By the end of 1971, France had launched seven spacecraft into earth orbit or beyond; Italy, four; Japan, three; the People's Republic of China, two; Australia, one; and the United Kingdom, one.[20] All the current members of the "space club" of nations that had an independent capability of launching satellites into earth orbit were in place, except for India, which later joined the group. Whereas other nations' activities had been largely overlooked in the American press, the arrival of Halley's comet in 1986, with Russian, European, Japanese, and no American spacecraft to greet it, thrust our international competitors into the national spotlight.

First, consider the Soviet Union, still by all accounts the major competitor of the United States. In the 1960s, both nations were heading for the moon. Officially, the Soviet Union still denies that it was trying to get to the moon, but the evidence indicates otherwise. James Oberg, an authority on the Soviet space program, has shown that the Zond series of lunar probes conducted in the late 1960s were indeed part of a lunar program. The "super booster" rocket that the Soviets were trying to build in the 1960s would have been useful only for a lunar launch. Finally, the strongest corroboration is a report by Library of Congress analyst Marcia Smith of a Soviet response to a query about the nature of one of their spacecraft. The Australian government wanted to know what *"Cosmos 474"* was, as there was some danger that this satellite might hit Australian territory in 1981. *Cosmos* is a standard Soviet designation for unmanned satellites, and the name gives no indication of the satellite's mission. The Soviets responded that it was a "lunar cabin."[21]

During the 1970s and 1980s, the Soviet Union concentrated on a space station, while the American space program diversified and built the space shuttle. Although the Soviets suffered some early setbacks, including the death of three cosmonauts who suffocated because of an airlock malfunction, succeeding generations of *Salyut* space stations were launched into orbit and visited for extended periods of time by Soviet cosmonauts. In some cases, humans spent over six months in space in the Soviet space stations. The activities in these stations apparently covered a broad range, but details have not been extensively reported in the West. No one knows exactly what the Soviet budget for space science is, but most estimates indicate that they spend somewhat more on space activities than the United States does.

A third major player in the game, which ranks well behind the United States and the Soviet Union in terms of total space expenditures, is the European Space Agency, or ESA. In the 1960s, various European countries formed two consortia, one to develop launch vehicles and one to do space research. These consortia merged in 1975 to form ESA. In 1987, ESA consists of thirteen countries (Austria, Belgium, Denmark, France, West Germany, Holland, Ireland, Italy, Norway, Spain, Sweden, Switzerland, and the United Kingdom), with Canada having status as an associate member. In 1984, ESA spent $0.85 billion, compared with NASA's $7.5 billion.

Many member states of ESA have their own space programs as well. For example, Canada developed the remote manipulator arm that has gone up on various shuttle flights. France has played a major role in the Ariane program and in remote sensing. Germany rented the cargo bay on one space shuttle flight for materials-processing experiments.

ESA has forgone the expenses of manned space flight and has concentrated its efforts on space science, applications, and the development of its own launching vehicle, the Ariane rocket. ESA and the CNES (the French space program; CNES stands for Centre National d'Études Spatiales) have established a quasi-private corporation, Arianespace, to manage Ariane. However, relationships between the two institutions and the French government are convoluted, as the CNES retains 34% ownership in Arianespace. Launches are made from a facility in Kourou, French Guiana, South America. Ariane has failed four times in eighteen attempts at launch. The most recent failure, in May 1986, caused Arianespace to redesign the third stage and not to launch for six months. Before 1986, Arianespace and the space shuttle were fiercely competing for the opportunity to launch satellites for private industry, and Ariane was winning an increasing share of the market.[22]

Japan is also a competitor. Although it is usually ranked fourth among the space powers, the country that brought you the Honda and the Toyota is forging ahead quite rapidly.[23] Both launch vehicles and a well-focused program in space science and applications are part of the Japanese space program. The Japanese H-2 rocket has the same capacity as ESA's Ariane 2, and ESA's lead over Japan in the launch vehicle area is indicated by the current availability of Ariane 2 as opposed to the expected availability of the H-2 in 1992. Two separate Japanese agencies develop applications satellites (for communications and meteorology) and scientific satellites. Current discussions of the Japanese space program refer to the possibility of a manned spaceflight program, possibly in cooperation with other countries.[24]

Two other countries, China and India, are currently developing and launching their own rockets. They, too, are competing for commercial satellite launches. China recently scored a point by landing a contract with an American company for launch services. Teresat, an American company based in Houston, has signed a contract with the Great Wall Industry Corporation to launch a communications satellite using the Long March 3 rocket from a site in central China in 1988.[25] Most Americans think of China as the land of the bicycle, and there are conflicts between old and new lifestyles in China. For example, a limousine carrying a team of American aerospace managers from the Martin Marietta corporation became entangled with two horse-drawn wagons in Peking traffic, and one of the horses kicked in one side of the car.[26]

India, Australia, and the many countries that are members of multinational communications satellite consortia complete the roster of nations that are participants in space exploration. India, a third world country like China, has a space program which is in many respects rather similar. India has not sought to develop its own rockets in quite the way that the Chinese have, and will thus suffer more because of the setbacks in the American shuttle program. Australia played a major role in space twenty years ago; it was the third nation to launch a satellite into orbit from its own territory.[27] Now it is trying to get back into the space business as a satellite builder. Completing the list of countries participating in space exploration are the dozens of countries that are members of several large international consortia that have been put together to manage the communications satellite industry. At the present time, the role of these countries is basically limited to being consumers of space technology, most particularly communications.

The various countries in the international space arena are in a curious position of simultaneously competing and cooperating with each other. ESA and NASA probably represent an extreme case. On the one hand, they have been competing viciously for the rights to launch commercial satellites on their various space vehicles. But the two groups cooperated in building instrumentation for one of the most complicated space satellites ever launched: the *Space Telescope*. Cooperation sounds, perhaps, harder than it is, for the people within NASA and ESA who do the cooperating are in space science, and the people who do the competing are in the launch vehicle section of the respective agencies.

*     *     *

NASA was an agency put together to go to the moon, and its managers and Congress had seen its role as primarily an instrument of national pride and power. The major themes that affected the agency during the *Apollo* program were, in retrospect, quite simple compared to the complex political context it finds itself in now. Then it was a simple blunt instrument of global politics, doing battle with the Soviet Union in a race to the moon. The astronauts played a central role as cosmic gladiators or national symbols. NASA managers and the public, too, could focus their attention on a single megaproject rather than being faced with a multifaceted space program which attracts the interest of diverse communities.

The contemporary space program occupies a much more complex world. The different communities that participate in the space program—space scientists, engineers, enthusiasts of the manned space program, supporters of international cooperation for its own sake, the military, and commercial interests—each perceive the space program in a different way. NASA has to balance the competing demands of these various constituencies, many of which seem scarcely aware of the existence of the others. A similar change from simplicity to complexity has taken place in the international arena. What was a two-nation race to the moon has become a multinational arena in which different nations, or consortia, bring their own sets of priorities to their own space programs. NASA is simultaneously competing and cooperating with a number of countries in a number of different areas.

In such a world, American "preeminence" would seem difficult to achieve, if *preeminence* means that America is to be the best in the world in everything. During the 1970s and 1980s, various national space programs pursued divergent goals. For example, the ESA consortium deliberately decided not to pursue manned space flight, but to develop its own expendable rocket-launching capability and to cooperate with other countries on a variety of space science projects. The Soviet Union concentrated on space stations and long-duration space flight. The United States forged ahead on the broadest front, supporting a wide range of space science activities and developing the space shuttle. Even before the *Challenger* accident, America was not able to be the best in the world in everything that happened beyond the earth's atmosphere.

# CHAPTER 3

# Highways to Heaven

The sight of a successful rocket launch, with an important cargo rising upward into space atop a pillar of flame, inspires almost everyone who watches it. Only thirty years ago, it was almost impossible to bridge the few hundred miles that separate the earth's surface from the edge of space. In a few decades, the technology of space transportation had developed to the point where, at least before *Challenger,* it had become almost routine. The space shuttle was seen as being a fancier version of a jet airplane rather than the daring technological leap that it really is.

But one of the key barriers to the widespread use of the space environment by science and industry is the difficulty of launching stuff and people into earth orbit. The *Challenger* accident highlights some of the limitations of the space shuttle and obscures its technological superiority over expendable rockets. Many technological developments of the last decade or two offer hopes (no guarantees) of making it much easier and cheaper to get into orbit and to get around to different orbits once one is there.

There may, someday, be highways to heaven, true "highways" in the sense that they are relatively inexpensive to travel on (though they will never be as easy to use as the interstate highway system). In years past, new roads opened the way to new settlements by overcoming natural barriers; in the United States, these obstacles were mountain ranges. The discovery of the Cumberland Gap opened the way for the settlement of Kentucky and, later, the Middle West. The Oregon Trail, which made use of the South Pass through the Rocky Mountains, was the entrance way to the Far West. The toll for the highway to outer space is still quite steep, a major deterrent to the widespread industrial use of the space environment. But it could become cheaper.

## A ROAD MAP TO SPACE: DIFFERENT TYPES OF ORBITS

The geography of outer space is rather weird, from a terrestrial perspective. Most of us are used to thinking of distances from one place to another that are measured in some kind of distance unit, miles for Americans and kilometers for most of the rest of the world. Within a particular urban area, there is a rough correspondence between the distance between two places and the amount of effort that it takes to get from one to the other. It's true that, in a few geographical areas, there are natural or artificial barriers, like rivers with few bridges across them or urban areas with perpetually impassable traffic jams, but these are unusual cases.

Although one can write down a distance in miles that separates two locations in space, this distance often has very little to do with how hard it is to get from one place to another. There is no air resistance in space, so that once something is given the shove it needs in order to get to the moon, it will just coast on until it gets there. Thus, the best way of estimating the difficulty of getting somewhere in space is the amount of push an object needs to get there. To get from one place to another, its velocity has to change. (The speed of an object measures how fast it moves, and the concept of velocity also takes into account the direction in which it moves.) The change in velocity required to get from one place to another is known to engineers as $\Delta V$. ($\Delta$ is delta, which is the fourth letter in the Greek alphabet and the conventional mathematical symbol for change.) This number, $\Delta V$, is the best measure of distances in space.

For example, when a satellite is put into earth orbit, it has to cover only a few hundred miles of distance to get above the earth's atmosphere, but it has to accelerate to a speed of 8 kilometers per second (8 km/sec) before its orbital velocity will carry it around the earth. The speed of 8 km/sec corresponds to 18,000 miles per hour, or 30 times the speed of sound. The few hundred miles that have to be crossed to get to space are almost irrelevant compared to the tremendous speed required. The Concorde supersonic passenger plane can cross a few hundred miles in about 20 minutes, but it travels at only $\frac{1}{30}$ of the speed needed to reach earth orbit.

How does an orbiting satellite stay up? A satellite orbiting the earth stays in orbit because it falls around the earth, rather than toward the earth. Its motion along its orbit is sufficiently rapid so that the earth's gravitation pulls its path into a circle (or some other geometrical shape) rather than pulling it toward the ground. Satellites in low earth orbit are close to the earth and feel a strong tug from the earth's gravity. Consequently, they have to zip around in their orbits very fast in order to avoid falling into the atmosphere and burning up. A satellite farther up feels a weaker pull from the earth's gravity, and it can and does move slower.

A satellite orbiting the earth several hundred kilometers from the earth's surface is in the nearest location in space, called *low earth orbit* (or LEO). (A list of acronyms [LOA] is at the back of the book.) LEO is the first step toward anywhere else in space, but it is a bigger step than you might think. If you were considering a trip to the moon, for instance, you might think of reaching LEO as taking only one step beyond your front door. The peculiar geography of space means that, in terms of effort, LEO is nearly three-quarters of the way there. In order to reach the moon, you have to escape from the earth's gravitational field, going on an escape trajectory. But it turns out that once you are in LEO you need only an additional 3 km/sec of velocity change to reach this escape trajectory, which will take you past the moon. LEO is, in terms of distances, less than 1 percent of the way to the moon, but once you have reached LEO, you have done 70% of what you need to do in order to reach the orbit of the moon.

The velocity needed to get to most locations in space near the earth, but beyond LEO, is about the same that's needed to get to an escape trajectory. One special location beyond LEO is the geostationary orbit (or GEO), where the communications satellites circle the earth every twenty-four hours. A satellite in geostationary orbit remains fixed from any particular location on the earth's surface.

Now, if you actually want to land on the moon's surface, rather than just get to the moon's orbit, you have to do something else. When you follow an escape trajectory toward the moon, you're traveling pretty fast when you pass the moon. If you want to stop and land on it, another velocity change is required, a deceleration of about 2 km/sec. But if you don't want to stop at the moon and simply want to go on to Mars, you need to work only a bit harder, producing a velocity boost of 3 km/sec. Missions to the outer Solar System vary in their requirements for velocity changes, depending on where you want to go and what you want to do when you get there, but velocity changes of 2–3 km/sec beyond what's required to escape earth orbit are reasonably typical for missions to the inner planets.[1] Considerably higher velocities are required to get to a planet in the outer Solar System or to escape from the Solar System entirely, if you do it by brute force. However, it is possible to get a "gravity assist" by flying quite close to another planet, whipping around it, and changing your speed. Table 1 indicates the velocity changes required to get to various destinations.

It's not just the magnitude of the velocity change required that separates the problem of getting around in space into two rather different tasks, that of getting to low earth orbit and getting from low earth orbit to anywhere else. There are a number of propulsion systems that are much more efficient than chemical rockets; however, they work only in space. These engines provide very low thrust, gentle shoves rather than the tremendous push of a huge rocket engine. Low-thrust engines cannot be used to get to low orbit because a sudden, tremendous push is needed in order to get off the launchpad. But at the

Table 1.   Effort Required to Reach Various Destinations

| Trajectory | Velocity Change Required (km/sec) |
|---|---|
| Earth to low earth orbit (LEO) | 8.6 |
| LEO to geostationary orbit (GEO), escape trajectory, or lunar orbit | 3.8–4.1 |
| Lunar orbit to lunar surface | 2.2 |
| Earth escape trajectory to Mars | 2–3 |
| GEO to Apollo asteroid | 1 or less |
| Earth escape to escaping from Solar System | 12 |

present time, these advanced technologies are only in the stage that NASA planners delicately call "concept formulation," a bit more than mere artists' conceptions but far from being a working propulsion system.

## PRESENT TRANSPORTATION SYSTEMS

Let's turn from dreams of the future to what's actually available now. We basically have one way of getting to space: rockets. Rockets work by mixing a fuel and an oxidizer together, burning them at high temperatures, and ejecting the exhaust gases out the back. Rockets are built in stages, so that a large engine that powers the initial stages of ascent is dropped off once it is no longer needed rather than being carried around as excess baggage. The faster the exhaust shoots out, the better the rocket works, so that there is a premium on developing high-temperature engines. It sounds simple, but building a piece of machinery that actually works at a temperature of thousands of degrees is a rather complex technical challenge.

Two types of rockets have been used in the space program. The simplest to build are the solid fuel rockets. The fuel and oxidizer are mixed together and stuffed into a long pipe with an opening at one end. Once the fuel is ignited, it burns to the end; a solid rocket can't be turned off. Riding a vehicle propelled by a solid rocket is a bit like driving a car with the accelerator pedal firmly fastened to the floor; in each case, it's full speed ahead until the fuel runs out. Because of their relative simplicity, solid fuel rockets are quite inexpensive, and so they were chosen to provide much of the thrust needed to lift the space shuttle into orbit.

In a liquid-fueled rocket, the fuel and the oxidizer are fed into the rocket engine by a complicated network of pipes. The most efficient rockets burn hydrogen and oxygen to make water. However, these elements are gases at normal temperatures, and gases make poor rocket fuels because they can't be stored compactly or pumped efficiently. Rocket engines that use these fuels must be quite complicated, because liquid hydrogen and liquid oxygen have to be pumped through the maze of plumbing to make the rocket work. Liquid-fueled rockets are much more controllable than solid-fueled rockets because varying the rate at which fuels are fed to the engine varies the power in much the same way that an accelerator pedal controls the speed of a car. Before the space shuttle, the only rockets used for manned space flight were liquid-fueled ones. If something went wrong, the rocket could always be turned off, and the crew could bail out.

The apparent reliability of expendable rocket technologies was shattered by a trio of accidents that followed the *Challenger* tragedy. On 18 April 1986, a Titan rocket carrying a military payload exploded just seconds after lift-off from the Vandenberg Air Force Base on the California coast. The Titan, an unmanned rocket, was reportedly carrying a KH-11 spy satellite.[2] These spy satellites usually work in pairs, and it is believed that the Titan disaster left only one of these in orbit. This rocket failure was particularly catastrophic because the launchpad suffered serious damage, too, which made it difficult to launch a replacement Titan quickly.

Scarcely two weeks later, on 3 May, a smaller Delta rocket was launched from the Kennedy Space Center. Disaster struck again, when the Delta lost power, spun out of control, and was destroyed by space-craft controllers. The Delta was carrying a $57.5-million weather satellite destined for geosynchronous orbit. The National Oceanic and Atmospheric Administration (NOAA) was then down to one weather satellite to monitor storm systems in the Western Hemisphere. After that, NASA was apparently left with no way to get anything into space. The gloom lifted somewhat when, later in the year, a Delta launching of two military satellites was successful, at last.

But for the Western world, there was still Ariane, the French rocket that at one time seemed to be merely a Gallic toy, a way of unnecessarily demonstrating that France was not the fifty-first state in the United States, dependent on an American partner for access to space. NASA's troubles had left Ariane as the only available commercial launcher, and companies that had counted on the space shuttle were lining up to buy one of the precious launch opportunities on the Ariane's manifest. But on 31 May 1986, the third stage of the Ariane failed to ignite, and in a now familiar scenario, the rocket and its payload had to be destroyed. (If a rocket carrying explosive fuel threatens to fall on people, controllers have to blow it up, unfortunately destroying the payload in the process.) The Ariane failure was particularly distressing because it was the fourth in eighteen launches and the second one that involved the third stage. The Western world was left without a launch capability.

Despite these rocket failures, one of the results of the *Challenger* disaster will be an increased reliance on expendable rockets for the near future. In 1984, the U.S. Air Force finally won a long bureaucratic battle with NASA and ordered ten Titan rockets for use in launching heavy military payloads. (NASA, like trucking companies all over the world, had sought to get as much cargo as it could on its truck, the space shuttle.) In the aftermath of *Challenger*, the Air Force ordered another thirteen Titans and announced plans to buy at least twelve smaller "medium launch vehicles." NASA itself will be using expendable rockets to launch a number of payloads, particularly scientific ones. American aerospace companies currently manufacture a number of rockets that can launch payloads of different sizes. Before 1986, the companies sold these rockets only to government customers like NASA, NOAA, and the U.S. Air Force.

One consequence of *Challenger* may be the emergence of an American private-launch-vehicle industry. A number of entrepreneurs had sought to compete with both NASA and Arianespace for the opportunity to launch payloads—mostly communications satellites—for paying customers. Before *Challenger*, NASA had done its best to defend its turf and had tried to squelch these companies.[3] The entrepreneurs' task was made particularly difficult because NASA and Arianespace, both of which have access to government funds, became engaged in bidding wars in their competition for commercial payloads.

When President Reagan announced his support for the construction of an orbiter to replace the *Challenger*, he also directed NASA to get out of the business of launching commercial satellites, clearing the way for the private launch industry. The same company that sells twelve medium launch vehicles to the Air Force may well use rockets of the same type to put commercial satellites in orbit. The shift to a private launch industry will be gradual, however; it is still not clear just what will happen to the forty-four contracts that NASA currently has for space shuttle payloads. NASA administrator James Fletcher stated that the agency will try to fulfill its commitments, and many of these forty-four payloads need to be launched on the shuttle because of foreign policy, national security, or other reasons. Whether the private industry can succeed also depends on the number of satellites that companies want to launch. If there are only ten paying customers per year, Ariane may be able to accommodate all of them, making it difficult for an American private launch industry to get started.[4]

Rockets remain a reasonably reliable way to get into orbit, but the technology is really quite old. The next logical step in propulsion technology comes naturally from thinking about what one does in building a rocket. A great deal of effort and quality control is put into manufacturing an extraordinarily complicated plumbing system, which is used for a few minutes, and then the whole thing is just thrown away into the ocean. In principle, a lot of money could be saved if the rocket could be reused. This thought was the basis for the space shuttle, a transportation system that was supposed to slash the cost of getting into orbit.

## THE SPACE SHUTTLE

How did the nation get itself into a situation where a single accident could cripple the space program for a decade? The consequences of *Challenger* will affect a wide variety of people who directly or indirectly rely on routine access to space. The space program is no longer the national frill that critics accused the *Apollo* program of being. International banking, global and domestic businesses, and arms control agreements all hinge on the ability to have the right kind of hardware in the sky. Why is the space shuttle not the inexpensive space truck that it was supposed to be? The answers to these questions, rooted in history, should be important guidelines as the country decides where to go next in space.

In selling the space shuttle in the unfavorable climate of the early 1970s, NASA had to make a number of compromises that have haunted the program ever since. Perhaps it could have been a cheap space truck. But the shuttle, as it was built, is not a truck. It has many useful characteristics, but truly low cost is not one of them.

The first critical decision that affected the future of the shuttle program was made even before the program had been sold to Congress, as NASA had to decide just how big the space shuttle was going to be. Different engineering groups were coming up with designs that were described by Alex Roland as ranging from "a Chevy to a Cadillac."[5] The "Chevy" designs would have been evolutionary outgrowths of the aerospace technology of the time and would have had a modest capability. They would have carried a relatively light payload of 10,000 pounds or so into earth orbit. Such a capability could not have handled something heavy like the *Apollo* spacecraft, which weighed about 100,000 pounds (including the lunar modules), or the Skylab space station with a similar weight. However, it could have handled many of the communications satellites of the time and could have been ready quickly and inexpensively. The Cadillac design resembled the present space shuttle in its lift capability of 65,000 pounds. But it was to be completely reusable (unlike the present shuttle) and would have cost an estimated $12 billion in 1972 dollars.

Another constraint on the shuttle program was economics. The Office of Management and Budget (OMB) insisted that the shuttle be "cost-effective" in some sense. Never before had a space program been subjected to this type of test, and the reasons for its imposition are still not clear. The Vietnam war was consuming the federal budget, and perhaps the Nixon administration had no special interest in the space program. NASA responded to the OMB's request and issued a contract to Mathematica, a consulting firm, for a study. The very existence of this study, preliminary as it was, forced NASA to continue to justify the shuttle as a good economic investment throughout the shuttle program.[6]

Because the shuttle was justified as the cheapest way to get into orbit rather than as a national commitment to a new form of space technology, the shuttle budget was chipped away at in the usual budgetary haggling. A skeptical Congress and Office of Management and Budget set a firm ceiling of $7.5 billion in 1972 dollars as the total cost of the shuttle program. NASA was told that cost overruns could not exceed $1 billion in 1972 dollars.[7] Again, this was the first time that a firm cost ceiling was placed on a space program. NASA can (and has) cut the costs on some space missions by cutting out some pieces, but this is impossible with the shuttle. An earth observation instrument with one of several cameras missing can still do useful work; a space shuttle with one engine missing certainly cannot fly.

With these firm cost limitations, NASA found itself stuck between a rock and a hard place. Realistically, the budget would allow only for the inexpensive "Chevy" design. But the U.S. Air Force, one of NASA's prime customers, couldn't live with a Chevy. The Air Force wanted to launch spy satellites that weighed far more than the 10,000-pound limit of the inexpensive design. They also wanted a shuttle that could maneuver some after it reentered the earth's atmosphere, so that it had some flexibility in selecting a landing site. Under political pressure from the Air Force's friends in Congress, NASA agreed to produce a shuttle that the Air Force could use. The shuttle had the dimensions of the Cadillac design but was only partly reusable, requiring a new external fuel tank and extensive reassembly of the solid rocket boosters for each flight.

The congressional parsimony led to a space shuttle program run on a shoestring. Although the *Challenger* tragedy is generating much criticism of the shuttle design, NASA and its contractors did a remarkable job in getting the shuttle to fly at all under these circumstances. It was delayed for about three years because of problems with the main rocket engines and with the tiles—features of the shuttle system that ended up working reasonably well. But because the shuttle is only partly reusable, one of the major operating costs of the shuttle—the need to pay the ground crew at the Kennedy Space Center to refurbish it before it flies the next time—is far higher than had been anticipated. It's quite possible that many of the design compromises that were made during that time affected the shuttle's eventual ability to fly safely and frequently.

If you read some of the space literature of the early 1970s, the space shuttle seemed like a dream come true. Even discounting some of the hyperbole that seems, unfortunately, to have become standard fare in some of the literature about the space program, the promise of the shuttle was tremendous. Regular and inexpensive access to orbit would allow one to do a number of things in space that would otherwise be impossible: repair satellites, retrieve satellites that are in the wrong orbit, and, most important, try out a number of new ideas without having to invest millions in them.

No one anticipated the tremendous cost of operating the shuttle, of paying tens of thousands of people to work at the Kennedy Space Center for months to get each shuttle ready for launch. Before *Challenger*, NASA was still trying to reduce the cost of each flight by putting great pressure on the launch schedule. Because many of the costs of operating the shuttle are fixed, the more shuttles that are launched each year, the less the cost of preparing each shuttle is. The flight rate climbed relentlessly, with nine shuttles flying in 1985 and fifteen projected in 1986. Before *Challenger*, I had heard reports of an era when the "shuttle would be up there all the time." Allowing for some mild exaggeration, this description would presumably apply to a fleet of five or six shuttles, each of which would be turned around in a month or two.

Another source of launch pressures was the competition between NASA and ESA's Ariane rocket for commercial customers. Although neither NASA nor ESA probably charged customers the full launching costs (including the costs of developing the rocket, maintaining a worldwide network of ground stations, and so on), both agencies were very concerned about being competitive, in cost terms. Still a third source of pressure came from NASA's reputation as a "can-do" agency. When one of NASA's customers requested a modification to a payload or a modified launch date, the agency followed tradition and allowed such requests in spite of the additional effort required to comply with them. The presidential commission investigating the *Challenger* accident, headed by William Rogers, took exception to NASA's flexible attitude and cited it as a cause of the extreme pressure on the work load of the space shuttle program.[8]

Pressures to maintain a frenetic launch schedule have often been cited as a cause of the *Challenger* disaster. Individuals at NASA have vigorously denied this argument, but I find it hard to read the Rogers Commission report and to come to any other conclusion. Evidently NASA, in a misguided effort to make the space shuttle into the space truck that it is not and can never be, created an atmosphere in which launching fifteen shuttles per year, rather than safety, was the paramount consideration.

Has the shuttle program been a success? It was sold on the basis of making it very cheap to send stuff into orbit, as a "space truck."

When you read some books written before the shuttle actually started flying, some of the promises of freight charges of $100 a pound or even less to send stuff into near earth orbit sound like hollow prophecies indeed.[9] It turns out to be surprisingly slippery to figure out exactly what it costs per pound to send stuff into orbit; it depends partly on the extent to which the development cost and the full cost of operating the Kennedy Space Center are included. A reasonable estimate is that in 1970, a Delta rocket could put something into low earth orbit for about $3,500 (in 1985 dollars); the cost of using the shuttle to do the same thing was quoted, before *Challenger,* as ranging from $1,000 to $2,000.[10] After *Challenger,* with a slower launch rate, the cost will go up. Clearly, we are a long way from $100 a pound.

However, as even the skeptics like space historian Alex Roland admit,[11] the shuttle does provide a number of in-orbit capabilities that are simply impossible with conventional rockets. The most dramatic of these is repairing and retrieving satellites in orbit. In April 1984, on the eleventh shuttle mission, the crew grabbed the Solar Maximum Mission satellite, fixed its balky guidance system, and put it back in orbit. Later that year, Dale Gardner and Joseph Allen rescued two communications satellites, which were stuck in a wrong, useless orbit near the earth.[12] The opportunities for doing space science in orbit are much greater on the shuttle than they could be with the use of expendable rockets. Space scientists often have difficulty appreciating this, because 1986 was the first year in which the shuttle was supposed to fly lots of space science experiments. Instead, 1986 was the year in which scientists sat, frustrated, grounded, and wondering whether their project was going to survive at all.

## MODIFIED SPACE SHUTTLES AND THE *ORIENT EXPRESS*

Thus, the shuttle is not the cheap highway to space that it could have been. But the space shuttle design is based on rather old technology. Since the shuttle was designed in the very early 1970s, we have discovered a number of new lightweight but strong materials, not to mention the enormous advances in computer technology. A space shuttle, designed now, would be fundamentally different from the one designed fifteen years ago. Although no one can say for sure how much it will cost to use new technologies to get into orbit, it seems likely that these new technologies can lower the cost significantly.

The National Commission on Space, which recently issued a far-reaching report on the next fifty years of the space program, pointed out that there are two essentially different, but nevertheless complementary, approaches to cost reduction.[13] One approach involves taking existing systems and tinkering with them. A second, more risky approach involves using brand-new technologies, new ideas, and new launch systems. NASA has been supporting research on new space technologies since its inception, but the level of support dropped from a high of nearly $900 million (1986 dollars) during the *Apollo* days to only $150 million at present. Such a small budget particularly limits the ability of aerospace engineers to investigate brand-new technologies.

A good illustration of the first approach of using existing technologies lies in some of the extensions of the space shuttle that NASA is currently exploring. The Shuttle Derived Launch Vehicle (SDLV) would be an unmanned shuttlelike vehicle that could carry two or three times as much payload into orbit as the shuttle could. The solid rocket boosters and a modified external tank would be coupled to a non-recoverable cargo carrier and a recoverable propulsion and avionics module. The avionics module would contain the costly items needed to guide this craft into orbit: computers, guidance systems, controls, and so on. Another way to modify the shuttle is to build a heavy-lift launch vehicle by replacing the solid rocket boosters with more powerful, bigger, fully reusable liquid rocket boosters. Such lift capability would probably be needed if the Strategic Defensive Initiative (SDI, or Star Wars) system is actually deployed. United Technologies felt sufficiently confident of the need for the SDLV that it spent its own money on a design, which it claims could be built in three years.[14]

These new shuttlelike craft might modestly reduce the cost of getting to low earth orbit. One significant component of the cost of a space shuttle launch is the cost of keeping people at landing fields around the world to accommodate one of the many Byzantine abort modes built into the shuttle system. If a single main engine is lost more than 3 minutes 51 seconds after launch, but before 5 minutes 20 seconds, the shuttle can do a transatlantic abort landing somewhere in Africa. From 5 minutes 20 seconds to 5 minutes 33 seconds after launch, the abort mode is to go once around the earth and land at Edwards Air Force Base. The unmanned shuttle-derived vehicle would cut costs both because pilots would be replaced with cargo and because people around the world wouldn't have to be kept on standby. The heavy launcher would still incur operational costs similar to the shuttle's, but the effort that goes into a shuttle launch would result in more pounds of cargo in orbit.

More and different space shuttles will be built by other countries. The French space agency CNES is well along with the development of a mini-space-shuttle called *Hermes*.[15] Unlike the space shuttle, *Hermes* would not launch satellites. It can carry six crew and a cargo of up to 4.5 tons into orbit, in contrast to the 32-ton lift capability of the shuttle. It would be used for a number of purposes, including ferrying crew to a space station, carrying experiments that require human presence in space, refurbishing or repairing satellites, and possibly doing military missions.

The Soviet Union is apparently working on a space shuttle of its own, though little is known for certain about it. *Aviation Week* reported that it is much like the American shuttle in size and shape, though different in important details. The main engines are not reusable and are located on the external fuel tank, so the shuttle is really more like an upper stage launched by a giant expendable rocket. Satellite photos at the Tyuratam launch site show a number of support structures like launchpads and runways, adding to speculation that a Soviet shuttle flight is imminent. There is considerable disagreement regarding when the Soviet shuttle will actually fly, with the U.S. Defense Department predicting flight in early 1987 and other experts, like James Oberg, seeing a Soviet shuttle flight after 1990, if at all.[16]

Much more speculative are the new technologies that could possibly make it very inexpensive to get into orbit. The reason that the space shuttle's operating costs are so high, compared to the operating costs an airline incurs when it flies a jet like the Boeing 747, is that launch preparations are so extensive. A 747 flying across the Atlantic lands, refuels, gets a new crew, and is on its way back to the United States in a few hours. A space shuttle has to be tipped up on its rear end, attached to a giant fuel tank, hitched up to a couple of solid rocket boosters (which themselves have to be put together from a bunch of pieces recovered from the ocean), loaded according to a continuously changing cargo manifest, and then launched in a very complicated, time-consuming, and expensive process. Although the space shuttle orbiter looks like an airplane, it's basically the upper stage of a rocket.

The dream is that of a plane that would take off horizontally, accelerate through the sound barrier, past the Mach 2.2 speed limit of the *Concorde*, and possibly all the way to the Mach 25 needed to reach orbit. (An aircraft traveling at Mach 2 is flying at twice the speed of sound.) Such a plane would require a type of engine, a supersonic combusting ramjet (called a *scramjet*), that doesn't exist yet. Both the American Defense Advanced Research Projects Agency and British Aerospace have been working on designs for such an aircraft, and they at least have some names. President Reagan dubbed the American version the *Orient Express* because it could be used to carry passengers from New York to Tokyo in an hour or two by flying up above the atmosphere (and avoiding the creation of a sonic boom). Another name used for it is the *hypersonic transport,* which emphasizes that it will travel much faster than the supersonic transport. The British call theirs HOTOL, for Horizontal Takeoff and Landing.

Can this dream become reality and, if so, how soon? So far, only design work has been done. The next step is a "proof-of-concept" test, which would show whether a scramjet will work in the way that it's supposed to at relatively low speeds. The wind tunnels that are used for testing aircraft parts melt at speeds above Mach 8, so flight tests would be necessary to show whether such an engine can really provide the boost needed to reach earth orbit. President Reagan mentioned this plane in his 1986 State of the Union message. If adequate ground testing is successful, a flight test by 1992 might be possible. But designs, artists' conceptions, little models shown at aerospace fairs, and the other trappings that accompany a project in its early stages are not a working aerospace plane, and the next few years should demonstrate whether this particular new technology will be able to live up to the high hopes generated for it.[17]

## BEYOND LOW EARTH ORBIT

In principle, rather different rockets can be used to get beyond low earth orbit, because they no longer have to push against atmospheric drag and the strong force of the earth's gravity. But in practice, chemical rockets are still the preferred way to get from low orbit to a higher orbit, usually the geosynchronous orbit used by the communications satellites. A number of companies have developed various upper stages that can be carried into orbit aboard the space shuttle and can then be used to boost satellites even further. Depending on the goal, a choice can be made between McDonnell Aerospace's Spinning Solid Upper Stage, Boeing's Inertial Upper Stage, or Orbital Science Corporation's Transfer Orbit Stage. These are all solid rockets, preferred for shuttle use because an accident leading to the premature ignition of the solid fuel is viewed as rather unlikely. After *Challenger,* NASA canceled its efforts to modify the liquid-fueled Centaur upper stage for use on the shuttle because of the danger to the human crew.

A reusable orbital transfer vehicle (ROTV) is planned for the relatively near future—the early 1990s, according to the National Space Commission's timetable, but realistically speaking, the late 1990s. This vehicle would be used to retrieve satellites either in other low orbits or in geosynchronous orbit. The ROTV could, for example, go and get a communications satellite that needed repair and bring it back to the shuttle or space station, where the astronauts could either fix it on the spot or return it to the ground. The ROTV would use much less rocket fuel in bringing payloads back from geosynchronous orbit if it could use aerobraking to slow it down. It would dip into the upper atmosphere of the earth, deploy some kind of a lightweight parachute to slow it down, and cut the parachute loose when it had slowed to orbital velocity. Related to the ROTV, but nearer at hand, is the orbital maneuvering vehicle (OMV), which would be used to deploy and retrieve payloads that were in low orbits but some distance from the space shuttle itself.

## USING NATURE TO GET AROUND

Using a rocket for propulsion from one orbit to another is simply using brute force. To go from the earth to low orbit, a great deal of force is necessary, but to go from one orbit to another, a small force applied over a long period of time can be just as effective. There are a number of imaginative ways of exploiting the laws of nature in order to make it easier to get around the solar system. One of these, gravity assists from orbits around massive planets, has been used extensively already. Some other ideas currently being tossed around are further from reality.

When a spacecraft makes a close approach to a massive body like a planet, the planet's gravity acts to change the direction of the spacecraft's orbital path. Relative to the planet, the spacecraft comes in at some speed, zips around the planet, and leaves at the same speed, but not necessarily in the same direction. Because the planet is moving in the solar system, it is possible to use the planet's motion to push the spacecraft in the desired direction, in a maneuver that is called a "gravity assist."

An important and substantial use of gravity assists was by the *Voyager* mission, which went past the outer planets. When *Voyager* went by Jupiter, the giant planet's gravity transformed its outward motion along Jupiter's orbital path, speeding its way toward Saturn. A similar boost from Saturn propelled it toward Uranus and later Neptune. Without such gravity assists, it would take much longer to explore the outer solar system. Without gravity assists, a trip to Neptune might take as much as fifty years, but *Voyager* made it in twelve years with a little help.

But must rockets be used at all to get beyond low orbit? One of the neatest, cleverest, and simplest means for getting around in space is the tether, which is a fancy name for a big long piece of string (or, more accurately, some ultrastrong material like Kevlar). Two masses are put in orbit at different distances from the earth. If they were not connected to each other, the lower one would orbit faster and the higher one would orbit slower. But they are connected, and they fly at some sort of average speed. When the tether is suddenly released, the upper mass will be traveling faster than required for a circular orbit and will move to a higher orbit. If the two masses are spun and then disconnected, an even greater boost results, with the separation between the two masses after half an orbit being up to twenty-five times the tether length in favorable cases.

Tethering sounds like black magic, but it's only celestial mechanics. What makes it feasible and useful now is the development of strong, light materials like Kevlar, which allow the manufacture of a tether that's long enough to give a significant boost. The concept has been tested, in a rather different context, in a balloon experiment, where a chemical analysis of the upper atmosphere was made by a gondola lowered fifteen miles below the balloon that carried it. In a vaguely similar joint Italian-American experiment, an instrument on a sixty-mile tether will explore the earth's ionosphere and upper atmosphere in the next few years; before *Challenger,* this experiment had been scheduled for 1987. Shorter tethers were tested during the *Gemini* missions in the 1960s.

Many valuable things can be done with tethers. For example, the space shuttle could be connected to the external fuel tank, and the tank could be used to boost the shuttle into a slightly higher orbit. The big, orange external tank (given the wonderful name of ET) almost, but not quite, reaches earth orbit during a shuttle launch. A minor modification of the shuttle would allow the ET to reach orbit. If the shuttle and the ET are connected with a twenty-five kilometer tether and the two of them are set spinning around each other to build up some speed, the ET can be released at just the right time, falling into some convenient ocean and boosting the shuttle to a higher orbit. The shuttle itself, when ready to go home from a mission to the future space station, could be lowered on a tether and released appropriately, boosting the space station into a higher orbit. The shuttle could carry less fuel, as this maneuver would help it on its way home. The space station would not need to be boosted up in order to counteract atmospheric drag, which will tend to make the station come back into the atmosphere. If the tether includes a long metal wire, the effect of the earth's magnetic field on this long wire would be to generate electrical power.

Another elegantly simple way of using nature for a push around in orbit is solar sailing. A huge, gossamer-thin sheet of aluminum-coated plastic will tend to be pushed around by sunlight, or perhaps by the solar wind. A perfectly reflective mirror will develop a thrust perpendicular to the sail and toward the shady side, permitting some maneuvering. A solar sail can be used to move outward or inward in the solar system. Going inwards is a bit subtle. If the sail is used to slow down orbital speed, the vehicle is sent on a plunge inward to the inner solar system. Slowing down again once the vehicle has reached the proper level will put it into a circular orbit. The solar sail is an old idea, first seriously considered by the Soviet space pioneer F. A. Tsander in 1924.

Tethers are real, or at least close to being real. Solar sailing may remain the stuff of science fiction and technological speculation for the foreseeable future. The report of the National Commission on Space is a very far-sighted document, mentioning a lot of futuristic technologies that may or may not work. The *Orient Express,* HOTOL, tethers, aerobraking, electromagnetic mass drivers, ion engines, robot mining systems, and other largely untested visions of the future are all discussed in a fair amount of detail. The conspicuous absence of solar sailing indicates the obscurity in which it lives in American thinking. My technical source on solar sailing was a rather old Soviet publication.[18] Perhaps it deserves more attention.

A new type of rocket engine may also play a role in getting outward from low earth orbit. If rocket exhaust can be made to travel faster, a more efficient rocket can be built because it takes less mass in fuel to give the payload the same kick. A chemical rocket works by burning something in a chamber with a hole in one side, and the pressure of the hot gases causes them to go out the hole. Limitations on the temperature of the burning gas produced by the types of fuel that can be used and the materials that can be put on the wall of the combustion chamber limit the exhaust velocities that can be achieved. These limitations can be overcome if electricity is used to accelerate particles to high speeds. The particles are ions, or electrically charged atoms; engines that are currently being tested use the heaviest atoms that are practical, meaning mercury or some inert gases. The thrust generated is really quite small (a 26-kg engine generates a thrust of a thousandth of a pound), but it is sustainable over long time periods.[19] If solar cells were used to power an engine like this, they could run for years. Such engines have been built and tested.

The use of electricity to propel spacecraft is also behind the concept of the mass driver, one of the key ideas in Gerard O'Neill's scheme for building a huge space colony in one of the stable points in the earth–moon system, as far away from the earth as the moon is, but sixty degrees ahead of the moon in its orbit. Superconducting magnets lift a vehicle above an aluminum track because the track does not let magnetic lines of force penetrate it freely. An electric motor accelerates the payload to high speeds. The Japanese National Railways tested a magnetically levitated train using a similar scheme in 1972. If there comes a time when it is necessary to lift large quantities of mass off the lunar surface, the mass driver may be built.

The different technologies discussed in this section are at differing stages of development. Any particular technology has to go through a number of stages, which lie between the first germ of an idea and a working device. The particular stages vary from industry to industry; I have borrowed a number of terms from NASA planning documents and have added a few others to come up with the list in Table 2.

Table 2.    Status of Advanced Propulsion Technologies[a]

|  | Ion Engines | OMV | Tethers in LEO | Tethers to GEO | ROTV | *Orient Express* | Solar Sailing |
|---|---|---|---|---|---|---|---|
| Ideas |  |  |  |  |  |  | ● |
| Concept formulation |  |  |  |  |  |  |  |
| Concept definition |  |  |  | ● | ● | ● |  |
| Proof of concept tests |  |  |  |  |  |  |  |
| Preliminary design |  | ● |  |  |  |  |  |
| Detailed design and development |  |  | ● |  |  |  |  |
| Flight tests | ● |  |  |  |  |  |  |
| Operational system |  |  |  |  |  |  |  |

[a] Abbreviations: OMV—orbital maneuvering vehicle (one low orbit to another); ROTV—reusable orbital transfer vehicle (LEO to GEO); LEO—low earth orbit; GEO—geostationary orbit.

Any technology starts with an idea. One then has to develop the idea somewhat, to work out the numbers, and at some point, if appropriate, to perform a proof-of-concept test. Will a scramjet really work? A proof-of-concept test might, under appropriate circumstances, require the tester to build a smaller model of the device. A proof-of-concept test for the idea of a microwave oven might mean putting a small vessel containing water in front of a microwave generator to see if it got hot. Sometimes a proof-of-concept test is unnecessary, if the basic physics behind the idea is clearly understood, but usually the test is needed. Until a proof-of-concept test is done it won't be known whether an idea will work at all. When it is known that the concept works, one can proceed to designing an actual device, which then must be flight-tested before you know how much it will really cost to operate.

The table summarizes the stages of development of various technologies. Although all of them may seem futuristic in an era when space travel means ascending into the heavens atop a pillar of flame, some of them are better developed than others. Where I could, I took the status of a particular technology directly from NASA's *Space Systems Technology Model*.[20] The ion engines are built and small models have been tested; they could be used in a real spacecraft with some confidence, though someone would have to be paid to develop a detailed design of an engine that was large enough to do something with. The orbital maneuvering vehicle and the use of tethers in low orbit are reasonably well understood. The other technologies are more speculative. Just because something like the *Orient Express* was included in a presidential State of the Union message doesn't mean that it will work.

*    *    *

Four failures in 1986 grounded the American and European space fleets. It is difficult to look beyond these accidents and tragedies to the future. The *Challenger* disaster in particular made it clear that the space shuttle was not a truck, although financial considerations forced it to masquerade as one from the moment of its birth. These accidents are horrible tragedies, particularly considering the loss of human life.

They also meant setbacks for the program, but they will not stop it in its tracks. The space shuttle and the expendable rockets of the Western world are beginning to fly again.

There are a number of technologies in various stages of development that could make the promises of the shuttle—low cost and routine access to space—a reality. Only the scramjet engine applies to the first step of getting into space, that of reaching low earth orbit. Once one is in low orbit, then, in principle, one has the option of using some really efficient propulsion systems instead of chemical rockets. If we could reduce the cost of getting into space to the level of several hundred dollars per pound, this reduction would generate enormous changes in the level of interest in space and the variety of things we could accomplish there.

# Summary of Part I

The *Challenger* explosion in January 1986 was the most serious setback to the American space program in its thirty-year history. It tragically killed seven people, smashed NASA's apparent invincibility to bits, led to lots of muckraking in the popular press, and showed that the space shuttle from earth to orbit was dramatically different from the Eastern Airlines shuttle carrying airplane passengers between Washington and New York City.

*Challenger* came at a time when the American space program was facing a number of other challenges. The space program had grown very multifaceted. Overseas competitors had maintained and strengthened their efforts. The Soviet Union's space activities had progressed at a steady pace through the 1970s, largely involving activities in earth orbit but also including some planetary missions like the one to Halley's comet. The European Space Agency, Japan, and a number of other countries had developed significant space capabilities of their own. Instead of a simple two-horse race, Americans were confronted with a multinational arena in which we were sometimes cooperating and competing with another nation at the same time. The public's perception of NASA was increasingly at odds with reality.

Despite the *Challenger* disaster, the opportunities beckon. A number of new technologies offer some hope—not promises, but hope—for opening the new high road to space in the way that major mountain passes opened the road to the American West. It's hard to look beyond the gloom of the *Challenger* disaster and contemplate these new frontiers. But anyone contemplating what we should be doing in space in the next fifty years has to look beyond the short run—beyond the next five to ten years, when launch opportunities will be scarce. We can go into space, and we may be able to get there quite inexpensively in the twenty-first century. The next question arises: What do we do in space once we're up there? The next two parts of this book will focus on past space accomplishments and future plans.

# PART II

# The Practical Uses of Space

Many of the most exciting moments in the space age have been based on looking down on the earth, rather than looking up to the stars. Astronauts consistently report that looking out the window is one of the most enjoyable ways of spending time in orbit. Very early in the space game, in the early to mid-1950s, the first photographs of the earth's curvature from space gave us a glimpse of what could be seen from the "high ground." Our first views of the whole earth, obtained in the late 1960s, generated hopes that our perception of humanity would change in some fundamental ways. Would we realize that we really live in one world, and that the national boundaries that divide one group of people from another are mere artifacts with no underlying reality? Could humans understand that we live on a tiny spaceship orbiting in the cosmos, and that we can never throw anything away but only put it somewhere else?

The hopes of peace lovers and environmentalists have not been realized; apparently, human nature is intractable enough so that the mere sight of a green globe suspended against the stark blackness of interplanetary space is not enough to change human ways developed for generations upon generations. However, the orbital vantage point has found another, more practical use as the key to the first commercial uses of space. Space satellites can function as the hub of a communications network, an excellent platform from which to photograph the earth, or as part of a global system for search and rescue on the open seas. Some uses of earth-watching satellites are rooted in pure science as well, for the global view we can obtain from space presents scientists with a golden opportunity for understanding how our planet works.

# CHAPTER 4

# Communications Satellites

The classic example of the large-scale use of space for practical purposes involves communications. Orbiting satellites are a key part of many communications networks, both civilian and military. People living in remote areas, far beyond the reach of telephone wires, can talk to the rest of the world by satellite at quite modest cost. These satellites receive and retransmit anything from telephone conversations to television programs. The communications satellite corporation, COMSAT, has made a profit since 1967. Dishes for direct reception of TV signals dot the landscape in many areas of the nation.

Since 1964, the number of global telephone circuits has increased from a few hundred to tens of thousands. The insatiable desire of people to talk to each other, to watch TV programs, and to send information from one computer to another halfway around the world shows no signs of slacking off. I believe that the communications industry can meet the challenge of providing ever stronger ties to link our global village together. In the year 2000 and beyond, the comparatively small communications satellites that we now have may well be replaced by giant antenna farms transmitting hundreds of signals along very narrow pencil beams of radio radiation.

For most people outside the communications industry, the impact of these changes will be very subtle. Slowly and gradually, many of us are beginning to think of places across the ocean as places that really aren't so far away, or so alien, at all. The network news organizations can bring us live pictures of events around the world, and so Olympic triumphs and hostage tragedies are no longer things that happen to people who are just names in a newspaper thousands of miles away. As long as the industry can keep the busy signals off the phone lines, the average person will become aware of this technological revolution only when he or she realizes that a phone call to England really involves very little effort.

## THE GLOBAL VILLAGE

Communications is the first and, so far, the only area in which the space program has actually earned a financial profit. Although profit isn't the universal dirty word that it was ten years ago, some people may still wonder why profit is important in the space program. The motives of human adventure and curiosity are powerful ones and may—to some of us—even justify megaprojects like space stations and missions to Mars. But as our ambitions become larger, the cost of space activities becomes greater, and it is important to demonstrate that, in some areas at least, the space program has produced a substantial return on the investment we have put into it. The net earnings of the COMSAT corporation are not the best way to measure the gains that humanity has got from the space program. However, it is an easily understandable way to demonstrate to a large number of enterprising people that putting hardware in orbit isn't just game playing, no matter how much fun the game may be.

Generations of Americans have been brought up with the best telephone system in the world. A hundred years ago, it was "Hello, Central? Get me Kalamazoo!" Now, by pushing a bunch of buttons, almost any American or Western European can reach anyone else in the world, instantaneously, with no waiting. Evening news programs can contain live reports from around the world. When Prince Charles married Lady Diana, the entire world attended the wedding. The blood and gore of Vietnam made their way quickly to American television sets, bringing the tragedy of war home to us in a graphic way that had not been possible before.

The business world, too, is affected by the communications industry. Worldwide networks of computers sharing data via satellite make it possible for multinational businesses to exist. I find it really hard to imagine the world a century and a half ago, when it took months for letters to go from Los Angeles, which was then a small Indian pueblo and Spanish mission, to New York.

Because this worldwide communications network functions so flawlessly, and because using it is so cheap and easy, most of us are probably unaware of the tremendous technological and organizational achievement that has transformed the earth, from a communications standpoint, into a global town. The system hardly ever breaks down and doesn't cost very much, and so most of us don't even know it is there. But think how difficult it would be to set up a network like this without complex technology. Currently, *Intelsat 5A,* a communications satellite first introduced in 1980, can carry twelve thousand telephone conversations at once. Nearly a dozen such satellites can be in use at one time, and so the global communications system, as it is currently configured, can carry something like a hundred thousand phone conversations at once around the globe. Suppose you were to try to install such a communications capability using the technology of seventy-five years ago, when all phone conversations went by wire. You would have to bury tens of thousands of cables under the oceans, and to keep fixing them when they were cut by fishes and other marine critters. It simply couldn't be done. As recently as 1964, there were only 128 telephone circuits in submarine cables linking the United States with other countries.[1]

## HOW COMMUNICATIONS SATELLITES WORK

Arthur C. Clarke, a far-sighted young British Royal Air Force (RAF) radar officer, invented the communications satellite concept in 1945, the concept underlying our global communications system, well before the space age began. Clarke later became well known as a popular science writer, as a science fiction writer, and as the creative genius behind the science fiction movie, *2001.* Clarke was one of the early members of the British Interplanetary Society, a group of early British rocketeers who met to exchange ideas, and so he was thinking about space even then. His critical role in the genesis of the space age was in inventing and popularizing a way in which space could be used for practical purposes.

What seems to be the first written description of a commercially viable space venture is provided in a memorandum to the Council of the Interplanetary Society, dated 23 May 1945, a document that is now in the archives of the National Air and Space Museum. At TELECOM '79, a conference of the communications satellite industry, Clarke spent the better part of a morning autographing posters consisting of this memo. The guts of the present system, the twenty-four-hour orbit for satellites, is in it. Clarke's friends in the Interplanetary Society encouraged him to go out and publish his ideas, and once he received clearance from the RAF censors, in July 1945, he submitted a slightly expanded version of the memo to *Wireless World,* which printed it in October 1945. Later on, he described this idea in more detail in some of his books.

In his reminiscences,[2] Clarke pointed out that the fact that a low-orbit satellite zips around the earth fast and that a high-orbit satellite meanders around more slowly had been known for hundreds of years. It was first published in 1687 in Isaac Newton's *Principia Mathematica,* the physics treatise that developed and synthesized most of our current ideas about how things move when they are pushed. What Clarke did in 1945 was the epitome of creative genius: putting two pieces of information, ideas from completely different fields, together in a unique way.

It was making the connection between the twenty-four-hour orbit and a possible practical use that required some vision, and it is vision that distinguishes the leaders from the plodders. Clarke knew that all communications systems have some central switching stations, which receive signals from one place and send them somewhere else. In 1945, all these switching stations were on the ground, giant switchboards in basements in big cities, with armies of telephone operators. What if one of these switchboards could be put up in the sky, in orbit, where anyone with the right kind of antenna could have access to it without stringing wires all over the globe?

By itself, this second idea seems like an irrelevant dream. Why put a switchboard in orbit? But Clarke, seeking a way to use space for something practical, foresaw the unique possibilities of a space-based communications system. A worldwide system, with a few orbiting switchboards and lots of hardware on the ground, would be most attractive if the machinery on the ground was cheap. But just think about what the sun does during the course of a day, or what the moon does, or what the stars do. They move across the sky in rather complicated paths, paths that change with the time of year. A fancy, expensive antenna would be needed to track the orbiting switchboard as it whirled around the earth, traveling rapidly across the sky. Was there any way for a space communications system to use cheap, easily built ground stations?

The solution to this communications dilemma came from the pages of Newton's *Principia,* as translated by Clarke's creative young mind: Put a satellite in the right orbit, and perhaps you could get around the need for an antenna that has to track it as it moves across the sky. If it's in an orbit that is too low, it circles the earth faster than the earth rotates, and it moves toward the east when you look at it. If it's in an orbit that's too high, as the earth's natural moon is, it moves very slowly, and it moves toward the west. (The moon, like the sun, rises in the east and sets in the west.) And so there had to be a middle ground, where the satellite wouldn't move at all, from the perspective of an antenna fixed to the earth's surface. The laws of orbital mechanics, first set forth by Newton in 1687, tell where this special orbit is. A satellite 22,300 miles (35,880 kilometers) from the earth's center takes exactly 24 hours to orbit the earth once. An antenna on the ground would take exactly 24 hours to spin around the earth's axis because the earth rotates. Thus, from the perspective of someone on the ground, the satellite wouldn't be moving. If the satellite were in orbit over the earth's equator, it wouldn't move north or south during the course of a day either. Thus, once you pointed your antenna at any particular communications satellite, you wouldn't have to move it again, ever, as long as the signals you wanted to get came from that particular satellite. This special orbit is actually called a *geostationary orbit,* abbreviated by NASA, which loves acronyms, as GEO, for *G*eostationary *E*arth *O*rbit. Sometimes you hear people talk about the Clarke belt, just another name for the GEO.

By the time Clarke wrote many of his space travel books in the early 1950s, he had fleshed out his vision so well that his concept of a communications satellite system became more complete and salable. He envisioned the basement warrens of communications centers being transformed into orbiting space stations, where armies of telephone operators would live, work, and sleep in geostationary orbit. If you wanted to call your cousin who was touring in Outer Mongolia, you would pick up your phone, which would be hooked up to your rooftop antenna, send a signal up to the orbiting switchboard, and a signal would be sent to the antenna perched beside your cousin's yurt.

The system almost, but not quite, worked out that way. The explosive growth of the computer and electronics industry overtook Clarke's original concept, and human operators are no longer needed, in space or on the ground, to handle the vast majority of telephone calls. As a result, people don't have to live in the orbiting switchboards, which is just as well. (Any time people are put in space, it costs a lot more money, not to mention the psychological costs of such isolation.) The first communications satellites were much simpler ones in low earth orbit. *Echo 1* and *2,* launched in the early 1960s, were giant aluminum-covered plastic balloons in low earth orbit. They were the most visible of all satellites because their large size and reflective surface made them easy to see from the ground. I still remember trying to puzzle out the constellations as a child and coming across these bright ''stars,'' which soon revealed their nature as orbiting satellites when they moved off toward the east, progressing on their ninety-minute circles around the earth. In July 1962, *Telstar 1,* the first civilian satellite to receive, amplify, and retransmit signals, was launched into low orbit. But a year later, *Syncom 1,* the first geostationary satellite, marked the beginning of the use of the geostationary orbit, a trend that has continued ever since.

The transformation of Clarke's idea into reality was an exceedingly tough job. John R. Pierce of Bell Laboratories found out about Clarke's idea and persuaded Bell Labs engineers to do some of the gory calculations needed before a credible proposal for a communications satellite could be made. Clarke credits Pierce with being as much the father of the communications satellite industry as Clarke himself is. An additional difficulty was in selling the idea that space could be used for something practical, and here, Clarke's writing skills came to the fore. He was one of a few of the rocket pioneers who had the talent and the inclination to write books and magazine articles publicizing the idea of space travel, seeking to make it respectable as well as fascinating. Clarke reminisced, "My efforts to promote and publicize the idea may have been much more important than conceiving it."[3]

In the 1960s, Marshall McLuhan first introduced the term *global village* to describe how electronic communications could transform the world from a bunch of isolated societies into a single, close-knit human family. Neither whole-earth pictures nor a flotilla of communications satellites has, so far, completely overcome the barriers that past history, language differences, and cultural and religious traditions place in the way of a fully united humanity. The technology can only build and pave the streets of the global village. The world's people must choose whether to walk on them or to remain shut up in individual countries, like fearful city dwellers locking themselves in apartments after dark. As time goes on, more and more people are taking advantage of satellite communications. Individuals, communities, and nations are taking the first few hesitant steps along these global highways in the sky, the radio links provided by communications satellites.

Before satellite communications, there was little use of even taped video material from locations elsewhere in the world. If an American TV network wanted to show film from Queen Elizabeth's coronation in 1952, they had to fly the film across the Atlantic and could not use it until hours or days later. (The telephone cables available at the time would not allow the electronic transmission of video.) Fifty years ago, in the 1930s, millions of Ukrainians died of starvation—deliberately induced or accidental—in Soviet Russia. No one cared until recently. One can't just blame the absence of information on the closed Soviet society, because blurred, out-of-focus pictures of corpses on the streets of Kiev did appear in American newspapers. These fuzzy pictures had little impact.

In contrast, in the mid 1980s, the use of video clips or live broadcasts from widely separated locations has become commonplace. In some cases at least, these efforts at communications have brought people closer together. Video clips of Ethiopian kids with their bones and bellies sticking out stimulated American relief efforts. The global rock concert "Live Aid" raised $50 million to help, partly because of American concern but largely because technology enabled audiences around the world to experience the concert simultaneously.

Global communication has been cited as an influence on policy decisions, particularly in a democracy like the United States. The network news coverage of the Vietnam war, which brought bloody battlefield scenes into America's living rooms, has been credited with developing the popular hostility that brought an end to this far-flung military venture. Would hostage crises be such a major preoccupation if we couldn't see live interviews with the hostages and their captors every night? In both the 1979 Iranian and 1985 Lebanese crises, many commentators argued whether the TV network news should give the story so much coverage, but there was no debate on its impact. The level of interest in the Olympics increased dramatically in 1964, when *Telstar* first brought live coverage of the sporting events into people's living rooms. Now we can all share in the triumphs of the athletes. Humans around the world can genuinely admire the performance of Mainland Chinese gymnasts as long as they can overlook the biased commentary of some sportscasters.

## COMMUNICATIONS SATELLITE TECHNOLOGIES

The current global communications system has, at its heart, several dozen orbiting communications satellites, sometimes called *birds*. Most of these are owned and operated by Intelsat—the International Telecommunications Satellite Organization—and are used for overseas phone calls and television broadcasts. Intelsat is owned by member nations, in proportion to the amount of traffic that they generate. In 1985, Arabsat, the first international consortium of countries other than Intelsat launched its satellite on a shuttle mission, with a Saudi Arabian prince participating as an astronaut. Arabsat is a regional system covering twenty Arab countries, with headquarters in Riyadh.[4] A number of geographically large countries (such as Canada, Indonesia, and the USSR) have launched their own communications satellites, and others lease circuits on the Intelsat system. Individual companies, such as RCA and Satellite Business Systems (SBS, owned by Aetna, Comsat General, and IBM) in the United States, have their own satellites. The U.S. military operates a communications satellite system. Future NASA scientific satellites will no longer communicate directly with ground stations but will relay their signals through a tracking and data relay satellite (TDRS, pronounced *Tee-driss,* if you want to use the acronym). Clearly we've gone a long way beyond the three satellites that Clarke first envisioned.

Each satellite contains a number of transmitters, or transponders as they are called. The transponder sends a signal down to the earth; the part of the earth that can receive the signal from a particular transponder is called its "footprint." In some cases, as with the TV channels in the United States, the footprint is quite large, covering the entire country. But in other cases, clever antenna designers can accommodate the wishes of particular customers who might, for example, want a footprint to be confined to a particular country. The satellite receives signals from antennas on the ground (the "uplink") and then rebroadcasts them back to earth (the "downlink").

The ground components of the system, called "earth stations," are antennas that send signals up to the satellites on a very narrow beam, as well as antennae that receive the signals. In the case of most phone communications, the antennas are owned by a phone company like AT&T or MCI, which then redistributes the messages through the domestic telephone network. In 1982, the largest global system, Intelsat, included some three hundred earth stations in some 146 countries. Private individuals can also own their own earth stations, usually for use as television receivers.

## TELEPHONES

In terms of such things as total traffic carried, most of the impact of communications satellites has, so far, been on ordinary telephone communications. In 1964, there were only 678 telephone circuits that crossed the world's oceans. Making a phone call overseas was a very big deal, and the cost was hundreds of dollars in today's money. Twenty years later, in the mid-1980s, forty thousand voice circuits girdle the globe, sixty times as many as there were earlier. Most important for those of us who use the telephone, the cost has dropped to only 5% of the cost of an overseas phone call in the 1960s.

Communications satellites have done more than just make long-distance phone calls cheap. There are some parts of the world where it would be essentially impossible to set up a phone system at all. Take the state of Alaska as an example. Books like John McPhee's *Coming into the Country* can perhaps give you a feel for how large and remote it is.[5] Although some might find it romantic to homestead hundreds of miles from civilization, life in such places can really prove quite a hardship without contact with the rest of the world. And when one considers trying to integrate this state with the telecommunications network of the rest of the country, problems become extreme with a state where telephone lines would have to extend hundreds of miles across forbidding mountain ranges and icy rivers. You cannot drive to Alaska's capital, Juneau, from elsewhere in the state. You simply could not string a network of telephone lines across Alaska, though you could probably link the two major cities.

The answer for Alaska was communications satellites. In the 1970s, the state decided that it would use this technology to connect every remote village in the state to a statewide telephone network, also giving every Alaskan in a village access to television. Many of these villages have only twenty to thirty people. Extending any kind of telephone wire to such a village would clearly be uneconomical.[6] In Alaska, satellites now carry something like 70% of the communications within the state. Not only are satellites cheaper than stringing telephone wires, but the small, isolated villages simply would never have telephones if it were not for the satellite capability.

The Inuit (or Eskimos) in the eastern Canadian Arctic have taken a similar step in connecting the fourteen communities in and around Baffin Island by satellite. Letters could take as long as five weeks to go from one community to the next. The Inuit have a tradition of reaching decisions by consensus, and it was impossible for these fourteen communities to act in a united, effective way in dealing with the territorial government. The Baffin regional council spent $4,300 for a computer for each village. Ron Mongeau, head of the Baffin Regional Council, is quoted as saying that this technology will have the "long-term impact of changing the sociology of the North,"[7] presumably by allowing the Baffin Islanders to retain some of their traditions amidst the onslaught of modern civilization.

Third World countries have taken advantage of this technology in much the same way that Alaska has. These countries do not have their own space capabilities; nevertheless, they have joined the space program by buying the satellites and paying countries with a launch capability to put them in orbit. A country like Indonesia, containing thousands of islands, would have to spend lavishly on undersea cables in order to bring telephone service to each island.

## TELEVISION

A more visible impact of communications satellites has been on television, and on the big and not-so-big dishes that are now appearing in suburban areas as well as in homes in the wide-open spaces. The use of satellites in television is really nothing terribly new. Network television programs have been distributed nationwide over satellites for years, where your local TV station picks up the network feed from the satellite and retransmits it into your area. One reason that we're limited to so few broadcast networks (like ABC, CBS, NBC, and PBS) is that there are very few available VHF television channels (those are channels 2–13), and in crowded parts of the United States, there's room for only a few VHF TV stations. (*VHF* means very high frequency, referring to a particular frequency of the electromagnetic wave that carries the TV program.) The remaining part of the dial on your TV set, ultra high frequency (UHF, channels 14–82) covers a frequency of radiation that doesn't carry as clearly or as far as VHF, and that therefore has proved less successful at penetrating the home market.

Radio and TV signals are a form of electromagnetic radiation, traveling electric and magnetic fields that push electrons around in your antenna and produce an electric current that, when amplified, becomes the program that you watch. Signals from different stations are distinguishable because they have a different frequency or, in other words, the electrical fields flipflop more times per second. Frequency is measured in hertz, or cycles per second, and radio and TV signals cover wavelengths that vary from tens of kilohertz (1 kilohertz = 1,000 hertz = $10^3$ hertz), through the megahertz range (1 megahertz = 1 million hertz = $10^6$ hertz), up to the gigahertz range (1 gigahertz = 1 billion hertz = $10^9$ hertz). For those who are interested, the frequencies of some types of radio and TV radiation are given in Table 3.

Table 3.    Frequencies of Radio and TV Signals

| Band | Frequency | Use |
|------|-----------|-----|
| AM radio | 500–1600 kilohertz | Commercial radio |
| VHF | 60–88 megahertz, 104–140 megahertz | Commercial TV, channels 2–13 |
| FM radio | 88–104 megahertz | Commercial radio |
| UHF | 300–3000 megahertz | TV, channels 14–82 |
| C-band | 6 gigahertz (uplink) 4 gigahertz (downlink) | Communications satellites |
| Ku-band | 14 gigahertz (up) 11–12 gigahertz (down) | Some present—and many future—communications satellites |

The cable TV industry works in much the same way that the broadcast industry does, with one important difference. Your friendly local cable TV company, just like your TV station, owns a number of twenty-foot-diameter antennas that pick up signals from the various satellites. But instead of rebroadcasting these signals in the various TV channels, the cable company sends the signal out through wires that may or may not pass by your home. These cables are not limited to carrying one signal in the way that a TV broadcast is, and so, if you get cable TV, you can watch as many different channels as your cable company is willing to give or sell you. In principle, you could get any of the 107 currently broadcast channels, networks that can give you programming that runs from Jerry Falwell to Playboy. Most cable systems pick a dozen or so channels to offer you, some for an extra charge. The limitation on cable TV is that, for the system to work for you, your home must be passed by a cable. The fraction of American homes that are passed by cable TV wires has grown explosively from a very small one at the end of the 1970s to nearly half by the mid-1980s.

Both cable and network TV systems have an intermediate link between your house and the satellite: a local TV station or cable company. Until recently, these links justified their existence because the signals from the satellite could only be picked up by a large dish-shaped antenna. These antennas used to be quite expensive, costing tens of thousands of dollars. In contrast, the spidery TV antenna on the top of your house that picks up the signals rebroadcast by your local station costs ten or twenty dollars (if you put it up yourself) or a few hundred dollars (if someone else installs it for you). It made sense to put up with the mild inconvenience and expense of using an intermediate link, as few people were willing to spend the better part of a year's salary on a TV antenna.

What made the difference in the mid-1980s was the development of technology that made much smaller and cheaper antennas to pick up the satellites directly. These antennas look like dishes. They reflect all of the radio waves that hit them to a focal point, where a little gadget is suspended above the center of the dish. Amplifiers and cables and other electronics then pipe the signal to your TV, and then you see "Dynasty" or "Dallas" or whatever. In the 1970s, you used to need a big dish to collect a strong enough signal from the orbiting satellites. Specifically, in 1978, the Federal Communications Commission wouldn't license a ground station unless it was sixteen feet across. In the 1980s the amplifiers and other electronics that take the received signal, boost its power, and deliver it to your TV set improved considerably. Another plus has been that the transponders, the devices on the satellite that broadcast the signal, are now more powerful, broadcasting 8½ watts of power, not just five.[8] By the mid-1980s, someone living in the middle of the United States needed only a seven-foot dish to produce a decent signal.[9]

Consequently, a big change has taken hold over rural America. It used to be that, in a remote area, there wouldn't be enough TV viewers around to support the cost of a broadcast TV station. A station that is watched by only a few hundred people can't charge very much for advertising and can't recover the capital costs of building the big receiving antenna and all the other gear that are needed to start a local TV station. But now, someone who wants to watch TV doesn't need an intermediary: the local TV station or cable company. Earth station costs plummeted from a hefty $6,000 in 1982 to about $2,000 in 1985. At first, people in the industry thought that the market would be confined to remote areas. But the business has grown, and a trade magazine claimed that one million people owned backyard or rooftop antennas in 1985.[10]

This technology seems to fulfill a variety of needs for different people.[11] Ed McMahon of the "Tonight Show" (also host of "Star Search") cited the vast variety of available entertainment, far more than the channels commonly offered by cable networks.[12] Others enjoyed the opportunity to see the "uncensored" feeds directly from the networks; you don't just see an anchorman materialize in front of you, but you can watch behind the scenes while he yells at another reporter, tells dirty jokes, and does anything banal enough to seem absolutely fascinating when a TV personality does it. People with international interests, such as students at Columbia University's Russian Research Center, can watch foreign newscasts directly. Last summer, after a conference in Kiel, West Germany, I spent some time in Denmark, where one can see news broadcasts originating in Denmark, Sweden, East and West Germany, and Russia. As my knowledge of Russian, Danish, and Swedish is virtually zero, I couldn't understand much of what was said, but I still found that seeing the same news through different patriotic filters makes our own prejudices seem very clear.

But enthusiasm for rooftop earth stations seems to go beyond just a desire for variety. Many people are frustrated by these intermediate links in the TV business, arbiters of the public taste that unilaterally decide just what's to go out over the airwaves. Buy your own dish and no one will decide whether or not you can watch the Playboy channel, Russian news, Spanish-language programming, or whatever. The independent, individualistic streak in the American national character has always chafed at being spoon-fed. Buy your own earth station, and you are no longer beholden to the top brass of the media who have decided that the American public can't take anything stronger than the intellectual pabulum that is the staple of prime-time TV.

And so the communications satellite business has wormed its way into our culture, becoming as much a part of our lifestyle as hot dogs and automobiles. Reach out and touch someone: you couldn't do it without those birds in geostationary orbit, which permit everyone in the country to call their mothers on Mothers' Day. What next?

# THE FUTURE

The communications industry is planning for explosive growth. INTELSAT, the world's phone company, planned that international phone traffic will double from 1983 to 1988 and double again from 1988 to 1993. Historically, demand has grown by 25% per year for ten years, which corresponds to a doubling in capacity every three years, so the planners are anticipating a mild slowdown. The domestic communications satellite industry, which only began in 1976, is anticipating similar growth.

The phrase "doubling in five years" actually seems rather bland, but think what it would mean in the absence of technological change. If these trends hold up, it means that, for each of the satellites in orbit in 1983, one more would have to be orbited by 1988, and another two more by 1993. Consider all of the hardware currently in place in the business, built up over the twenty years since *Telstar* first flew in 1963. All that stuff would have to be built again in the next five years, and then again in the next five. That's how exponential growth, growing by a given percentage each year, works.

It sounds like a big job, but in fact it's even bigger than it might seem. Communications satellites can't be bunched too close together because a ground station has to be able to tell the difference between a signal from one satellite and a signal on the same frequency from another satellite. Radio antennas can't focus radio waves absolutely precisely, and the smaller the antenna, or the higher the frequency of the radio radiation, the more difficult it is to focus the radio radiation. Because all communications satellites must be in the same geosynchronous orbit, circling the earth over the equator once in twenty-four hours, there's a limited amount of space or orbital arc to put them in. In heavily populated parts of the world, slots on the orbital arc are hotly contested, and they are allocated to individual countries by the World Administrative Radio Conference (WARC). So, even though space is limitless, there is a part of space, the geosynchronous orbit, that is as precious to the communications industry as midtown Manhattan real estate is. The illustration shows a number of communications satellites tightly bunched together in the sky over Texas. The two

Seven satellites in geosynchronous orbit are visible in this photograph by Paul Maley taken from a site in southern Texas. The streaks are made by stars as the earth rotates under them; the dots are the satellites, which don't move relative to the earth and remain stationary relative to the camera. (Copyright by Paul D. Maley.)

satellites that are right next to each other in the picture are occupying the same slot in geosynchronous orbit, operated by the same company. Their transmissions must be coordinated so as not to interfere with each other.

For the next ten years at least, some modest extensions of current technology will probably enable us to cope with the gargantuan appetites of the world's communicators. Earth stations are being built better. There are currently plans to decrease the spacing between satellites from the current four degrees to two degrees. (Two degrees is about four times the size of the full moon, or the diameter of your thumb if you hold it at arm's length.) A two-degree spacing will allow us to double the number of satellites. Still more capacity can be added by having the satellites receive and rebroadcast signals at different frequencies. Although the number of different radio frequencies that can be used is limited—you don't want taxicab and police dispatchers to disturb your watching the World Series—there are some frequencies that are allocated to the satellite business that are still not fully used. Specifically, many communications satellites, and all of the ones carrying TV programming, operate in what's called the *C-band,* with a downlink at 4 gigahertz. Future plans call for an increasing number of satellites to use the higher frequency Ku-band with a downlink at 10–12 gigahertz. Capacity can grow not just because one has an extra frequency band to work with, but also because different channels can be crammed a lot closer together in the higher frequency band. So, if the projection that use will double every five years for the next decade or so holds up, these modifications in existing technology can carry us through the mid-1990s at least.

These minor technical changes will not be completely painless or unnoticeable, particularly to someone who owns his or her own earth station. Any irregularities or holes in an earth station antenna must be smaller than the wavelength of radio radiation being picked up. C-band radiation, the kind currently in use, has a wavelength of 7.5 centimeters, or 3 inches. A wire mesh dish with 1-inch holes has irregularities that are smaller than the wavelength of the radiation, and it works well in the C-band. But if all the TV signals are carried on the Ku-band, which at 11 gigahertz has a wavelength of 2.7 centimeters, or a little over 1 inch, that same mesh dish will be so-so at best, according to current tests.[13] Mesh dishes have become quite popular because they look nicer and the wind blows through them, so they don't have to be so rigidly built to hold up in a windstorm.

An additional problem with using the higher frequencies is that rain can absorb some of the radio signal that comes down from the satellite.[14] The cable and broadcast industries can probably figure out a way around this, as rain doesn't completely wipe out the signal and a major operator can invest in a backup way of getting the signal through. But again, a homeowner with equipment that just barely worked when it wasn't raining would be out of luck. Developing nations are also afraid that, when it's their time to get some slots in the orbital arc, they will be stuck with only the higher frequency channels. Many developing nations are in the tropics, where it rains heavily and where a high-frequency communications channel might work far worse than a lower frequency channel. Their problem is that they may not be interested in spending millions of dollars of hard currency on satellites right now, but they don't want to be shut out of prime low-frequency assignments in the orbital arc forever.[15]

A big issue for the future is scrambling. In early 1986, a mainstay of cable television, HBO, decided to scramble its signal. An earth station owner has to buy a black box to decode the signal and see a real program instead of bleeps. An irate earth station owner, calling himself Captain Midnight, jammed the HBO signal on the evening of 27 April 1986, interrupting HBO's programming with his own irate complaints about HBO's decision to scramble its signal. This issue generated a fair amount of political controversy in 1986, and it remains to be seen whether the protests will bring results.

One of the early hopes of the cable industry was that it would produce universal access to the broadcast end of the TV industry, so that anyone with a bright idea and a few bucks could put a program on. This hasn't happened yet. Although local cable companies do have outlets for locally produced programming, very few people seem to watch it, and interest has never mounted for it. I think this trend is likely to continue, for the near future at least, not because the technology doesn't allow local programming to exist, but because people's tastes in TV programs have evolved to the point where the cost of programming is very high. Rough ballpark estimates of the cost of TV programming run around $100,000 per hour. When I first saw that figure, I found it hard to believe, but I think I can now understand how the costs escalate. Recently, I was a guest on the "Newton's Apple" science TV program, broadcast by PBS. Host Ira Flatow, a dozen technicians, and I spent a full day in a TV studio to prepare a six-minute segment. We ran through our segment on comets time after time so that the final product could be the polished, apparently perfect performance that the public expects. When you pay fifteen people for a day's work and get six minutes of air time out of it, you have spent real money.

An additional concern of people in the communications satellite business is competition from fiber optic cables. These long strands of glass can carry many telephone conversations as modulated light waves. Communications satellites will always be the only way to go for people living in remote areas, and they will probably remain an important factor in the transoceanic communications market. However, much of the satellite traffic between major cities on the same continent could, in principle at least, be lost to cable. The threat of technological competition from a ground-based system has stimulated a good bit of research on ways to make satellite communications less expensive. If fiber optics takes hold, the competition for slots in geosynchronous orbit could slacken.

## SPACE COMMUNICATIONS AND PLAIN LIVING

But what difference is all this going to make to the average person? How will our lifestyles be affected? As an example, take the videoconference, which may replace meetings in some circumstances. A number of companies have established a network of rooms where groups of people can get together and meet face to face without spending hours on airplanes. Alexander Giacco, president of the Hercules Chemical Corporation, can use videoconferencing to talk to many of his managers, who can no longer hide from headquarters.[16] J. C. Penney built videoconferencing rooms in shopping malls with Penney stores, partly for their own use but also for rent to paying customers.[17] The Aetna and Travelers insurance companies in my hometown of Hartford, Connecticut, led the move toward videoconferencing, and United Technologies and Emhart (which also have headquarters in the Hartford area) have used the systems as well.[18] Reports on the future of videoconferencing are mixed, with aerospace trade publications like *Aviation Week* being quite optimistic and more general publications such as *Forbes* remaining guarded.[19] I think that *Forbes* is being too cautious; I've seen this technology used in too many different places. But it still has its limitations; I suspect that it's particularly useful for those meetings where participants have had previous face-to-face contact.

Many workers find themselves part of a national network of people doing similar things who communicate only by phone. People selling a similar product out of different company offices, librarians who staff interlibrary loan departments across the country, coauthors of scientific articles, committees, and many other groups "meet" only by phone. It's a new custom. Ten years ago, many friends were very surprised that I had collaborated in several research projects and coauthored articles with a number of other astronomers, none of whom I had met face to face until our paper was published. Now this type of coauthorship seems almost routine.

The world is thus becoming more democratic in the sense that one no longer needs to be part of the elite, located in a major national center, in order to be involved in work that has a national impact. In my own work, I find this quite important. My colleagues and I are all part of "invisible colleges" that are tied together by the phone lines and by Federal Express. A growing number of workers are telecommuting, that is, spending most or all of their working day at home, using a computer to send their work to the main office. Perhaps one of the most extreme examples of telecommuting in my business is "remote observing," where an astronomer who is assigned time on a telescope located on a remote mountaintop does not physically travel to the telescope but remains in his or her home office, watching the data come in on a screen and talking to technicians on the mountain. Friends of mine who have tried this remote observing tell me that this kind of telecommuting works surprisingly well.

*    *    *

Communications satellites are the only piece of the space program that make money. They work because they are in a special orbit—in which a satellite orbits the earth at the same rate as the earth turns on its axis. Thus, they appear to be stationary from the perspective of someone on the ground. Inexpensive ground antennas allow anyone access to this global communications highway in the sky, and backyard dish antennas have sprung up across America as millions of TV viewers have tapped into the system.

Someday, this new technology will have a fundamental impact on how we live. Already, the impacts are there, even if they are not obvious. They are an integral part of the information society that we are rapidly moving into.

# CHAPTER 5

# The Orbital High Ground

## Weather Watching, Spying, and Star Wars

An orbiting satellite can photograph, send radio signals to, or shoot at objects anywhere on the part of the earth that the satellite can see. This unique vantage point not only is the key to the communications satellite industry but also offers a number of other practical applications. Anyone in the world who can receive TV signals from anywhere, even people in the boonies who can pick up only the tiniest TV stations, can see satellite pictures of the earth on their evening weather forecasts. These pictures and other global weather data are a key part of weather forecasting, not just visual trimmings added to the evening news. Satellite-based navigation systems can pinpoint the location of any military or civilian vehicle in the part of the world that the satellite sees —half the globe from geosynchronous orbit.

But the phrase "high ground" in the title of this chapter usually has a military meaning; a place with a commanding vantage point, where defenders since time immemorial have held out against vicious attack, dropping objects on enemy heads. The civilian uses of space have their military counterparts in the fields of communications, meteorology, and navigation. The space program was started for the purpose of photographing enemy territory from orbit. We would probably not have any arms control treaties at all were it not for the spy satellites that can monitor the other side's activities without requiring on-site inspection. The most controversial proposal for using the orbital high ground is the Reagan administration's Strategic Defense Initiative

(SDI), popularly known as Star Wars. Somewhat related to SDI, yet different, is the antisatellite weapons program, which has also experienced accelerated development in recent years.


## WEATHER WATCHING

Two and a half years after *Sputnik,* when the American space effort was barely under way, the United States took the lead with the launch of the first meteorological satellite, named *TIROS* (Television and Infrared Observation Satellite). The most obvious product of weather satellites like TIROS and its successors is a photograph of the world, showing the storm systems more distinctly. Up to then they were visible only as wiggly lines on weather maps. The picture on your local evening TV weather forecast is now familiar and expected, illustrating one of the many ways that space activities have insinuated their way into our daily lives. These photos and other ways of using satellites to make global meteorological measurements have played a vital role in advancing the science and art of weather forecasting.

Many other satellites quickly followed the first TIROS. Acronyms changed with dizzying speed in the early 1970s as a variety of satellites with different capabilities were developed and as the weather satellite program searched for a bureaucratic home. ESSA satellites were named for the Environmental Science Service Administration, a now-defunct government organization. We then had ITOS (Improved TIROS Operational Satellite), in 1970, renamed NOAA-1 after the National Oceanic and Atmospheric Administration, the agency that then took over the operation of these satellites and still runs them. Another name you may run across is NIMBUS, a designation that, in most if not all cases, is reserved for a satellite that is a test bed, where instruments intended for future satellites used for weather observation and remote sensing, the two areas discussed in this chapter, are evaluated. (Calling it a ''test bed'' means that NASA does not guarantee that the instrument will work.) Currently, the weather satellites are called GOES, for Geostationary Operational Environmental Satellite.

## Weather Satellites and Weather Forecasting

Just what do these weather satellites do? The answer to this question is more complicated than you might think. The weather satellites aren't just weather-forecasting machines. Some background information about the general problem of weather forecasting is needed if you are to understand just where the weather satellites fit into the bigger picture.

It was much easier to plot the route of the *Apollo* spacecraft to the moon than it is to forecast tomorrow's weather because the force acting on a traveling spacecraft—gravity—is much simpler than the multitude of forces and processes that govern the weather. The condition of the earth's atmosphere is influenced by heating, cooling, evaporation, rain, snow, sleet, hail, wind, ocean waves, the interaction between wind and ocean waves, mountain ranges causing updrafts, and more. Although many of these processes that drive the global weather machine are quite complex and poorly understood, the force of gravity is described by a very simple mathematical expression.

It's not that weather forecasting is impossible, only that it can be quite complicated. Even describing what a good weather forecast is is not easy. Take the question of whether it will rain tomorrow. In some parts of the world, making such a prediction isn't hard at all. If you live in a desert climate like Tucson, Arizona, you will be right almost all the time if you predict "no rain." If you understand the climate cycles, you can do even better. Someone who predicted "no rain" for southern Arizona, and who predicted "chance of rain" for the rainy month of August, would be right 90% of the time—considering that a forecast of "chance of rain" is correct whether it rains or not.[1] Consequently, a "90% correct" batting average is rather poor in Tucson.

Before the space age, a weather forecaster gathered a few observations from weather stations on the ground and used them to make a rather crude weather map. Intuition, a few looks out the window, and other tricks in addition to the map were the basis for a guess of what the local weather would be. Weather prediction was an equal mixture of science and black art.

In the past twenty-five years, two related developments, satellites and computers, have brought much more real science to the weather-forecasting business. Satellites can provide pictures of the weather all over the earth, not just where weather stations happen to be. Computers can take these billions of observations, use what limited knowledge we have of how the sun, air, land, and sea interact, and calculate what will happen next. If the jet stream is pushing weather systems from west to east, and if it's raining in Chicago on Tuesday, you can calculate when the rain will move eastward to New York. (But even in this example, it isn't quite that simple. The effects of the Great Lakes and the northern end of the Appalachian mountain chain on weather systems can add significant complications.)

In some parts of the world, even this simple type of weather prediction was impossible before satellites. For example, storms that hit Britain blow in off of the Atlantic and could strike without warning unless one is lucky enough to hear a report from a ship in the Northern Atlantic. When a computer is used to do things right, satellites become even more vital. The computer forecasts require information on the entire earth, and satellites are the only way to provide it.

And is the space-age weather forecasting any better than relying on the old black art, on using the feeling in your bones to tell when it will rain? (There may be some scientific basis to that old tale, as arthritic joints often feel more painful when the air pressure is low, acting as a kind of barometer.) The answer seems to be that, in some respects, the improvement is significant, and that, in other areas, it is only marginal. Will it rain tomorrow? Modern forecasters can beat someone working with climate averages (as in the Arizona example above) by 24%–30%.[2] This improvement is really rather modest because, in some parts of the world, you can beat the climate averages just by looking at a weather map.

Satellite observations may play a larger role in the accuracy of long-range forecasts. This is one area in which weather prediction probably has a great economic return. A correct forecast of whether it will rain tomorrow will save some people from wet heads, canceled picnics, rained-out baseball games, and soggy parades. But a forecast of the rain trend for the next month, or two, or three, can help farmers save their crops (and everyone's grocery bills) because they can irrigate or harvest them at the right time. Temperature forecasts would allow utilities to have enough fuel to handle the 5 P.M. surge when every office worker goes home and turns on the air conditioner full blast, without having to waste money stockpiling tons and tons of coal as a contingency measure. A somewhat dated estimate by meteorologist Jack Thomson, quoted by Congress's Office of Technology assessment,[3] gives a figure of $13 billion (1971 dollars, equal roughly to $40 billion 1985 dollars) for the total economic cost of bad weather, reflecting such things as crop losses, lost workdays, and so on. About $0.5 billion (1971 figures, equivalent to about $1.5 billion in 1985 money) could have been saved if adequate 30- to 90-day forecasts had been available.

One and a half billion dollars is a great deal of money—about one-quarter of NASA's annual budget. Is this amount, which is only an estimate, based on fact? I suspect that when we do develop a reasonable 90-day forecasting system, the savings could be even bigger. Take the electric utility business as an example, just to see whether these numbers make sense. The cost that the electric company pays to produce (or buy) electric power varies according to a particular company's mix of power sources: hydroelectric (from dams), coal, oil, natural gas, nuclear, and others. When a power company has an unexpectedly high demand for electricity, it has to buy more expensive power, either by scrounging for fossil fuels in the marketplace, by buying it from some other power company that has a surplus, or by firing up some spare generators that can produce expensive electricity in an emergency.

Better long-range weather forecasting could save utilities and their customers money by allowing them to rely more on cheaper power sources. Americans use 600 billion kilowatt hours of electricity every year. If weather forecasts could cut the power cost by 10%—this is just my estimate, but I think it's a reasonable one—we would save about half a penny on each kilowatt hour, or $3 billion per year (about half of NASA's annual budget), from this sector of the economy alone.[4] Add in other weather-dependent sectors like agriculture, transportation, and tourism, and we come up with big bucks from better weather prediction. The gains would probably be spread out so much through the economy that it would be virtually impossible to identify the savings, but if one industry alone can produce a billion-dollar dividend from space, and by itself come close to justifying NASA's budget, we can see why even the most persnickety bean-counters could find long-range weather forecasting interesting.

I picked the utility business as an example because the dollar savings can be estimated fairly easily, but in many ways, the biggest payoff from good long-range weather forecasts will be in human rather than dollar terms. The global village created by satellite communications has brought the starving millions in Africa into American living rooms. I suspect that most of us would find the elimination, or at least the alleviation, of the problems of large-scale famine to be a morally laudable goal, more exciting than cheaper electricity. Satellites can't eliminate hunger, as shortfalls in the harvest are only part of the African hunger problem. Political systems that discourage efficient farming (or any farming at all), high birthrates, and poor transportation and distribution systems play a role, too. But better long-range forecasts will certainly help.

## Storm Warnings: Vital Weather Data from Space

There is one area of weather forecasting in which satellite data are vital. In many parts of the world, sudden, severe storms that are born and grow over oceans could, before the space age, strike continental shores without warning. Read just about any British yachting book, particularly those written before satellite forecasting had much of an impact, and you read of stories where a quiet Sunday afternoon sail turned into a nightmare when a gale that no one knew about blew in off the North Atlantic Ocean. Fishing off the west coast of Europe is a hazardous occupation for much the same reason. Space has been used to produce accurate and timely storm warnings for Europe in a way that was impossible before 1957.

The shape of the United States is rather unusual in that most of us live in areas where the storms that affect us have hit somebody else first. Thus, for Americans, severe weather could have been predicted to some degree before satellites came along. Most people, particularly in the tropics, are less fortunate. Storms come in from the sea, and the warning signs of hurricanes that appear in cloud patterns are confusing at best. Even in the United States, the sophisticated modern weather satellites can spot the formation of a squall line that will, later in the day, spawn tornadoes and hailstorms. The precise tracking of hurricanes can also allow for better storm warnings.[5]

## Predicting Weather Prediction

Can improved technologies make some of the dreams of reliable short- and long-range weather forecasts come true? I've mentioned three major areas of weather forecasting: short-range prediction, long-range prediction, and severe storm warnings. In short-range prediction, I don't see much in the way of dramatic improvement because the weather patterns that determine whether it will rain, snow, or do neither in a particular place are small potatoes on a global scale. We'll need to understand the overall circulation of the atmosphere first. With severe storm warnings, again, I don't see much in the way of changes. Satellites have meant a great deal, but they're doing now just about all that they can do.

The greatest potential for some fundamental improvements in weather prediction, improvements assisted in a crucial way by measurements from space, is in the area of long-range forecasting of general global trends. Although it's true that there hasn't been much improvement in the recent past, the past is probably not a good guide to the future. In many ways, especially when it comes to understanding the oceans, we're just beginning to learn what measurements we can make from satellites if we're clever. We can do more than just take pictures of clouds; using available technology, we can measure global temperatures, winds, ice and snow cover, vegetation on land and sea, and ocean currents. Technological advances will gradually increase the precision of these measurements and will gradually lead to forecast improvements. Thirty-day weather forecasts, which can be seen in some newspapers, may well become more than the amusing irrelevancies that they seem to be now. I think we'll be able to do more than look at the widths of the brown stripes on woolly bear caterpillars in order to determine whether it will be a cold winter or not.

A cloud on the horizon, at least for the immediate future, is the political status of the weather satellite program. NASA's charter and general outlook dictate that it should be involved in exploration and research, not in the operation of a network. The problem seemed to be resolved when the National Oceanic and Atmospheric Administration (NOAA) took over the operation of the satellites, but then it became a political football when the Reagan administration tried to cut back the NOAA system from two satellites, one serving as a backup, to one. Given prior cutbacks, NOAA apparently had to reduce service to Alaska and Hawaii.[6] As in many other areas of the space program, NASA and other government agencies are facing challenges in finding a suitable organizational structure for a program that involves space but that, for one reason or another, shouldn't be run by NASA. In such a situation, a program can find itself without a guardian in a sea filled with budget-cutting sharks, and it can be curtailed again and again until it almost disappears beneath the waves. The existing monitoring program will undoubtedly continue, but additional resources must be used to create a usable archival data base and to pay the salaries of scientists to analyze it.

## NAVIGATION AND RESCUE ON THE OPEN SEA

Communications satellite technology and the jet airplane have meant that the Atlantic and, to a smaller extent, the Pacific oceans are no longer the barriers to travel that they once were. The vastness of the ocean, which makes the task of the oceanographer very difficult, is scarcely apparent to the transatlantic airplane passenger. When you're inside an airplane your surroundings are much the same, whether you're waiting at the departure gate or winging your way across the cold, frigid North Atlantic. The extent of the world's oceans is much more apparent to someone crossing by boat. (I was lucky enough to cross the Atlantic on a passenger ship with my family when I was fourteen years old, and I still remember an impression of endless waves. Get up one morning, and you see waves. The next morning, the crew may tell you that you're five hundred miles further along, but you still see just waves. How do you know where you are? There are no road signs in the open ocean.)

But just as satellites can shrink the globe by providing a worldwide communications net, just as they can put meteorologists in a position to understand the global picture, so the satellite view of the entire earth can make the business of navigation and rescue much easier. In the mid-1980s, the U.S. Navy continued development work on its NAVSTAR satellite system, called the *global positioning system,* or GPS. It provides an inexpensive, very accurate method of determining the position of any ship or airplane (or automobile) anywhere on the earth. The GPS supersedes various earlier satellite navigation systems that were less accurate but still serviceable. Most of the satellites needed for the GPS were in orbit prior to *Challenger,* but a few more remain to be launched to allow the system to work anywhere on earth at any time.

Satellite navigation works because you can measure (directly or indirectly) the length of time that it took a radio signal to get from the satellite to you, and thus determine how far away you are from the satellite. Light and radio waves travel fast—they can circle the earth seven times in a second—but modern electronics can measure time intervals that are much shorter. If you know exactly when a satellite is transmitting a radio signal, and you can time your reception of it, you can tell how far from the satellite you are. If you can compare the times that signals from two different satellites arrive, you can tell how much closer you are to one than to the other, and you can obtain some idea of your position. Do this with a sufficiently large number of pairs of satellites, and you can locate your position relatively easily. There are also ways in which you can measure how fast you are traveling relative to the satellite (by using what's called the *Doppler effect,* where the frequency you detect is increased or decreased by your motion relative to the transmitter).

The general idea of measuring radio signals to determine one's position at sea was first implemented during World War II with transmitters located on the ground. These vintage navigation systems could determine a ship's position only within a mile or so. Worse yet, they simply didn't cover large parts of the earth's surface. As with communications satellites, it is the global vantage point of an orbiting satellite, compared with the smaller range of a transmitter located on the ground, that makes the difference. You can see a satellite in high orbit from half of the earth's surface.

Determining your position from these signals can't be done manually—you can't use headphones and a stopwatch to take advantage of NAVSTAR. With the NAVSTAR system, the signal from each satellite contains codes that provide information on the exact shape and size of the satellite's orbit. A microcomputer has to unscramble the signals and codes, using the difference between the arrival time of signals from different satellites and the frequency of the transmission as received on the vessel or airplane to show where you are. If you're on a military vessel, you can take advantage of the precise positioning system (of course, it's called the *PPS*), which can determine your position to within 15 meters (about 50 feet) horizontally and 30 meters (about 100 feet) vertically. If you had a big enough house, you could almost determine whether you were in the front hall or the back bedroom. Most of us, meaning nonmilitary users, have to be content with the standard positioning system (SPS) where the errors are 100 meters (330 feet) horizontally and 162 meters (500 feet) vertically.

Reports of military demonstrations of satellite navigation indicate that even these publicly stated estimates of the capabilities of the system may be underestimates. Undersecretary of Defense for Research, William Perry, testified in 1981 about a demonstration of a helicopter landing at the Yuma, Arizona, proving grounds. The pilot landed the copter within a few feet of a big *X* marked on the runway, using only the global positioning system to give his position both vertically and horizontally. Such a system, had it been available for general military use, could have saved Operation Eagle Claw, the disastrous attempt to rescue the Iranian hostages in 1980. Three helicopters had to be abandoned, one because its gyro and radio navigational aids failed and two because of equipment failure perhaps caused by the flight path, which took the helicopters through *haboobs,* desert dust storms. The GPS would have allowed the helicopter pilots to go around the *haboobs.*[7]

It's comparatively easy to use the global positioning system. The cost of buying and installing a receiver and interpreter for the military system ranged in 1985 from $5,000 to $45,000. As I write this, receivers for the nonmilitary system cost about as much, but the price will probably drop as a mass market develops. At one time, Congress required civilian users to pay a fee of $370 per year to use the system (you had to rent a decoder for that amount to pick up the signals), but even this minimal fee was eliminated in 1984 after the Korean Airlines disaster in August 1983, when a civilian airplane veered hundreds of miles off-course, encroached on Soviet airspace, and was shot down by the Soviet military; nearly three hundred innocent airline passengers were killed.

Space-age services to mariners and pilots go beyond navigation. Suppose you're on a boat or aircraft somewhere, and despite the storm warnings, you end up being lost at sea. If you're near a coastline, someone may hear your desperate SOS message and come to help, but if you're in the remote polar regions, or in the open sea far from land, the chances that someone will hear your message in time to be able to provide help would seem slim indeed. However, the weather satellites can come to the rescue. They have, along with all their other gear, a system called *SARSAT* (*S*earch *a*nd *R*escue *Sat*ellite), which can pick up the distress signal and relay it to a rescue coordination center. Despite the deep freeze in U.S.–Soviet relations in the mid-1980s, America and Russia are cooperating in the SARSAT program, along with Canada, France, the United Kingdom, Sweden, and Norway. At present, the only satellites with the necessary equipment are *NOAA-8* and two Soviet weather satellites; plans exist to extend the system to more satellites.[8] In the first months of operation, SARSAT saved forty lives.

A rather clever idea has made the SARSAT scheme even more useful. Because of the Doppler effect,[9] you can tell whether a satellite is moving toward or away from a distressed vessel that is transmitting an SOS. Since 1972, all U.S. ocean-going vessels have had, by law, to carry an EPIRB (emergency position-indicating radio beacon), and since 1970, all aircraft carry an ELT (emergency location transmitter), which transmit signals at frequencies of 121.5 and 243 megahertz. The satellite can measure the frequency of the distress signal, see how it changes as the satellite orbits the earth, and so detect the position of the downed vessel rather precisely. *NIMBUS 6* tracked the *Double Eagle I,* the first balloon to cross the Atlantic and provided critical position information when the balloon fell into the ocean near Iceland and the crew needed to be rescued.

There are still a few bugs in the technology, with perhaps the biggest problem being false alarms. Something like 97% of the calls for help are not real distress signals, in comparison with an 8% rate of false alarms for U.S. fire departments. Some of the false alarms make good stories.[10] A Swedish woman was cleaning house when someone—she or a child—inadvertently pushed the transmit switch and found out about it only a few hours later when helicopters started looking for a plane crash near her house. Another false alarm led the Civil Air Patrol to a helicopter with traces of marijuana concealed beneath a haystack.

## SPY IN THE SKY

The Cold War was at its peak in the 1950s and the early 1960s. Russian tanks rumbled through the streets of Budapest in 1956 to invade Hungary, and Western Europeans feared that the same tanks would soon be invading Bonn and Paris. The Berlin Wall, built in 1961, added to the siege warfare atmosphere. A former boss of mine built a fallout shelter in his house when he moved to Delaware in 1962, just before the Cuban missile crisis. Some basically reasonable people carried battery-powered radios around all the time so that they would be the first to know about the feared Russian invasion. Many regarded a nuclear confrontation between the United States and Russia as a certainty; Nevil Shute's book *On the Beach* graphically illustrated the outcome of World War III: the extinction of humanity.

The horrors of Hiroshima and Nagasaki were not as distant a memory then as they are now. Many leaders had great hopes that the nuclear arsenals could be dismantled, but negotiations through the 1950s always foundered on the same rock: verification. The ragged history of arms control agreements has made it clear that both sides need to be able to ensure that the other will live up to the terms of a treaty. One of the few cases of successful voluntary compliance with a treaty involved the Geneva Protocol of 1925 banning the use of chemical and biological weapons, which was generally adhered to during World War II, though there have been allegations of subsequent Soviet violations.[11] But this treaty involved weapons which have never been of major strategic importance, unlike nuclear bombs.

So we need to be able to monitor the activities of the other side if we are to be able to make a sturdy arms control agreement. Before the dawn of the space age, however, verification of Soviet compliance with an arms control agreement was essentially impossible. The United States could set up radar installations in countries on the Soviet border, but we could never be sure that these installations would always exist. For example, the fall of the shah of Iran in 1978 led to the loss of two American monitoring sites near the Soviet border. These sites were strategically located, quite close to the Soviet Kasputin Yar launch site and the Baikonur cosmodrome (which are just west of the Caspian Sea and east of the Aral Sea, respectively).

The radar installations can provide only limited information because they can only intercept radio communications and track the paths of rocket launches. Flights over the Soviet Union by long-range aircraft like the U-2 were also possible, but these involved a violation of Soviet airspace and were counter to international standards as well as to Soviet law. The operational difficulties of this method of information gathering were graphically illustrated when U-2 pilot Francis Gary Powers was shot down in May 1960 and had to withstand a public trial. In the late 1950s, many words were spoken on the importance of on-site inspection, but this proved unacceptable in the Soviet Union, which does not allow its own citizens, much less foreigners, to travel freely about the country. I'm not sure that the U.S. military would be particularly open to the idea of letting Russians wander around our military bases either.

Space has come to the rescue. Space reconnaissance is invariably first on the list of modern ways in which we can verify an arms control agreement.[12] Since 1960, over two hundred American reconnaissance satellites have flown over the Soviet Union, taking detailed pictures. The Soviet Union has launched over twice that number to spy on the United States. The existence of these satellites was officially denied, or not confirmed, until the late 1970s, but it was widely known well before then that satellites could perform vital tasks like counting airplanes. In the 1970s, arms control negotiators worked out three major arms limitation agreements: the Antiballistic Missile (ABM) and Strategic Arms Limitation Talks (SALT) treaties of 1972, and the later SALT II treaty of 1979, which was, before 1986, adhered to but never ratified by the United States. I doubt that any of these treaties could even have been seriously thought about without the space verification capability.

The central role of satellites is illustrated by some of the language of the SALT II agreement. This unratified but influential treaty defines aircraft types in terms of functionally related observable differences (FRODs), things you can see in satellite reconnaissance photos that indicate whether a particular plane "can perform the mission of a heavy bomber." (The arms control community apparently can't resist acronyms either.) Bomb bay doors on an airplane might, for example, be a FROD. In some cases, the treaty required a nation to put FRODs on particular types of aircraft in order to allow verification procedures.[13]

Reconnaissance satellites are generally launched into very low orbits. Because there's still a bit of atmosphere in these low orbits, atmospheric drag causes them to spiral slowly into the ground. The first of the current generation of satellites, officially named *Keyhole-9* (*KH-9*) but popularly dubbed *Big Bird,* was launched in 1971 but had a lifetime of only fifty-two days. The short lives of these satellites, along with the need to have a lot of them if you are to have anything approaching comprehensive coverage of the Soviet landmass, is the reason that both sides have launched so many of them.

These satellites take pictures of military installations and can return the pictures in one of two ways. In the 1960s and 1970s, satellites returned film in at least two reentry vehicles, which were recovered in midair by aircraft flying out of an air base in Hawaii. The current *KH-11* satellites convert the pictures into digital radio pulses and radio the images down to the ground.

The key capability of any reconnaissance satellite is its resolution, roughly defined as the size of the smallest object it can photograph. The exact resolution of current systems is highly classified, but it has been variously quoted in the open literature as somewhere between 3–4 inches[14] and 6 inches,[15] on an average day. The clarity of a particular picture depends on the weather. Typically, air currents and pollution in the atmosphere limit the resolution to about 3–4 inches from a distance of 100 miles, though on a perfect day, 2 inches is possible, and on a lousy day, the limit is 6 inches.

The resolution needed to verify an arms control agreement depends on what the photo interpreter is trying to do. Typical arms control agreements specify that the other side is to construct no more than a certain number of weapons of a particular type, and so what is usually wanted is a count of airplanes or missiles. Satellite reconnaissance plays a role that goes beyond arms control agreements, providing in some cases valuable information about such things as the civil war in Nicaragua or the antics of Colonel Qaddafi's Libyan military. In budget hearings, the U.S. Congress heard of various target resolutions for various tasks, illustrated in Table 4. It seems clear that most of the major types of missile carriers can be identified rather clearly from earth orbit. Although in principle missiles could be hidden in silos, continuous coverage would probably allow someone to spot anyone putting the missiles into the silo as long as they were not built underground, too.

Satellite imagery is undoubtedly very valuable to arms control, but it is not a magic solution to all problems. A satellite picture needs to be interpreted by an expert who recognizes various types of aircraft

Table 4.    Interpretation Tasks[a]

| Target | Detection[b] | General Identification[c] | Precise Identification[d] | Description[e] |
|---|---|---|---|---|
| Bridge | 20 ft | 15 ft | 5 ft | 3 ft |
| Rockets and artillery | 3 ft | 2 ft | 6 in | 2 in |
| Aircraft | 15 ft | 5 ft | 3 ft | 6 in |
| Missile sites | 10 ft | 5 ft | 2 ft | 1 ft |
| Nuclear weapons components | 8 ft | 5 ft | 1 ft | 1 in |

[a]Excerpted from U.S. Senate, Committee on Commerce, Science, and Transportation, *NASA Authorization for Fiscal Year 1978, Part 3* (Washington, DC: U.S. Government Printing Office, 1979), 1642–1643, as quoted in Blair and Brewer; Richelson (see Notes 14 and 15 of this chapter).

[b]Detection: Location of a class of units, objects, or activity of military interest.

[c]General identification: Determination of general target type.

[d]Precise identification: Discrimination within target type of known types.

[e]Description: Size/dimension, configuration layout, components/construction, count of equipment, and so on.

and so on. Judgment, analysis, and verification from other sources also play a role. In 1975, Pentagon officers suspected that the Soviets were firing laser beams at our satellites, trying to blind their infrared sensors. It was only after comparing data from many places, including defense and civilian weather satellites, that investigators realized the true cause: gas fires from natural gas flaring off high-pressure gas pipelines in Russia.[16]

There are a couple of other ways in which space plays a role in the verification process for arms control. Nuclear weapons tests can be partially monitored by other means, principally monitoring the radioactivity of the atmosphere and detecting the earthquakelike vibrations of the earth from a seismometer. But space sentinels play a role. An exploding nuclear bomb emits a short blast of gamma rays, X rays, and neutrons. Six pairs of Vela satellites, launched between 1963 and 1970, watch the globe, searching for clandestine nuclear tests. They may have spotted one on 22 September 1979, when a flash of light came from one of three places in the ocean: off Prince Edward Island, Canada, or possibly South Africa, or possibly Antarctica. This event could possibly have been a flash of light from a bit of space junk close to one satellite, or it could have been a nuclear test. The Vela satellites had relatively primitive sensors, and the origin of this event will remain obscure forever. The GPS, the U.S. Navy's set of navigation satellites, will have an improved set of sensors, and according to congressional testimony as quoted by J. Richelson, "will allow the detection of nuclear weapons detonations anywhere in the world at any time and get its location down to less than a [deleted]." Richelson went on to speculate that "a reasonable guess for the deleted word is *mile*."[17]

Another aspect of space surveillance is based on looking up into space to keep track of the other side's satellite launches and activity in antisatellite (ASAT) weapons systems. For a long time, the ground system was a set of huge Baker-Nunn optical cameras located at Edwards Air Force Base in California; Mount John, New Zealand; Pulmosan, Korea; Saint Margarets, New Brunswick, Canada; and San Vito, Italy. These 3-ton, 10-foot-tall behemoths can photograph an object the size of a basketball orbiting near geosynchronous orbit (as long as the object is not too black). This system produces pictures on film, and analysis of the film takes a long time. A more automated system called *GEODSS* (Ground Based Electro-Optical Deep Space Surveillance system) is being installed, consisting of two 40-inch computer-controlled telescopes at each station. The stations are in New Mexico, Maui, South Korea, and Portugal, and on the island of Diego Garcia in the southern Indian Ocean. The GEODSS's computers are smart enough to check each image of the sky against a star map and against a list of known space objects, and so it can detect new satellites. It will also undoubtedly find a number of astronomical phenomena involving stars that vary significantly in the amount of light that they emit.[18]

## VISIONS OF STAR WARS

On 23 March 1983, President Reagan completely changed our approach to the military use of space by setting forth a vision of a world in which nuclear-tipped ballistic missiles might become worthless junk:

> What if free people could live secure in the knowledge that their security did not rest upon the threat of instant U.S. retaliation to deter a Soviet attack, that we could intercept and destroy strategic ballistic missiles before they reached our own soil or that of our allies? . . . I call upon the scientific community in our country, those who gave us nuclear weapons, to turn their great talents now to the cause of mankind and world peace, to give us the means of rendering these nuclear weapons impotent and obsolete.[19]

The Strategic Defense Initiative (SDI), popularly known as Star Wars, has been established in the wake of President Reagan's speech. Reagan's vision was apparently based on a number of new technologies that have been under development in the defense department for a long time: particle beam weapons and laser weapons. Laser beams travel at the speed of light and can zap a target halfway around the earth in less than a tenth of a second. Particle beams travel only a little bit slower. In principle, these weapons could destroy enemy missiles much faster than the more traditional antimissile defenses could, where an ordinary rocket is used to shoot down another rocket.

Much has been written on SDI, and the purpose of the next few pages is not to give a detailed account of the program. Rather, my purpose is to provide a brief overview of the dimensions of SDI and the meaning of a "defensive shield," as well as to highlight some of the outstanding issues in the public debate. A more comprehensive and relatively balanced treatment of SDI can be found in a recent Office of Technology Assessment (OTA) report, as well as in Jeff Hecht's book *Beam Weapons*.[20] Shorter treatments can be found in an excellent pair of articles by Gerald Yonas, supporting SDI, and Wolfgang Panofsky on the other side.[21]

Since 1983, there has been an extraordinarily dramatic and very public acceleration of military research on technologies related to ballistic missile defense. SDI has combined the divided, small-scale research programs of the various services into one organization, and money has been thrown at these programs at an accelerating rate. Despite all the criticism of SDI, this program has tripled its budget in three years, a remarkable achievement in an era of great concern about deficit spending. In the fiscal year ending in September 1984, the Pentagon spent $468.7 million on laser weapons, $51.5 million on particle beams,[22] and assorted amounts on other SDI-related programs to bring the total to $991 million. As is shown in Table 5, the Reagan administration submitted a series of requests that would represent nearly a fivefold increase in three short years, culminating in the fiscal year 1987 request for $4.8 billion. Depending on one's viewpoint, Congress either lagged behind or resisted throwing money away, appropriating somewhat less than the Reagan request. Still, spending on SDI has more than tripled in the three years between fiscal 1984 and fiscal 1987. The

Table 5.    Government Spending on the Strategic Defense
Initiative (in Billions of Dollars)

|  | 1984 | 1985 | 1986 | 1987 |
|---|---|---|---|---|
| SDI request | 0 | 1.8 | 3.7 | 4.8 |
| Congressional appropriation | 1.0 | 1.4 | 2.8 ⎫ | 3.5 |
| Department of Energy research related to SDI | 0 | 0.2 | 0.3 ⎭ | |

$3.5 billion slated for fiscal 1987 is a lot of money, more than double the Government's spending on basic science through the National Science Foundation ($1.5 billion), and almost half of NASA's total budget ($7.5 billion).[23]

But what is SDI? What are we throwing all this money into? Basically, two versions of SDI are being discussed in public, and the differences between the two versions are important. They are not necessarily self-contradictory, but when one tries to resolve the two approaches to SDI, it becomes clear that SDI is not just a magic technological solution to the arms race but part of a much more complicated picture.

In many speeches, the President still promotes visions of an impenetrable nuclear umbrella that will shield America from its enemies. I have heard this called the *Astrodome defense,* a defensive system that would ward off attacks from ballistic missiles in the same way that the Astrodome keeps rain off the Houston Astros. In June 1986, Ronald Reagan addressed the graduating class at Glassboro High School in New Jersey:

> Let us leave behind, too, the defense policy of Mutual Assured Destruction—or MAD as it is called—and seek to put in its place a defense that truly defends. Even now we are performing research as part of our Strategic Defense Initiative that might one day enable us to put in space a shield that missiles could not penetrate—a shield that could protect us from nuclear missiles just as a roof protects a family from rain.[24]

Other administration officials continue to echo this theme. Defense Secretary Weinberger was quoted in July 1986 as saying, "It is not our missiles we seek to protect but our people, and we must never lose sight of that goal."[25] Much of the public debate on SDI, as well as the conversations at the summit at Reykjavik in October 1986, has drawn on the imagery of this leakproof shield. And it does need to be leakproof, for the Soviets currently have over ten thousand nuclear warheads. If even 1% of these get through, they could kill ten to twenty million people and cripple the country.[26] Furthermore, American leaders would have to be confident that Star Wars would work the first time it was tried in a real battle situation.

But when one talks to the scientific community, even those involved in the SDI program (who are the only ones I will quote in this paragraph), an entirely different picture emerges. Gerald Yonas, Chief Scientist of the SDI Organization, contradicts the Astrodome defense: "Nobody believes in a 100% leakproof defense. Nobody believes in 100% anything that's ever worked on military systems."[27]

A staff report to Senators Proxmire, Johnston, and Chiles echoes a similar theme:

> During our briefings by General Abrahamson and his program managers, there was never any discussion of an impenetrable defense shield that would protect all Americans from nuclear war. Rather, the SDI program is aimed ultimately at creating successive layers of ballistic missile defenses, effective enough as a whole to deter the Soviets from attacking in the first place. General Abrahamson made it quite clear that the objective of SDI is *deterrence*.[28]

Louis C. Marquet, director of DOD's Directed Energy Office, said that "the goal of the program is not to build a perfect defense system. The goal of the current Strategic Defense Initiative is to bring to the table the technical issues."[29]

So, on the one hand, we have a hopeful vision of SDI: President Reagan's Astrodome defense. On the other is the vision of SDI as a deterrent. Perhaps I can reconcile the two visions of SDI to some extent. One picture presented by SDI proponents is that the deployment of a defensive scheme would be accompanied by deep reductions in nuclear forces. If each of the superpowers had fewer nuclear warheads, a defensive system that was only 80%–90% effective would be very useful indeed.

Just before SDI began, Freeman Dyson presented a vision of such a world in his book *Weapons and Hope*.[30] It's a "live-and-let-live" concept, pulling back from the Utopian vision of banning the bomb entirely but recognizing the absurdity of our present world, where both sides are armed to the teeth to win an all-out nuclear conflict that neither wants. The vision is that each side has a few hundred nuclear missiles, sufficient to deter any nuclear attack by a Third World aggressor or by the other side. A defensive system could play a very important role in such a world because the few missiles that got through it (and there *will* be a few) would not destroy American or Soviet civilization, although the tragedy would be immense. The defensive system would take care of the possibility that a mad, charismatic leader like Libya's Qaddafi, Iran's Khomeini, or Uganda's (former) Idi Amin might steal a missile or two and hold the world hostage, or the possibility of an accidental launch of a few missiles by either side.

The live-and-let-live world sounds rather nice; American Presidents since Truman have felt uncomfortable with the doctrine of mutual assured destruction. But how do we get there from here? The development of a defensive antiballistic missile system is only a small part of the task of making the transition from the current world to the live-and-let-live world. As the Office of Technology Assessment report explains, "Most explanations of the SDI policy approach by Administration officials tend to emphasize a scenario in which the Soviet Union agrees to deep reductions in all kinds of offensive nuclear forces."[31] In other words, deep cuts in the nuclear arsenals of the two superpowers would have to be negotiated before SDI could be an effective shield.

Is it reasonable that the two superpowers would agree to the massive arms reductions that would be required? The past history of arms control negotiations is not encouraging. The bold offers discussed at the Reykjavik summit in late 1986 may be difficult, even impossible, to implement concretely. The few arms control agreements that we do have between the United States and the Soviet Union were very hard to achieve, and none of them involved really deep cuts in existing weapons systems. Indeed, some observers take the very cynical view that the arms control agreements are basically agreements in which the superpowers have formally agreed not to construct weapons that they don't want to construct anyway, and in which one meaningless bargaining chip has been exchanged for another. Such cynicism undoubtedly overstates the case, but it does highlight the difficult path that arms control has had to follow.

We can see from a close look at the Star Wars program that the administration's view of the future of SDI is much more complex than the simplistic picture that often emerges in presidential speeches. Just suppose that SDI did succeed in the purely technical sense, that throwing a trillion dollars at scientists, engineers, and aerospace companies did produce a workable defensive system deployed in space. This step, by itself, would not give us anything like a nuclear shield to protect people. An effective (but not impenetrable) nuclear umbrella would be produced only by arms control negotiations. SDI is not the easy technical solution to the arms race that it is often portrayed as being. Its supporters see it as part of a solution. But it's not the whole answer.

## The Reality of the Strategic Defense Initiative

Both of the two visions of the role of SDI apply to the far future, to a world that is quite different from the present one. The Astrodome defense requires a technology that does not exist and may never exist. Using SDI as a deterrent requires the negotiation of extraordinarily deep cuts in nuclear forces. Visions of the future, whether utopian or not, must be kept in mind when we consider what should be done about SDI. Still, an evaluation of our present perspective requires a more detailed consideration of just what kind of celestial battle stations are visualized by SDI planners. What could a defensive system designed for SDI do, in technical terms? Is it feasible? What purpose could it serve if deployed under the present circumstances? What issues should we keep in mind when debating SDI?

The ballistic missile defense (BMD) envisaged by SDI is extremely complex, and a schematic that illustrates various phases is shown in the accompanying illustration. Let's follow what an enemy missile would face along its trajectory from launching to the target. What's really new in comparison with previous antiballistic missile (ABM) programs is the idea of destroying missiles in their boost phase, during the three to five minutes after launch when the rockets are firing, when the missile is easy to detect, and before inexpensive decoys can be deployed. The speed of laser beams is critical to success at this stage because other weapons could not reach the target in time. Directly after the boost phase is the postboost phase, when upper stages and "buses" from ballistic missiles coast upward through the atmosphere. Much of the technology that can destroy missiles in the boost phase is applicable to this second phase, except that the missiles are no longer rocket-powered, so that heat or light sensors can no longer home in on them.

More than ten minutes after a missile launch, interception of the ballistic missile becomes considerably more complicated. In a world where antimissile systems are deployed extensively, the offensive side

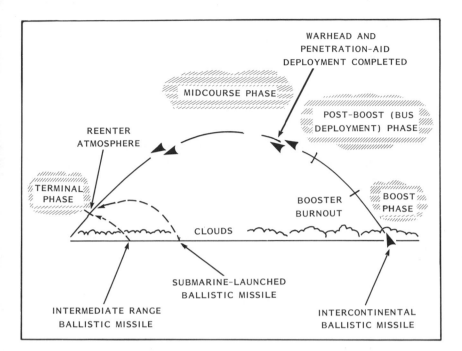

The defensive scheme envisioned by SDI planners is a multilayered one. Missiles can be intercepted during the boost phase, the postboost phase, the midcourse phase, or the terminal phase. (Source: D. Walker, J. Bruce, and D. Cook, "SDI: Progress and Challenges," unclassified staff report submitted to Senator William Proxmire, Senator J. Bennett Johnston, and Senator Lawton Chiles, 17 March 1986.)

will employ a variety of penetration aids. These can be dummy warheads, aluminum wires, or other devices to confuse the radar seeking to identify the real warhead. Each of the missiles could deploy dozens of decoys, a deployment that would result in a million targets in a full-scale attack. The technological challenge here is separating the bomb carrier from the decoys. This midcourse phase, when the warhead and decoys follow a trajectory above the atmosphere, lasts about twenty minutes. The final stage in which interception is possible is the "terminal" phase, the final minute when the warhead reenters the atmosphere on the way to the target. The idea is that even if a few missiles get through the first layer of this multilayered structure, succeeding layers will shoot them down.

Thus, the ballistic missile defense envisioned as a result of the SDI program is a complex, multilayered one. It can defend us against only certain types of enemy attack. In attempts to shoot down missiles launched from submarines positioned a few hundred miles off the American coast, or against intermediate-range missiles fired at European cities from the Soviet landmass, the boost phase and midcourse phase opportunities for interception would be much more limited. In each of these cases, the boost phase would be much shorter (because the missile wouldn't have as far to travel), and the missile would never leave the earth's atmosphere. SDI would also be much less effective against cruise missiles, low-flying automated airplanes that hug the terrain and are difficult to detect by radar.

But the technological challenge of designing a defensive system is even greater than it might seem on the basis of the above discussion. Building a laser or beam weapon that can zap a target ten thousand miles away is only part of the job. The interception task that must be accomplished during each of these phases is in itself quite complex, involving a number of different steps. Jeff Hecht, in *Beam Weapons,* did a particularly good job in drawing a distinction between a weapon that can destroy anything that happens to be in the firing line and a weapons system that is militarily useful. Hecht enumerated the many different tasks in using a beam weapon. In order to make a ballistic missile defense work, one must:

- Identify the target, which will not be sitting at the other end of a firing range with a giant bull's-eye around it.
- Track the target until the weapon is ready.
- Point the weapon at the target.
- Focus the beam on the target.
- Compensate for atmospheric effects on the beam.
- Maintain the focus as the target moves.
- Make sure that the energy in the beam is deposited on the target and not reflected away.
- Verify that the target has been destroyed.[32]

An additional factor is that the weapons system that accomplishes all of these tasks must work the first time it's used, must be ready at all times, and must be operable by ordinary military personnel, not people with Ph.D. degrees in physics or electrical engineering.

The complexity of the interception task is the reason that scientists, including those in the SDI Organization quoted above, deny the possibility of an Astrodome defense that can completely block a full-scale attack. The entire present Soviet land-based missile force consists of 1,400 missiles. Suppose you were to try to defend against an all-out attack by these. If all 1,400 missiles were launched at once, you would have to do each of these eight tasks to destroy a missile, or a total of 11,200 tasks, within thirty minutes. But this calculation even understates the true challenge of providing an impenetrable shield in the current world. Because decoys can be deployed after the boost phase, a successful system would probably need to destroy most of the missiles in the boost phase, which lasts three to five minutes. To see how hard this is, suppose that you needed to destroy 90% of the missiles in the boost phase. With eight tasks required per missile (assuming you hit each one the first time), the SDI system would need to destroy 90% of 1,400 missiles in four minutes, and you would be doing 2,520 technologically challenging things per minute, or 42 per second.

The battle that would occur during the implementation of a ballistic missile defense would be fast-paced. During the time that it took you to read the last paragraph, you would have had to accomplish several thousand technological challenges. Aim! Focus! Zap! Verify! On target! Aim! Zap! Verify! Target! Pow! Blam! Sorry, you can't read these words fast enough to keep up with what would have to be done no matter how many speed reading courses you took, because you would have had to skip through my sequence of exclamation points in one quarter of a second. Computers would have to do it all. The pace of ballistic missile defense would make the final battle scene in the movie *Star Wars,* where Luke Skywalker takes about ten minutes to knock out one battle station, seem like slow motion. The defender has less than ten minutes to destroy all, or almost all, of the enemy missiles during the critical boost and postboost phases.

The speed of the battle would require a computer-operated defensive system. The computers themselves may be within reach of current computer technology, but the software, the instructions needed to tell them what to do, may be impossible to develop. One program would involve ten million lines of computer code. No one programmer could write it alone; a team of thousands of programmers would have to work for years, coordinating their efforts absolutely perfectly.

Can such a computer program be developed? SDI critics have their doubts. Computer programs often contain mistakes or "bugs" that have to be identified and fixed by testing. A critique of SDI's plans for battle management has called for a different type of computer system in which individual groups of battle stations would be managed by separate computers rather than one huge central one. Can such a computer network, quite different from virtually all computer systems developed so far, work? Would it be compatible with the SDI hardware? Can it be tested enough, without real battles, to be trustworthy? A number of software developers seriously doubt that the system would work at all.[33]

What role could SDI's missile defense play in today's world? If SDI is to be deployed soon, it will be facing ten thousand Soviet warheads, innumerable decoys, and the threat of an antisatellite weapons system. In the short run, there are a number of roles that SDI could play. Various people involved in the SDI debate have stated that such a system could:

- Strengthen deterrence by defending American land-based missiles.
- Hedge against a Soviet deployment of an advanced ABM system (such a deployment would violate the ABM treaty).
- Lead to the development of new technologies that could be used for other military or civilian purposes.
- Serve as a bargaining chip in arms control negotiations.[34]

In regard to the second role, the ABM treaty was signed in 1972, and it limits each side to one antiballistic missile site and forbids deployment or testing of additional systems. The primary difference between the ABM systems discussed then and the SDI is that SDI adds the critical capability of intercepting missiles in the boost phase, before decoys can be deployed. The Soviet Union has built its one allowed ABM site near Moscow; the United States began construction on its site but has since abandoned it. The Soviet Union, like the United

States, has been continuing to do research on ABM technologies, including beam weapons—the centerpiece of SDI. Their progress is quite unclear. One Soviet radar site has been cited as a possible violation of the ABM treaty, leading to suspicions that they may be preparing to ''break out'' of the treaty and deploy additional ABM systems.[35]

At the present time, SDI is a research program, but a very expensive one. President Reagan has lavished attention on it, and it has been widely publicized. Some of the critical questions raised about its ultimate goal have yet to be answered: How much will it cost? Can the other side overwhelm SDI with a cheap system of decoys and more missiles? How many space shuttle flights will be needed to deploy it?

Everyone who has written on SDI agrees that it is worthwhile to continue the research program. Few quarrel with the advisability of centralizing the various Pentagon programs on SDI technologies in one program. The debate concerns the way in which the program is conducted, the degree of urgency and visibility involved, its technological direction (Should mature technologies or more speculative ones be emphasized?), its role in arms control, its role in our overall nuclear strategy, and its budget. The future of SDI will remain a focus of congressional and public debate for a number of years to come.

## ANTISATELLITE WEAPONS

Related to SDI, but actually rather distinct from it, is the antisatellite (ASAT) weapons program. A satellite is much easier to attack than a ballistic missile. An intercontinental missile takes only about half an hour to get to its target, and it can be accompanied by a horde of decoys. A satellite remains in its orbit for months, years, or decades and behaves much more like the targets at the end of a practice range than the quickly firing missiles. Most existing satellites have been designed under quite severe weight constraints, so little or no effort has been made to make them less vulnerable to military attack.

Both the United States and the Soviet Union have had longstanding ASAT efforts dating back to the 1960s. Currently, the United

States is testing ASATs, while the Soviet Union stopped its tests in the early 1980s.[36]

An ASAT weapon is not really very complicated. A satellite is launched that maneuvers itself to get quite close to the target. The target can be destroyed in many ways. Debris from an exploding satellite can hit it, or electromagnetic radiation can heat up the target and fry it. A more sophisticated way of destroying the target is to use a homing radar device that will cause the ASAT to collide with the target. One easy way to destroy a satellite was discovered in 1962 by a high-altitude weapons test, which enhanced the radiation belts and destroyed several orbiting satellites. The United States successfully tested a homing device in the atmosphere in June 1984. The panoply of laser beams and particle beams that are part of SDI can also be used to destroy the satellite. But there is much more time to identify the target, home in on it, focus the weapon, shoot, and verify that the target was destroyed. The U.S. Air Force regards the Soviet ASAT system as being ''operational.''

Despite the apparent simplicity of ASATs and the vulnerability of satellites, military assets in space, such as communications, surveillance, and navigation satellites, are not as vulnerable as one might think. The Soviet ASAT system works only against satellites in low earth orbit, and it cannot reach to geosynchronous orbit, where the communications satellites are. Many of the military systems depend on a large number of satellites, and knocking out one will alert the other side and will only degrade, not stop, whatever that satellite was doing. Both sides have agreed that space assets are the legitimate property of the launching country; thus, destruction of a satellite would clearly be regarded as a hostile act. The military satellites are doing tasks that have been accepted as being peaceful, or at least legitimate.

A number of arms control advocates in Washington, most notably California Congressman George Brown, are concerned that too much ASAT development could lead to yet another expensive arms race, this time in space. In December 1985, Congress enacted legislation that banned ASAT tests in outer space. The intent of this legislation was to force the Reagan administration to the negotiating table. Some negotiating options short of an ASAT ban are being explored. One, referred to as a "rules-of-the-road agreement," would formally prohibit any satellite from maneuvering too close to another.

The rather limited role of ASATs would expand dramatically if a Star Wars type of defense were deployed. Most visions of a ballistic missile defense would require us to have a good deal of hardware permanently deployed in space, where it would be, in principle, quite vulnerable to ASATs. Planners have dreamed up ways of getting around this, such as a laser battle station that would "pop up" or be launched quickly from the ground in the event of an attack. However, it's hard to see how a pop-up weapon could be effective against missiles in their boost phase because there are only a few minutes available in which to destroy a missile.

\*     \*     \*

Although the communications satellite industry still dominates the commercial uses of space, there are a number of other ways in which space is used for practical purposes. The global perspective presented by the orbital vantage point is crucial to many aspects of weather forecasting, particularly in parts of the world where storms approach by sea rather than by land (as they do over much of the United States). Satellites can be used for navigation and for surveillance as well. The most controversial proposal for the military use of space is the Reagan administration's Strategic Defense Initiative, which is still only a research program in its very early stages (though an expensive one).

SDI and the closely related antisatellite programs have stirred renewed fears regarding the "militarization" of outer space. But the genesis of the space program indicates clearly that space had already been "militarized" in the 1950s, when scientific satellites were launched in order to clear the way for surveillance satellites. The coexistence between military and civilian space programs can be uneasy, particularly when both are competing for the same launch facilities and opportunities. However, it is a relationship that the country has lived with for nearly thirty years, with remarkably little friction between the two programs.

# CHAPTER 6

# Earth Science

## A Global View of a Green Planet

A few exploratory missions have demonstrated that observations from space provide a unique and tremendously powerful way of looking at our own planet, of measuring the temperatures and wind speeds all through the atmosphere, of mapping the invisible ocean floor, and of providing large-scale views of the slow, persistent ocean currents that govern our climate. A satellite that orbits the earth every ninety minutes can provide a global perspective that simply can't be had from a network of isolated, individual observing stations, or even high-flying aircraft. Satellite data have opened up a brand-new perspective on the world's oceans. Orbiting satellites that return pictures of the earth are indispensable to companies and researchers who map urban areas, monitor crops and pests, find new archaeological sites, and discover new geological faults and possible sites for new mines, to name only a few of the more dramatic uses of earth pictures.

Although we obtain new knowledge from these efforts, there is some practical payoff as well. Understanding how our planet works can help us understand and adjust to the large-scale weather patterns that can drastically affect our lives. As an example, the warming of the eastern South Pacific Ocean, called El Niño, produced flooding and droughts in twelve countries in its most recent appearance (in 1982–1983) and damaged billions of dollars of houses, crops, and factories. Further practical applications of the orbital perspective are produced by the capabilities of orbiting satellites to map large areas quickly.

With a few orbital pictures, one can quickly produce a chart of tidal currents in complicated estuaries; such a chart can guide efforts to control pollution.

Space has also given us quite a different perspective on our earth by discovering the environment around it. Before *Sputnik,* the earth was thought to be an isolated planet swimming in an empty sea of space. But the earth has a magnetic field that does more than just make compasses work. This field interacts with the flow of charged particles through the solar system to create a complicated magnetosphere. The magnetosphere is a giant cosmic laboratory for the understanding of the interaction of charged particles and magnetic fields, the field of plasma physics. In addition, the particle environment of the earth can have its effects on the design of spacecraft, which must work in an environment of speeding charged particles zapping delicate electronic devices.

The global perspective on our green planet has caused geologists, meteorologists, oceanographers, and biologists to realize that the earth, air, water, and life that they study are very closely related. The earth is one giant interrelated machine, and what happens in the ocean is closely related to what occurs in the air above it and the land near it. Strictly speaking, I shouldn't talk about "ocean science" from space. However, the orbital vantage point has meant the greatest difference to oceanographers because our previous perspective on the ocean had been limited to a few isolated observations made from surface ships. I'll begin an examination of this new field of "earth system science" with our new perspective on the seven seas.

## THE OCEANS

So far, most of the uses of space that I've written about have been activities that NASA has been doing for a long time. The first communications and weather satellites were launched in the early 1960s, just after the dawn of the space age. But even though this business of peering down at the earth is relatively old, it still holds some surprises. One new area of interest to both NASA and other government agencies, like the U.S. Navy, is the use of satellites to survey and understand the world's oceans. The prospects for mapping and understanding what are now tremendous, largely unknown blue-water expanses are very good indeed. At a convention in the mid-1980s, Jeff Rosendhal of NASA's Office of Space Science and Applications, told me that NASA's work on the oceans was anything but routine, old-time stuff that the agency had been doing for years. Bubbling over with excitement, he said, "For the first time, we're going to solve the problem of the earth's oceans."[1]

But just what does it mean to "solve the problem of the earth's oceans"? Why should we bother? Who cares about the briny deep? The answer is that we all care, or should. Major storm systems like hurricanes grow over the world's oceans, and year-long climatic disruptions are also oceanic phenomena.

Despite modern transportation technology, the oceans are vast, largely unmapped, trackless wastes. Eighty percent of the surface of the globe is covered with water, and the only humans to be found there are the crews of a fairly small number of ocean-going freighters and yachts. These ships provide fragmentary reports of data like wind speed, air temperature, water temperature, and wave height. In a three-day period in August 1978, fewer than fifty weather reports came from ships south of the Equator, between New Zealand and South America—an area ten times the land area of the United States.[2] We wouldn't know much about the North American climate with the same kind of spotty coverage, corresponding to only five weather stations for the entire United States.

So the first step in "solving the problem of the world's oceans" consists of simply finding out what's going on in these blue-water expanses. Orbiting satellites fly over the oceans again and again as they whirl around the earth, making one circuit every ninety minutes (if the satellite is in low orbit). NASA's *Seasat* (here the meaning of the name is, for once, clear—if we have *Landsat*s to observe the land, we have *Seasat*s to observe the sea) satellite, one of the most successful "failures" in the history of the space program, obtained more measurements of oceanic winds than had been obtained in all of human history, even though its mission was cut short, after 100 days, when it had a power failure. The *Seasat* mission opened our eyes to the opportunities for understanding the earth system with the aid of space observations.

## El Niño and the Effect of Oceans on the Weather

Still, who cares about how the oceans work? It isn't just those isolated freighter captains, crew, and passengers, and a few daredevils on sailing yachts who sail the ocean blue because they want to, who care. Ocean currents affect all of us in very practical, economic ways. In 1982 and 1983, people as far apart as Peruvians and Southern Californians were buffeted by a year of horrendous weather and sea conditions connected to the same global phenomenon, known as El Niño. Rainfall in normally dry Southern California tripled, and houses slid down cliffs into the sea. Fifteen-foot waves smashed into the Pacific shoreline, destroying expensive homes in exclusive districts like Malibu.

But the sufferings of Americans were mild in contrast to those visited on the citizens of coastal Peru. The annual anchovy catch, which supports many people, plummeted from twelve million tons to only two million tons. Those anchovies that did survive were overfished, and it will be a long time, if ever, before the fishery recovers. Moreover, Australians, Indonesians, and Indians suffered through widespread drought at the same time that Californians and Ecuadorians were doused by heavy rains.

We now know that each of these meteorological catastrophes was related to the condition of the oceans in the Eastern South Pacific. An ocean current, called the Humboldt current after its discoverer, carries cold, nutrient-laden water from the region around Antarctica up the western coast of South America. Every once in a while, around Christmas, the water becomes warm, as North Pacific water is blown toward the coast for a short time. Spanish fishermen have called this El Niño—referring to the Christ Child—because of its timing.

Recently, scientists discovered that the severe El Niños were part of a global climatological phenomenon known as the southern oscillation, a quasi-periodic, drastic change in the wind patterns over the South Pacific Ocean.[3] The usual El Niño lasts for a short time and doesn't do much to the local fishery, or to the weather elsewhere. A severe one develops from an inexplicable change in the placement of high- and low-pressure areas in the Eastern Pacific. Normally high pressure centered on Easter Island drives the winds toward the west. When the pressure difference becomes quite low, the trade winds reverse direction. Normally, the water west of South America is pushed westward, away from the coast and out toward the open ocean, making room for cold subsurface water to rise up. But in an El Niño, the warm equatorial water is pushed toward South America, eventually plunging downward. The rich cold water, loaded with algae, which are good fish food, no longer rises to the surface, and consequently, the fish starve in the warm, barren water from the equatorial regions.

The severe El Niño phenomenon, which occurred twenty-four times in the ninety years between 1890 and 1980, had been regarded primarily as a climatological curiosity, not of broader interest. During the El Niño of 1982–1983, we had weather satellites that could monitor the entire globe, and this particular El Niño was so bad that its effects in other areas were impossible to ignore. The suffering Peruvian fishermen were one of many groups of people affected by the worldwide bizarre weather.[4] The estimated damage produced by the 1982–1983 El Niño was $10 billion. If we could predict a severe El Niño, even somewhat in advance, some of the damage could be prevented. Maybe there is some subtle effect that occurs about six months before a severe El Niño, and satellite data are so comprehensive that we may well be able to spot this El Niño precursor and predict the severe weather and changed oceanic conditions.

## Space Measurements of Winds, Waves, and Water

How can satellites help solve the problem of the world's oceans? As is the case with weather satellites, it is the space-age capability of obtaining large-scale views of the entire earth, day after day, week after week, that is the key. Oceanographers have lagged behind meteorologists in using space, partly because it's harder to learn about the oceans than simply to take pictures of clouds. As weather satellites got better and better, doing more than just taking pictures, oceanographers realized that sophisticated satellite observations could provide essential data on temperatures, waves, and currents that had not been available previously.

Perhaps another stimulus was the reports of various astronauts of many and subtle changes in the color of the ocean water. Hundreds of years ago, Polynesian navigators in their outrigger canoes were sensitive to changes in the color of the ocean water; apparently, they used these changes as one of many pieces of information to guide tiny boats across thousands of miles of ocean. The Skylab astronauts, Soviet cosmonauts on Soyuz spacecraft, and astronauts on other missions saw the same subtle color variations. Perhaps some far-sighted ocean scientist realized that these were more than just a colorful, amusing sideshow to keep the astronauts from getting bored and saw the possibilities of obtaining crucial missing information.

In any event, it was in the late 1970s that oceanographers realized that satellites had the technological capability to provide oceanographically interesting information like water temperatures, wind directions, and ocean current speeds. Different instruments were then installed aboard spacecraft to provide this information. For example, temperature is measured by a device called a *radiometer,* which measures the amount of infrared (or sometimes ultra-short-wave radio) radiation emitted by the earth's surface. (Infrared radiation has wavelengths longer than visible light and shorter than radio waves.) The hotter something is, the more radiation it emits. Thus, by measuring how much radiation comes from a patch of ground or sea, one can figure out how hot it is.

In principle, it sounds so easy. In practice, it's harder—a lot harder—than it sounds. Clouds and fog can absorb infrared radiation, and one has to correct for that by measuring (usually) three wavelength bands of infrared radiation. One needs to build a lightweight scanner that can isolate the infrared radiation from a small patch of ocean surface. Hot land near a patch of ocean can emit lots of infrared radiation and contaminate measurements. If radio waves are being used, one has to be out of range of terrestrial transmitters that are using the same wavelength or one can pick up the voices of taxicab dispatchers and erroneously think that the sea is hot. Figuring out just how to use satellites to measure sea surface temperatures involves a long, continuing trial-and-error process. Scientists build a simple instrument, fly it, compare the measurements with measurements made on the earth's surface (or ''ground truth'') to see where the instrument works and where it doesn't, and design a better instrument.

The end results of all this are maps of the temperature over large regions of the earth's surface, such as those shown in the figure. The global-surface-temperature map illustrates temperature measurements worldwide. In regions such as the Eastern Pacific Ocean, ship coverage would be essentially nonexistent and there would be no way to determine temperatures, no way to unravel the mystery of El Niño.

## Winds and Waves

Satellites can observe the earth both passively, by producing images of the radiation that is emitted naturally, and actively, by using a radar instrument to bounce radio waves off the surface of the earth or the sea. A very clever use of radar has been to map the winds on the surface of the ocean. It's a bit tricky to do this, and all of the bugs

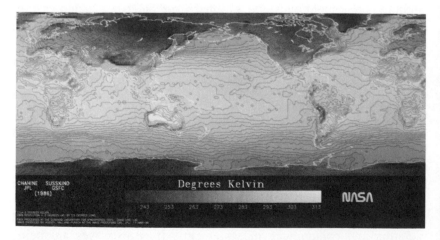

Temperature of the water in the world's oceans in January 1979, determined from both infrared and microwave data. The contours and shading indicate the run of temperature, which is known in some detail even over uninhabited regions like the South Pacific. The instruments referred to in the illustration are the High Resolution Infrared Sounder (HIRS) and the Microwave Sounding Unit (MSU), both on NOAA satellites. (Source: M. Chahine, Jet Propulsion Laboratory.)

haven't been completely worked out. There are a number of different kinds of waves on the surface of the ocean. The big ones, or swells, the ones that California surfers glide down, are produced by intense storm winds that may be some distance from the shore. Local winds produce tiny ripples, where the crests are an inch or so apart, and foam on the ocean surface. What happens is that very short-wavelength, high-frequency radio waves have a wavelength that's fairly close to the size of the tiny ripples, and so these radio waves are readily reflected by the ripples. The higher the wind speed (in principle), the more intense the ripples and the more intense the reflected radiation.

Thus, a satellite needs two instruments that, combined, can measure the wind speed. A radar points straight down and bounces short-wavelength radio waves off the tiny ripples. A radiometer measures the naturally emitted radio waves so they can be subtracted off. Calibrations of the wind speed measurements, where ships and buoys check the wind speed on the ocean (''ground truth''), demonstrate that errors are less than 2 meters per second (around 4 knots, or 4 miles per hour). Still more complex are devices (called *scatterometers*) that can measure the reflectivity of the sea from two different directions and infer the direction of the wind. We can determine wind directions within twenty degrees in many cases.

Again, the short-lived *Seasat* mission provided us with a brief, tantalizing taste of what these techniques could do. The illustration shows the wind field over the Pacific Ocean on 7 September 1978. The complex storm systems that rage around the Southern Ocean, ringing the Antarctic continent, are clearly delineated here. Think how much we might know about El Niño if the *Seasat* scatterometer had been operating in 1982–1983. A scatterometer like that flown in *Seasat* will be flown on the *NROSS* (Navy Remote Ocean Sensing System) mission, scheduled (before *Challenger*) for launch in 1989.

In addition to temperature and winds, satellites can measure a wide variety of other quantities. Ocean color is an indication of biological productivity, the abundance of algae in the sea. The sea ice coverage can be estimated relatively easily from photographs, though one has to be careful to distinguish between sea ice and clouds. A clever technique called *synthetic aperture radar* (SAR) can measure wave heights and patterns.[5]

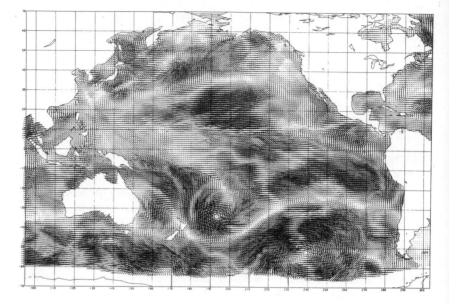

Marine wind field over the Pacific Ocean derived by Peter Woiceshyn of the NASA-supported Jet Propulsion Laboratory, from the *Seasat* scatterometer data for 6 and 7 September 1978. The lines show wind direction, and their length is proportional to the wind speed.

Satellites can measure the distance between them and the ocean surface, to extremely high precision, and can determine two important pieces of information. Gradual variations in the height of the ocean's surface are produced by ocean currents, which can be studied by this method (especially if a series of measurements are taken over time). In addition, the ocean surface tends to follow the shape of the ocean bottom, being higher over submerged mountains and lower over trenches. Thus, the *Seasat* mission produced a global map of the ocean bottom, available for the first time. NASA has planned a joint American-French mission, *TOPEX/POSEIDON* (TOPEX = Ocean *Top*ography *Ex*periment), to continue the *Seasat* work. This mission was originally scheduled for a new start in 1987 and a launch in 1989 (before *Challenger*); its planners may have to wait for a bit longer.

Satellite measurements of the earth system can be applied to the land and the atmosphere as well as the ocean. Scientists studying the land use many of the same sort of instruments that the oceanographers use. The upper atmospheric scientists face a slightly different challenge. The upper atmosphere, which includes the stratosphere, is not just a cold, dry, and scientifically dull layer separating the troposphere (the lowest layer of the atmosphere, where all the weather is) from the even colder reaches of interplanetary space. It contains the all-important ozone layer, which shields us from harmful ultraviolet radiation, and recent missions, such as the SAGE (Stratospheric Aerosol and Gas Experiment, launched in the early 1980s), have proved the way in which various chemicals, including the freon used in refrigerators and air conditioners, interact with the ozone layer. The use of freon in spray cans was banned in the 1970s as a result of fears that this chemical, once it got into the stratosphere, might eat up the ozone layer. Follow-on missions, such as the Upper Atmospheric Research Satellite (UARS) and various instruments to be used on space shuttle missions, will continue this work. Some immediate concerns about the ozone layer arose in 1986 with the discovery of large holes in the ozone layer in the polar regions; no one knows exactly why these holes are there or whether they are indications of any long-term effects on the ozone layer.

Has the increase in our ability to gather data about the oceans actually "solved the problem of the earth's oceans?" It certainly has produced a start, but as oceanographer John Apel summarizes, satellite oceanography "is certainly not the stand alone panacea for oceanographers that its most enthusiastic supporters have made it out to be."[6] The difficulty is that true understanding of the world's oceans requires more than simple fact gathering. Scientists need to understand, in precise quantitative terms, how interrelated forces like wind stress, solar heating, the interplay between land, sea, and air, and the global circulation of the ocean driven by the spinning earth all work together to make El Niños and hurricanes. One of the challenges for the next decade is the analysis of the data that currently exist by scientists, seeking to determine the underlying general trends that form the basis of a true understanding of the workings of the earth system.[7]

Because scientists now realize that the land, water, air, and living parts of the globe are all interrelated, plans for new spacecraft have been drawn up by interdisciplinary teams of earth scientists working together. Recently, such a group compiled a four-page shopping list of measurements that needed to be done on a regular basis. Soil moisture, cloud cover, wave height, ice cover, temperature, winds, lightning, are all measurements that ideally should be done globally on a regular basis. Specific missions like the two mentioned already, NROSS and TOPEX/POSEIDON, will address parts of the shopping list. The Upper Atmosphere Research Satellite (UARS) is designed to measure the complex dynamics of the stratosphere and other layers of the upper atmosphere. Here, as in many other areas of space science, it is not clear how many of these missions will survive the scarcity of launch opportunities that will result from the *Challenger* tragedy. In addition, earth scientists have called for the establishment of a centralized data base so that scientists in any part of the world can have easy access to the information that has been gathered already.[8]

## CHARGED PARTICLES IN SPACE NEAR THE EARTH

The outer layers of the earth's atmosphere blend with outer space in a region where the air is thinner than the best laboratory vacuum, but where the earth's influence is still felt. Before the space age, scientists visualized the earth and its thin atmosphere as orbiting through completely empty space. The earth is surrounded by a magnetosphere, the name given to the space environment near enough to the earth so that the earth's magnetic field rather than gravity dominates the motion of charged particles. The magnetosphere contains a large number of structures, invisible to the eye, that are produced as a result of the complex interaction of the earth's magnetic field with the solar wind. Various people call the study of these structures by different scientific names: magnetospheric physics, space physics, solar-terrestrial physics, or space plasma physics. I'll use the name space plasma physics because it's the most comprehensive. (A gas that contains charged particles rather than electrically neutral atoms and molecules is called a *plasma*.)

Although there was reason to suspect that the solar wind would interact with the earth, the first indication that such a complicated structure existed came with the discovery of the Van Allen radiation belts by *Explorer 1* in 1958. These belts are regions of space around the earth that are filled with charged particles—electrons, protons, and a few atomic nuclei—traveling at very high speeds. For historical reasons, the term *radiation* has been used to describe such charged particles. These belts near the earth should really be called the *Van Allen charged-particle belts*. The radiation belts exist because of a trapping mechanism produced by the geometry of the earth's magnetic field.

A plasma is different from an ordinary gas because its constituents are charged, and their motion is affected by electric and magnetic fields that coexist with the plasma in the same part of space. If you try to push ordinary, electrically inert gas across a magnetic field, it will move as though the magnetic field wasn't there. But try to push a plasma across a magnetic field, and the electrons and ions in the plasma will spiral around the magnetic field lines rather than moving in the direction you want them to go. The complex interaction between plasmas and magnetic fields is responsible for a wide variety of phenomena such as the acceleration of particles to very high speeds, electrical discharges such as the aurora borealis or "northern lights," and disruptions of radio communication.

## Why Study Space Plasma Physics?

Who cares about space plasmas? There have been three major themes underlying NASA's interest in this field. First, there is the intrinsic interest in this new phenomenon. The earth's magnetosphere may be invisible to the eye, but it is a highly complex structure that scientists seek to understand because it's there. The most visible effects of the magnetosphere are the auroras, an electrical discharge in the earth's upper atmosphere that produces a shimmering, hazy glow seen to the north in the Northern Hemisphere. Although auroras are most easily seen by people living in the Far North, in places like Alaska and Canada's Northwest Territories, they can sometimes be seen farther south.

Another motivation for NASA's plasma physics program has been gathering data about the environment in which satellites and people work. High-speed charged particles can do nasty things to the human body by smashing the molecules that make it up into tiny pieces. If the radiation dosage is relatively mild, the body can repair itself and suffer no immediate ill effects (though the long-term hazards of radiation are still being disputed). A heavy dosage of radiation can produce radiation sickness or even death. If a spacecraft is in an orbit that is subject to radiation, one can shield its contents be it electronics or people, from high-speed charged particles by building a craft with thick walls, so that the space particles hit molecules in the walls and slow down before they hit molecules in the people inside. Spacecraft designers require good maps of the particle environment around the earth and the way it changes with time so that they can design walls with the appropriate thickness.

Even a modest amount of radiation can wreak havoc on orbiting satellites, as I found out from my own first direct involvement with the space program in 1974. I had submitted a successful proposal to use a satellite called *Copernicus,* which had a one-meter-diameter telescope in it, to obtain some measurements of the ultraviolet radiation emitted by a few nearby stars. This work had to be done from space because these wavelengths of radiation are absorbed by the earth's atmosphere. I went to the satellite headquarters in Princeton to plan my observations, which I had thought would take about eight hours. I was surprised when Don York, who was in charge of dealing with young astronomers new to the space game, told me that using the satellite it would take a day or more to finish my job. Why? One reason was that the satellite's orbit sometimes passed through the "South Atlantic anomaly," a part of the radiation belt that comes close to the earth's surface because the magnetic field is weak there. These particles set off the telescope's photomultiplier tubes in the same way that the ultraviolet radiation that I wanted to measure would. Trying to measure my star when the satellite was in the anomaly would be about as easy as seeing the stars in the daytime with the naked eye. As a result, spacecraft controllers had to turn off the instrument I wanted to use when the satellite passed through the anomaly, and at other times, the satellite's orbit caused my star to set behind the earth, and so my eight-hour observation ended up taking much longer.

The third major reason that scientists are interested in space plasma physics is that space plasmas can tell us about how plasmas behave on earth. For thirty years, plasma physicists have been trying to figure out how to control the hydrogen fusion reactions. These reactions are the heart of the tremendous explosive power of the hydrogen bomb, what makes the sun shine, and a potentially limitless source of energy for human activities. But you can't confine and control hot plasmas like those in the sun's guts by putting them in ordinary bottles or other containers. One way to contain hot, ionized gas (plasma) is to confine it with "magnetic bottles," keeping the charged particles in place in the same way that the earth's magnetic fields trap particles in the radiation belts. It's turned out that leakproof magnetic bottles are a lot harder to make than Coke bottles. (Incidentally, the shape of the field lines in one kind of magnetic bottle is the same shape as on the old Coke bottle, with a thick middle and a skinny top and bottom.)

The study of plasmas in space can perhaps help us build a fusion reactor that will contain plasmas in the same way that the radiation belts around the earth are naturally confined by magnetic fields. If we can develop a reactor that can control the production of hydrogen fusion, the abundance of hydrogen and other light elements in seawater along with the tremendous amount of energy released in such reactions offers hope for a very cheap, essentially endlessly renewable energy source. Such a development would make an outstanding contribution to humanity.

Furthermore, many other cosmic phenomena, such as the production of energy in quasars (the most powerful energy sources in the universe), may be roughly similar in nature to the processes occurring in the earth's magnetosphere. Giant planets like Jupiter have stronger magnetic fields than the earth does, and their magnetospheres are somewhat different. Pulsars, the tiny, spinning remnants of the evolution of stars much more massive than the sun, also have magnetospheres.

## Radiation Belts, Geomagnetic Tails, and Other Weird Beasts in the Magnetosphere

In the early days of the space program, virtually every scientific satellite carried a tiny box to measure the number of charged particles in the region of space where the satellite traveled. Nearly a hundred satellites made such measurements, mapping out the charged-particle and magnetic-field intensities near the earth. In many cases, a satellite provided only a tiny bit of information, as early satellites sometimes failed after a few days. But all of these missions have built up a reasonably complete picture of where the particles and fields are, as shown in the sketch.

The solar wind, shown blowing from the left in the drawing, was (after the Van Allen radiation belts) the second major discovery of the space program. Its existence had been suspected because of its effects on comet tails, but no one directly measured it until the late 1950s and early 1960s when the Soviet lunar probes *Luna 1, 2,* and *3* and the American *Pioneer 4* and *Mariner 2* first ventured outside the magnetosphere, the volume of space where the earth's magnetic field dominates the shape of the charged-particle regions (in the figure, it is the

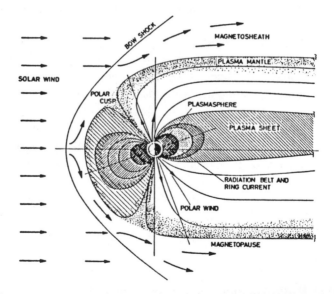

Schematic illustration of the various structures in the earth's environment. These structures are produced by the interaction of the solar wind with the earth's magnetic field. (From D. J. Williams, "The OPEN Program: An Example of the Scientific Rationale for Future Solar-Terrestrial Research Programs," in M. Dryer and E. Tandberg-Hanssen, eds., *Solar and Interplanetary Dynamics* [Dordrecht: Reidel, 1980]. Reproduced by courtesy of the International Astronomical Union.)

region inside the magnetopause). This wind first encounters the "bow shock" upstream of the earth, where a shock wave, or a sudden change in the pressure in space, deflects the solar wind. Particles that come close enough to the earth are caught by the earth's magnetic fields and are trapped for a while. In various complicated ways that scientists are currently studying, these trapped particles interact with each other and with the complex magnetic fields that govern the charged-particle flow in the magnetosphere. The various structures shown in the illustration result from these interactions, and exist around other planets with magnetic fields: Mercury, Jupiter, Saturn, and Uranus.

Scientists can describe the magnetosphere in excruciating detail; page after page of geophysics journals and conference proceedings are filled with pictures like the one shown here, which are suitably more complicated with more and more observations.[9] Our understanding of why the magnetosphere looks the way it does is still pretty general. Plasma, produced by the solar wind and ionosphere, is transported through the complicated structures in the magnetosphere and ends up in various stable places where it can sit for a while, like the radiation belts. The details that are so critical to a real scientific understanding are elusive, and the structure changes inexplicably. A relatively quiescent magnetosphere is suddenly disturbed by geomagnetic storms, where the intensity of the magnetic field at the earth's surface is suddenly disturbed by invisible, but detectable, events occurring in the magnetosphere.

Because so many satellites have measured the particle and field environment around the earth, various committees have repeatedly recommended that more attention be paid to the analysis of the data that are currently at hand. (Much of the planning of future space missions is done by committees of scientists assembled either by NASA or by other Washington groups like the National Academy of Sciences. In the case of space plasma physics, it is the academy's Space Science Board that has taken the lead.) Scientific insight here, as in other fields, can't come simply from collecting mountains of data. NASA has occasionally had difficulty supporting data analysis programs. Sitting in an office with a keyboard and screen and analyzing data, or looking out the window and thinking, is just not "the right stuff" by any stretch of the imagination. If we wish to understand the universe rather than just to marvel at it, data analysis and theory have to happen.

Space experiments in plasma physics have changed with the times. Although there are long-term monitoring programs that continue, NASA has used the space shuttle as an experimental probe, where plasma is injected into the magnetosphere from the shuttle and the effects on the magnetosphere are studied. Occasionally, NASA has launched a rocket to dump an "artificial comet" into the upper atmosphere of the earth, and the general public can sit back and watch the show (as long as it's clear, and as long as they're willing to get up at 3 A.M. if that's when NASA needs to launch its comet).

The highest priority major mission program for the future is the International Solar-Terrestrial Physics (ISTP) program. The European Space Agency (ESA), NASA, and Japan's Institute for Space and Astronautical Sciences (ISAS) have planned a six-spacecraft mission that, before *Challenger,* was scheduled to start in 1988 and continue through the mid-1990s. NASA will provide three spacecraft, ESA two, and Japan one. The various spacecraft will simultaneously measure the magnetic and electric fields, particle densities, and plasma characteristics in widely separated, carefully chosen locations in and near the magnetosphere. A recent ESA report on its space science plans for the next two decades identified space plasma physics, extended to include solar physics, as one of four high-priority "cornerstone" areas. It's quite possible that ESA may become the leader in space plasma physics in the twenty-first century. Because a recent National Academy of Sciences committee assigned this mission a high priority, and because of the participation by Europe and Japan, which are less affected by *Challenger* than NASA is, I believe that many of the parts of the ISTP program will come to pass before the year 2000.[10]

## LOOKING DOWN AT THE LIVING EARTH

So far in this chapter, I've described how satellite measurements have increased our understanding of the earth's oceans, atmosphere, and magnetosphere, the nonliving part of the planet we live on. Satellites can detect the effects of living creatures as well. Earlier, I mentioned the subtle and significant variations of the ocean color, noticed years ago by Polynesian navigators and by astronauts. These color variations allow us to track the constantly changing pattern of life in the open sea. There are a number of other ways in which satellite pictures can, in principle, tell us a great deal about life on our planet.

But how, exactly, do we use these satellite pictures? One fall day in 1985, I strolled across the University of Delaware campus, where I work, to seek the answer from Vytautas Klemas, director of the Center for Remote Sensing. Vic, as everybody calls him, has wrestled for years with the question of what can really be learned from satellite observations of life on earth and whether such a process is cost-effective. A slim, distinguished-looking, silver-haired man, he, along with his students, mostly studies the ocean currents in estuaries, where rivers meet the sea.

In his neat, well-organized office, Vic guided me through the maze of different satellite and image-processing systems to explain what he and his students do. Their activities cover an enormous range because Vic is very open to new ideas and to working with others who can make use of his expertise for their own purposes. One of his many articles describes a collaboration with Jay Custer, an anthropologist studying Indian sites in Delaware and neighboring states. Klemas, Custer, and Tim Eveleigh, a graduate student, demonstrated how pictures from space can help scientists to discover archaeological sites. (I've always wondered how archaeologists decided where to dig.) The project that I'll focus on here is a bit closer to the mainstream of Vic's work. It involves the use of satellite pictures to probe the danger from offshore oil spills and marine dumping.

In my opinion, the coast of the small state of Delaware ranks with the Big Sur and the Maine coast—especially the part east of Mount Desert Island—as one of the most unspoiled parts of the American shore. (I may be slightly biased because I've lived in Delaware for ten years, but I've lived other places, too.) Because the west bank of Delaware Bay, which forms the eastern boundary of the state, is so marshy, most of the settlement of the state is further inland. The shore road, Delaware Highway 9, passes through beautiful wetlands and goes by only an occasional house. These marshes are important as the habitat of rapidly disappearing species, as the place where many of the plants that the fish eat grow, and as a place for people to observe and enjoy. Whenever I can, I use Highway 9 on my trips to southern Delaware rather than the faster but much less attractive highway through the middle of the state. In a far-sighted move, the state government preserved the pristine nature of this coastline by passing the Coastal Zone Act of 1973, which forbade any future development on the shore.

Delaware Bay, which borders these marshlands, is a very busy place. Two major ports, the cities of Wilmington and Philadelphia, are upstream of the undeveloped shoreline, and medium-sized oil tankers pass through the bay every day. In the last ten years, there have been two major oil spills and countless minor ones. Complicated tidal currents in Delaware Bay can affect the dispersal of pollutants and oil slicks accidentally dumped into the bay. Space photographs of such a region, like the one shown here, can, in principle, do a great deal of mapping at one fell swoop. This photograph, showing various shades of gray in the bay, made Vic suspect that the boundaries between the various shades of gray represented fronts, or places where the tidal currents changed abruptly.

But how do you demonstrate that you can, in fact, use pictures like these to map tidal currents? Could Vic check his idea? In much of the remote sensing area, an essential activity is comparing satellite pictures with "ground truth," measurements of the same area taken on the ground. Vic commented, "In the beginning of any program like this, you need lots of ground data."[11] He started by comparing the satellite pictures with National Ocean Survey charts of the tidal currents and concluded that yes, indeed, the satellite pictures showed sharp "fronts" where the velocity of the current changed abruptly. The current charts were only a preliminary check, though, because they are available for only a few times in the tidal cycle. The satellite images can provide a picture of the currents at all times in the tidal cycle, and for the strong ("spring") tides as well as for the weaker ("neap") tides, and all times in between.[12]

In further studies of the ocean dumping problem, Vic and his graduate student William Philpot used satellite imagery to study what happens when acid wastes are dumped into the ocean. This stuff, the residue of a process used to make the titanium dioxide pigment in white paint, was deliberately dumped at a site 73 kilometers (about 45

Tidal currents in Delaware Bay, from an image with the Earth Resources Technology Satellite (ERTS) satellite obtained on 10 October 1972. (Courtesy V. Klemas.)

miles) southeast of Cape Henlopen, Delaware. Did the waste become adequately dispersed in the ocean before it reached the shore, if it did reach shore? How did the pollutant move during storms, which seemed to push it toward the shore? First, ground data were used as a check on the method. Surface drogues, floating buoys with radio transmitters in them, were released in the dumping area and were tracked from shore to check whether the satellite data could provide reliable estimates of the drift velocities. Then, the extensive archive of satellite data could be used to check on what happens to that waste under a variety of conditions. It turned out that when storms do carry the waste toward the shore, they dilute it very effectively, reducing the concentration by at least one million to one by the time it reaches shore.[13]

Do scientists need the data from space to study pollution? ''It's crucial," Vic told me. "You'd need a lot of ships to obtain a synoptic view of large areas. And ships are expensive, costing four thousand dollars a day and more." An orbiting satellite can easily obtain synoptic coverage of an area by mapping or photographing it day after day, week after week, and year after year on a regular basis. The need for such long-term coverage explains the need for a continuing series of earth-watching satellites.

There are many, many other areas in which satellite data can play a crucial role in disclosing the consequences of human activities on the earth. Particularly in Third World countries, where less developed road systems make large-scale mapping difficult, satellite techniques are much easier, quicker, and cheaper than either aerial or ground-based mapping. Because a satellite survey is repeated regularly, it's also very handy when one is looking for changes. Often, a state or county planning department would like to determine where the trees are currently growing on a piece of land that is proposed for residential development, not where the trees were twenty years ago when the last aerial photograph was taken.

Observations of human activities from space were first made by the *Apollo* astronauts in the 1960s, and scientists have been conducting this type of activity, under NASA sponsorship, ever since. A succession of various satellites has been used for this work, and they are now called *Landsat*s. The size of the smallest element you can see on the ground (called a *pixel*) has been gradually reduced, with the best resolution on the *Landsat D* being 80 meters (260 feet).

The future of this program was in doubt even before 1986. The Reagan administration, prodded, no doubt, by various claims that *Landsat*s can do such necessary things as geological mapping very cheaply, tried for years to sell *Landsat*—satellite, ground system, and all—to private industry. Hughes Aircraft (a subsidiary of General Motors) and RCA created the EOSAT corporation, which took over the operation of *Landsat 4* and *5* in early 1986. If EOSAT does not support the kind of research that can explore what more can be done with the *Landsat*s, who will? Can EOSAT compete with the French satellite, called *SPOT* (*S*ystème *P*robatoire pour l'*O*bservation de la *T*erre, or Satellite for the Observation of the Earth), which is being run by a subsidized, quasi-governmental French corporation and which has higher resolution? A difficulty that Vic pointed out to me is that ''We who work on the land portion of the earth observation program can't make grandiose statements about something like El Niño. For us, there's no big sweeping thing, just lots of little things.'' Vic and I have had the same experience as many space scientists have with the difficulty of obtaining funding for, or selling, packages of lots of little projects. It's a lot easier to find support for a project when you have one highly visible venture that can carry the rest of the program.

Programs like *Landsat* and particularly *SPOT* present the news media with an interesting opportunity because they can photograph anything, anywhere on the earth. *SPOT* was described by a reporter as the ''ultimate 'skycam' '' because it has a resolution of thirty-three feet, sharp enough to spot such interesting targets as military formations.[14] The media can have access to pictures of Russian military maneuvers in Siberia or troop maneuvers in Nicaragua. Satellite images provided some crucial information in the first few days after the Chernobyl nuclear reactor disaster in April 1986, a time when the Soviet Union was being particularly secretive about what had happened. Through *SPOT,* news organizations may now have access to the same type of information that has been available to the military for years.

*       *       *

Next to communications satellites, the large-scale, repeated view of the earth provided by orbiting satellites has been the closest thing to a ''practical'' use of space. As maps are the sorts of things that we are used to getting ''free'' from governmental agencies, it's hard to put a price tag on all of the results. Space images have a wide range of practical applications, including pollution monitoring, finding archaeological sites, land use planning, and reporting the news.

Oceanographers struggled for decades trying to make sense of isolated, spotty reports from ships, which were found only in a few well-traveled shipping lanes. Now, however, this and the next generation of satellites can for the first time fully map the watery part of our globe. The global phenomenon known as the southern oscillation, which El Niño is a part of, sends freakish weather conditions around the world, and we are just beginning to understand it. If we can understand this global phenomenon, we may be able to predict it. Scientists using satellites to observe the land have nothing quite like El Niño to hang their hats on, but the myriads of little projects, studying anything from predicting the results of oil spills to discovering archaeological sites, add up to a lot. Space observations have uncovered a completely new property of our planet: the magnetosphere. In addition to their intrinsic interest, studies of the magnetosphere can help us to design better space vehicles and to understand the behavior of plasmas in the laboratory.

The orbital vantage point, where a satellite can see a large portion of the globe at once, has been the basis for many space applications. None of these cases has demanded the presence of astronauts in orbit to actually do things. One other area, less well developed than earth watching or space plasma physics, does require astronauts in orbit to do experiments. Space is a laboratory in which gravity is absent, and a number of people hope to use this property of the space environment to make new materials.

# CHAPTER 7

# Materials Processing in Space

## Making Pie in the Sky, or the Next Industrial Revolution?

For years, people in the space community have dreamed and hoped for space industrialization. If space is to play a big part in our way of life, it has to involve more than people who do research for the government or work for the military. Communications satellites have been with us for almost the entire history of the space program, but no other multi-billion-dollar businesses have emerged from our activities in space. The Solar Power Satellite scheme seemed like a possible space industry in the mid-1970s, when long lines at the gas stations prompted a search for alternative energy sources. However, the energy crisis faded, a number of problems arose, and this scheme is now slowly sinking out of sight. Another possibility for commercialization is the earth observation game, where two private companies (the French SPOT and the American EOSAT) have just started to try to make a go of it in the mid-1980s.

Space manufacturing is another way of possibly making money in space. Its most enthusiastic promoters visualize it as even eclipsing the multibillion-dollar communications satellite business. The major opportunity for profit may come from manufacturing new materials in space and selling them on the ground, and the original name for this piece of NASA's program, *materials processing in space* (MPS), reflected this emphasis. A second area of emphasis is using the space environment to study the behavior of materials like fluids in the absence of gravity. To reflect this second area, which includes more in the way of concrete results, NASA has adopted the more comprehensive name of *microgravity science and applications* to describe its efforts, sometimes undertaken in joint ventures with other companies. The same close relationship between basic science, applied science, and manufacturing that was apparent in the earth-observation phase of the space program is apparent here as well.

## THE MICROGRAVITY ENVIRONMENT

Gravity is something we all live with. If you let go of this book, it will fall to the floor. It falls because gravity is pulling it toward the center of the earth. You don't fall because the rigidity of the earth itself, the building you're inside of (if you're inside), and the piece of furniture you're on support you and hold you up.

It's not that gravity vanishes when you're in earth orbit; it's that it works differently. When you're in orbit, you're falling around the earth. You're no longer supported by something rigid, and the earth's gravity pulls you toward the earth's center. But your orbital motion carries you sideways, and so the effect of gravity is to pull you around the earth. What produces the microgravity environment inside an orbiting space station or shuttle is that everything in orbit—you, the cup of coffee in front of you, the coffee in the cup, your pencil, and everything else around you—is falling around the earth in the same way. The coffee in your coffee cup is no longer pulled down toward the bottom of the cup as it is on the earth. The coffee and the cup are both in orbit around the earth. Thus, if you jostle the cup a bit, some globs of coffee might start wandering around the space shuttle cabin, making a mess.

Until the early 1970s, the microgravity environment of space was regarded primarily as a hazard to the comfort and health of the astronauts. Some experiments on some of the later *Apollo* flights provided the first indications that the microgravity environment could function as an interesting lab as well. On *Apollo 14,* a set of experiments performed in a furnace demonstrated that materials of different densities remained mixed as they solidified. On earth, the coffee grounds in a coffee cup sink to the bottom because they are heavier, but in orbit, they would remain suspended in the coffee, and you would drink them. For eating and drinking, microgravity is a nuisance, but for materials processing it can be an advantage. There are some distinct advantages to solidifying materials that remain well mixed, such as making interesting alloys that simply could not be produced on earth. Other experiments on *Apollo 14* and *17* explored the behavior of fluids in the microgravity environment, and primitive though they were, both paved the way toward a more ambitious program on the space shuttle and prompted some blue-sky expectations of a tremendous industry in space.[1]

Convection is another thing that can be turned off only by going into space. If you heat part of a liquid or gas sufficiently, the hot fluid will expand, becoming less dense than its surroundings. As long as the fluid is subject to gravity, this hot stuff will rise to the top of whatever container your fluid is in. This type of motion, where hot gas rises and cold gas falls, is called *convection*. Now that microwave ovens have appeared on the market, the plain old oven in my old-fashioned stove is called a *convective oven* because the hot element at the bottom heats the air in the oven, and the convective currents cause this hot air to rise and then transfer heat to the food inside. In some applications, such as separation of material mixtures, convection can disrupt nice, uniform fluid flow patterns, making the process much less efficient. Many of the microgravity projects use the space environment as a convenient way of turning convection off.

Why is this called the *microgravity environment,* rather than the *zero gravity environment?* The relative motion of any spacecraft and a drop of fluid suspended in it is never exactly zero. The force of gravity is slightly higher in the part of the spacecraft nearest to the earth, and the spacecraft rotates as it orbits the earth. These natural processes produce an acceleration of the spacecraft relative to its contents. Under normal conditions in low orbit, this acceleration is equivalent to $10^{-6}$ (one millionth) of the earth's gravity. Depending on the size of the spacecraft, astronaut motions may produce accelerations considerably larger than this, but still small enough so that they don't bother the experimenters in a spacecraft the size of Skylab or the space shuttle (for most experiments, a relative acceleration of $10^{-3}$ times the earth's gravity is acceptable). But in a small craft like the *Apollo* spacecraft, an astronaut who bumps into the wall during sleep could disrupt one of the pioneering experiments. Despite this danger, the early work did demonstrate the feasibility of microgravity work in space.

## WHAT CAN BE DONE IN A MICROGRAVITY ENVIRONMENT?

Many millennia ago, one of our Neolithic ancestors conducted the first materials-processing experiment when he (or she) put copper and tin together to make bronze. Ever since then, human efforts to refine metals from ores, to alloy them, to temper them, to weld them, to pound them into useful shapes, and to make gadgets like computers out of various materials have taken place in an environment in which gravity pulls heavy things to the bottom of containers. Frequently, gravity imposes limitations on what we can do. No matter how careful we are, gravity is always driving flows in a slowly cooling melt, limiting the size of the largest perfect crystal we can make of any substance. In principle, there are many new or better materials that can be made in the absence of gravity.

NASA's microgravity program has two major thrusts. In one, the aim is to send astronauts into space to make new types of materials, to bring the stuff back to earth, and then to use or sell it. A growing number of microgravity researchers have taken a different approach, using the microgravity environment as a research lab to try to discover just what gravity does to various processes on earth. Just as a terrestrial research lab can be used to study the effects of temperature, pressure, and other such variables on something like the operation of a light bulb, the space environment can be used to study the effects of gravity.

As an example of this second type of research, consider the light bulb. Night baseball is made possible by high-intensity lights that illuminate the field. The light bulbs used in these stadium lights are not ribbons of hot metal like the light bulbs in your house. Rather, they use an arc discharge in which an electrical current passes between two strips of metal (the electrodes) separated by a low-density gas. The electrical current interacts with the atoms and molecules in the gas, making light. People who design such light bulbs are therefore interested in studying such arc discharges.

General Telephone and Electronics (GTE) performed an experiment in the space shuttle as part of a much larger program to study ways of making better, more efficient high-intensity light bulbs. In an arc lamp, which works because there is an arc discharge in it, hot, light gas rises and comparatively cool, dense gas falls in convection currents. These currents shape the arc and give it its name because the gas right between the electrodes is hottest and rises, forming an upward bow pattern. The gas in the arc is usually a mixture of elements that enable the light from the lamp to mimic the color balance in sunlight. Various elements respond to the convection currents in different ways, and analysis of arc lamps shows that light from particular elements comes from different parts of the arc.

On the ground, the effects of convection are mixed in with other processes, and unscrambling them, figuring out just what it is that makes some light bulbs work better than others, is hard to do. As Paul Ritt of GTE described this work at a scientific meeting, the ability to "turn gravity off," in this case turning convection off, in a simple experiment could lead to "some substantial product improvements and even new products."[2] GTE is not planning to make light bulbs in space; rather, they are using the space environment so that they can understand better what goes on in light bulbs and thus make better ones in their earth-based facilities.

A rather unlikely company has involved itself in a similar venture. Think about a farm tractor. Heavy, bulky, with a big and noisy engine, it has tires that may well be taller than you are. If it's bright green, the John Deere company made it. Farm tractors in space? No way. But this company, which you might think of as the least likely candidate for the space program, is cooperating with NASA on a number of projects. Deere, unlike many smokestack industries, spends rather heavily on research, investing 4% of its sales in product improvements.[3] It saw some opportunities in the microgravity environment, using it to explore various details of tractor making in the same way that GTE explored the manufacture of light bulbs.

Cast iron is about one-quarter of the weight of a farm tractor. Therefore, it's important for Deere to understand how it behaves when it's solidified. Graphite flakes within the cast iron contribute in an important way to its properties, and Deere researchers are studying whether convection or diffusion was the primary factor influencing the way in which carbon atoms form graphite flakes or spheres. As with GTE's light bulb, the easiest way to turn convection off is to take a furnace into a microgravity environment provided by NASA.

In this case, the initial work didn't even require Deere to send equipment into space. NASA is willing to make its testing facilities available to companies like Deere, and one of the facilities is a KC-135 aircraft that flies on a path that takes it into free fall for a short period of time. While the aircraft is falling freely toward the earth, its contents are also falling freely, and within the aircraft, gravity is turned off. This aircraft is used to accustom astronauts to microgravity; it's earned the nickname of *vomit comet* because many people respond to microgravity by getting sick. According to author Alcestis Oberg, an experiment on one of NASA's KC-135's showed that diffusion was the primary process in making graphite flakes, and Deere thought that more extensive space experiments were unnecessary for this purpose.[4] However, NASA claims that the company is planning experiments on board the shuttle with a larger furnace.[5] This collaboration between NASA and a company that is definitely not in a high-tech industry illustrates how pervasive the space program really is.

Within this two-pronged effort of materials processing, on one hand, and research, on the other, activities in the microgravity program are very diverse. However, four types of process can more or less cover most of the materials studied and research that takes place. First, *crystal growth and solidification of molten materials* happen very differently in the absence of gravity. The hope is, for example, that large crystals grown in space can be used to make ultrafast electronic chips that could function as the brains and guts of the supercomputers of tomorrow. Closely related to crystal growth is the problem of making alloys in large quantities if the components of the alloy don't mix well. When this is tried on the earth the heavier component sinks to the bottom of the container it's put in. As a result, on the earth, the alloy will form only at the interface where the two materials touch.

A second area involves *containerless processing*. It's hard to produce ultrapure materials on earth because they have to be solidified in a container. A different application of containerless processing is making a perfect sphere by solidifying a liquid drop of molten material. In the microgravity environment, liquid drops form spherical shapes because of surface tension. On the ground, machining an absolutely perfect sphere is difficult, if not impossible. Although this is the one area where an actual commercial product has been produced in space, the scope of activity has not been large.

A third area, which at the moment involves more basic research and less production of materials, is the study of *fluid flow and chemical processes* in the absence of gravity. Would a candle flame snuff itself out if gravity didn't carry the burnt gases away from the wick? (This phenomenon was investigated as early as the Skylab mission. The aim then was to understand how things burn in space in order to design safe spacecraft.) There's a type of convection called *Marangoni convection,* in which changes in the surface tension at the edge of a fluid pull the stuff around. This type of convection is very difficult to study on earth because more ordinary convection overwhelms it. Whereas Marangoni flows a few millimeters in length are the best that can be done on earth, scientist-astronauts on *Spacelab 1* produced a flow 6.8 cm long.[6] Related projects involve the study of bubbles and thin films. A project flown in *Spacelab 3* attempted to create a scale model of the convective circulation on the earth.[7]

Within these three major areas are a large number of new phenomena. G. Harry Stine, a writer with one of the most optimistic views of the prospects of microgravity, listed forty-three different processes that are characteristic of the microgravity environment.[8] But it is the fourth major area, *processing of biological materials,* which has been the focus of most interest, because a project involving McDonnell Douglas and Johnson & Johnson offered hopes, before the *Challenger* explosion, of being the first really significant commercial venture in the world of microgravity. Depending on how long it takes NASA to get the shuttle flying again, this project may still work out. McDonnell Douglas scientists were exploiting the unique properties of the microgravity environment to purify a medically useful protein several hundred times better than could be accomplished on the earth.

## MAKING MEDICINES IN SPACE: THE
## ELECTROPHORESIS EXPERIMENT

Purification of drugs is one of the most important steps in the process of making them. A biological or chemical process produces a veritable witches' brew of proteins, fats, fatty acids, carbohydrates, and other stuff. One of the ingredients is the chemical needed in a medical process, and it must be separated from the rest. The chemical industry has developed a number of ways of separating different materials, but they don't always work as well as the industry would like. The technique of electrophoresis has been used extensively in chemical analysis. If you want to find out what proteins are in a mixture, you don't need to analyze very much of it. It's a different problem to separate out large quantities of these proteins so that you can sell some of them as drugs. Separation techniques that can deal with large quantities of material generally do not produce really pure material like medicines.

The drawing is a rough schematic of a continuous flow electrophoresis apparatus of the type that has flown in space.[9] The apparatus is a tall rectangular box, which is relatively flat in the direction perpendicular to the paper. The mixture of materials flows into the

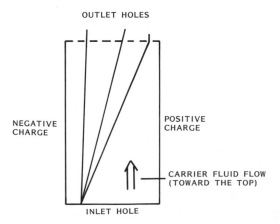

Schematic diagram of the continuous flow electrophoresis apparatus that has been used in space to purify biomedical materials.

apparatus in one small inlet hole at the bottom. Also flowing in at the bottom is a mixture of water, acid, and a buffering agent, which is called the *carrier electrolyte,* and which I'll simply refer to as the *carrier.* This carrier flows up through the apparatus and out the top, carrying the mixture along with it.

What makes the separation process work is that an electrical current flows through the system in a direction across the carrier flow (in this picture, horizontally). In the McDonnell Douglas cell, the acidity of the carrier is adjusted so that the proteins have a negative charge and are thus attracted toward the positive electrode at the right. Some proteins can be pulled through this fluid more easily than others. A small, compact protein can be easily pulled through the carrier. In this cell, it will be pulled off to one side and will exit through the top of the cell quite far off to the right in the diagram. A large, irregular protein, with many chains of amino acids sticking out in irregular directions, will have more difficulty making its way through the carrier fluid, and it will follow the carrier more easily, going on a straighter path to the top of the cell. The McDonnell Douglas cell has 197 outlet ports at the top of the apparatus, and the main idea behind it is that different pieces of the mixture will flow in different protein streams into different ports at the top.

It sounds neat and simple, but when it is put into practice problems develop.[10] The electrical current that passes through the flowing fluid tends to heat it. As the heating won't occur at a uniform rate in the apparatus because the protein streams will affect the heating rate, convection currents can be set up that mix the sample back together again. This problem can be solved by making the apparatus very small and thin, compressing the carrier between two flat plates that are only a millimeter apart. Then, there's very little carrier to heat, and squashing the carrier between the plates inhibits convection currents. This is all right if one is interested only in analyzing the mixture, but to produce chemicals to sell in quantity, the apparatus must be made big enough so that clinically useful quantities of the substance can be produced quickly.

These same problems also limit the concentration of unmixed material that can be extracted from this apparatus. If the concentration of protein is high, the heating of the fluid will be different inside and outside protein streams, and convection will mix it all back together. Another problem is that most proteins tend to be heavier than water. A concentrated glob of protein can, in some cases, sink back to the bottom of the apparatus, clogging up the input hole. Thus, there is a limit on what can be done on the earth. David Richman of McDonnell Douglas claimed that, in principle, a terrestrial apparatus could produce solutions that were 1% protein (by weight), but that the apparatus that flew in space produced, in practice, 0.25% concentrations of protein when it worked on the ground.

Theoretically, neither of these problems apply in space. Convection doesn't occur because there is no gravity to drive it, and it doesn't matter whether the protein streams become hotter or cooler than the carrier fluid. Sedimentation doesn't occur either. Just as coffee grounds float in coffee, the proteins will float in the carrier, not sinking back and covering the input hole. One can both make the apparatus bigger and get a higher concentration of protein out of the business end at the top.

Interestingly enough, few drug companies were able to take the leap of faith needed to invest resources in the idea I just described. Whereas aerospace companies are more inclined to gamble, only one aerospace company, McDonnell Douglas, had an active life-science group in house. They had participated in the *Mercury* and *Viking* programs, and when *Viking* ended in 1976, they started exploring electrophoresis. But McDonnell Douglas didn't want to take on the entire responsibility for the project because selling drugs involves knowing how to deal with doctors and with the Food and Drug Administration (FDA), which administers drug-testing programs. So McDonnell Douglas approached a number of drug companies, and according to one account, the typical response was that electrophoresis in space was ''not going to happen until 2040.''[11] It won't ever happen until someone is willing to take the risks. Johnson & Johnson's Ortho Pharmaceutical division did agree initially to test and sell the products made in space, leaving McDonnell Douglas free to explore other opportunities with other institutions. The joint venture acquired a name, EOS (Electrophoresis Operations in Space), which has another meaning as well. Eos was the Greek goddess of dawn, and space enthusiasts hope that the EOS project will be the dawn of a new era in space.

But how were these companies going to get into space? Would they have to pay each time to fly this apparatus aboard the shuttle to perfect the process? Or would they have to go through the usual proposal procedures at NASA, whereby everything they learned at great expense would have to be shared with all comers because government organizations can't keep trade secrets? NASA was very sensitive to the need to involve the private sector to a greater extent, and it developed the concept of a Joint Endeavor Agreement (JEA). This scheme allows NASA to work with potential paying customers in a way that protects the customers' investment in the groundwork. The general idea of a JEA is that NASA and the company will share the risks of a project. In this case, NASA will fly, for free, the electrophoresis apparatus while it is being tested and producing drugs to be used in animal and human clinical trials. During this period, the company can protect proprietary information as necessary. Once the material is marketed, the EOS project will have to pay the going rate to fly the apparatus in the shuttle or wherever they decide is best.

Until the *Challenger* blew up, progress on this experiment was really impressive. The first flight, on the fourth flight of the space shuttle in June 1982, demonstrated that the expected improvement in volume and purity of product would happen. Preflight calculations indicated that one could double the size of the apparatus, increasing the area of the input and output holes fourfold, and improve the concentration of protein from 0.25% to 25%, a 100-fold improvement. The expected 400-fold increase in productivity was realized in practice; the separation efficiency was improved 463 times. A year later, on the seventh shuttle flight, McDonnell Douglas tried increasing the voltage in the cell, and a 718-fold improvement over ground-based results was obtained. In 1984, they hoped to produce enough protein for the first tests on animals, but the equipment malfunctioned and the carrier solution became too hot, allowing bacterial contamination.[12] But in April 1985, one gram of pure drug was produced, and animal testing started.[13]

An illustration of how successful the Joint Endeavor Agreement scheme was in keeping the trade secret is that it was not until November 1985 that McDonnell Douglas released the identity of the drug that they were purifying in space. The drug is called erythropoietin, a human hormone that controls the production of red blood cells. Before this venture, it was available only in very small quantities for laboratory research and had not been used clinically.

Yet McDonnell Douglas's partner, Johnson & Johnson, seems to have got discouraged. A recent report indicated that they had decided instead to test erythropoietin samples made by Amgen, an earthbound genetic engineering firm.[14] Perhaps the uncertainties following the *Challenger* disaster, given that it's not clear at this writing how soon the electrophoresis apparatus could fly again, produced some second thoughts. And so, it's not now clear where this particular venture is headed. McDonnell Douglas's hopes had extended far beyond erythropoietin; they had a shopping list of something like twenty drugs that they had in mind for space manufacturing. One specific possibility has been announced: the production of beta cells in space as a possible cure for diabetes, done in collaboration with Washington University in Saint Louis.

## OTHER COMMERCIAL MICROGRAVITY VENTURES

The electrophoresis experiment often served as the symbolic leader of the fledgling microgravity industry. A number of other ventures have occurred as well. Most other companies are simply using the space environment to do research of interest to them, as John Deere and GTE did in exploring the properties of arcs in light bulbs and cast iron in farm tractor engines. 3M's joint endeavor with NASA, for example, involves growing crystals in orbit. Its initial emphasis is on basic research rather than on making products in orbit. Grumman, Batelle Columbus Laboratories, Union Carbide, and Westinghouse have all been cited as possible participants in similar projects. In mid-1983, 140 separate ventures had reserved space in NASA's Get Away Special program, where $10,000 buys space to fly a small, self-contained experiment on a shuttle flight.[15]

One venture has actually brought a product to market. John W. Vanderhoff, a chemistry professor at Lehigh University, worked with NASA scientists to design an instrument called the *Monodisperse Latex Reactor* (MLR). A plastic mixture contained in four furnaces, each about a foot tall, congeals and grows around tiny seeds to produce plastic spheres of absolutely uniform size. In one space shuttle experiment, these furnaces produced ten billion identical tiny balls. The National Bureau of Standards has stamped its seal of approval on them, certifying that the size of these spheres is ten micrometers in diameter, varying from this by only a tenth of a percent.[16]

But what in the world can anyone do with 10 billion plastic balls? No one needs all 10 billion, but quite a few people can use a mere 15 million. You can buy these in a 5-milliliter vial from the National Bureau of Standards for a cost of $384.[17] If you take a photograph of a substance through the microscope and put one of these spheres in your field of view, you can determine, exactly, just how big things are in the picture. This should be useful for people who want to know the sizes of particles in such things as paint pigment and the toner used in copiers. Other possible users include people investigating biological membranes who want to know the size of the organelles in the cell wall that are responsible for transporting food and waste products in and out of a cell.

NASA plans, on future flights, to make more latex balls in larger sizes. They have published "estimates" that claim that the market for larger 100-micron spheres could be $200 million to $300 million annually.[18] But as with other estimates of the market for space products, these are basically random guesses.

The idea of making spheres in space was behind one of the oldest proposals for space manufacturing: making ball bearings in space. Ball bearings have been in use for well over a hundred years. Some uses of ball bearings require bearings of exceptional uniformity or roundness, not because the bearing costs a lot but because the cost of a bearing failure can be large. In June 1973, Skylab astronauts tried cooling blobs of molten metal in space to see if they would make perfect spheres. The balls were not absolutely round, evidently because they cooled unevenly. I mention this story to illustrate that not every idea works out in practice.[19]

Besides the latex sphere and EOS projects, there is one other less mature commercial venture that involves making materials in space and using them on earth. A retired U.S. Air Force officer, a retired NASA manager, and a Harvard MBA joined forces to establish Microgravity Research Associates, with the backing of a dozen venture capitalists. Microgravity plans to grow crystals in space, starting with gallium arsenide. This substance is a semiconductor used in integrated circuits. These circuits are the brains of computers, calculators, and the microprocessors used in automobiles. The high-quality gallium arsenide crystals could, in principle, be used to make chips that would be much faster than the silicon chips used in most applications. Gallium arsenide can be made on earth, but the crystals are imperfect and the circuits are not as fast as they could be. This company hopes that gallium arsenide crystals grown in space will be the basis for the high-performance end of the world of high-tech products. There are people who would be willing to spend $100,000 to put a supercomputer on their desks, and they are the proposed market for Microgravity.

## THE SKEPTICS TAKE ON MICROGRAVITY

The potential for microgravity looks tremendous, on the surface. Enthusiasts point to multibillion-dollar pharmaceutical markets and even larger markets in the semiconductor industry. But the industry has just barely gotten started, and these market projections are mere guesses. It is sobering to read what was written on this topic ten years ago and compare it to what has happened. It's one thing to point out that there are forty-three different ways to fabricate materials in space that can't be used on the ground, but somebody has to be willing to buy them eventually. Currently, it costs a few thousand dollars per pound to launch something in the shuttle, and this cost applies to everything: the astronaut, the furnace, and the power supply, as well as the stuff actually made. For space processing to pay off, you have to find a product that you can sell for a million dollars per pound. There are a few drugs and semiconductors that are worth that much. There may be more products that simply can't be made on the earth.

NASA's hopes for a corporate rush to the space shuttle have not been realized. In most sections of this book, space limitations keep me from giving you more than a sampling of what's going on, but in this chapter, I may well have discussed every single microgravity venture that's been tried by private companies. For example, at the moment, only eight companies have signed Joint Endeavor Agreements for shuttle flights.[20] Many of these companies are already part of the aerospace community and should not be regarded as newcomers to the space program. Gail Bronson, writing in *Forbes* magazine, is even more downbeat: "Other than that [3M's and McDonnell Douglas's shuttle flights], a corporate marketing effort by NASA has produced zip."[21] Many companies are evidently adopting a wait-and-see attitude.

Why are they so reluctant to invest, if so many wonderful things can be done in microgravity? Because it's so costly to work in space, the space program provides an incentive to improve processes that work on earth—in electrophoresis, for example. A. R. Thompson of Britain's Harwell research laboratories dismissed the space experiments as "inevitably costly" and then wrote about his group's ingenious efforts to overcome the convection currents in electrophoresis by putting the carrier fluid in a skinny cylinder and then spinning the outer wall of the cylinder, inhibiting convection.[22] Others point out that it is merely difficult, not impossible, to grow large good crystals here on the earth.

Another reason for the corporate reluctance is rooted in the conservatism of the American financial system, combined with the high cost of space exploration. The ante for participating in the space poker game in a significant way, in the materials-processing business, can be quite high, in the range of hundreds of millions of dollars. Given all the uncertainties, bankers appear to be more willing to make more conservative investments of such substantial funds. Whereas Steve Jobs and Steve Wozniak could start Apple Computer in their garage, it's hard to start a space venture on a shoestring.

The bankers' reluctance to invest is understandable, given that some of the early claims of microgravity enthusiasts recalled those of the snake-oil salesmen of a century ago. Space program supporters face the difficulty that there is one way to sell things to Congress and to Presidents, and there is another way to sell things to bankers. The political arena understands, if not demands, visionary promises. No one knew whether we could land people on the moon when President Kennedy and NASA sold the idea in 1961. No one even attempted to come up with a "return-on-investment" figure for the space station in 1984. But bankers and private companies think in just those terms and distrust people who regale them with unsubstantiated claims of multi-billion-dollar markets. Consequently, even the enthusiastic supporters of this program pepper their speeches with comments like ''it's unclear to me as to whether it is really going to be commercially viable.''[23]

But it's not just bankers who suspect that the space microgravity program is occasionally oversold. A group of scientists flew an experiment on *Spacelab 2,* trying to grow large crystals of big biological molecules to test whether X-ray techniques could determine their structure. This experiment had limited success, partly because on that particular shuttle flight, the astronauts were chasing after a communications satellite that had failed to get into the proper geosynchronous orbit. The repeated firing of the shuttle's rocket engines seriously disrupted the microgravity environment. Gina Kolata of *Science* magazine, the leading American interdisciplinary scientific journal, wrote an article whose title, ''The Great Crystal Caper,'' mirrors the concern of many biologists. One of the reasons for the skepticism recalls similar criticisms of other microgravity experiments, namely, that the necessary ground-based work has not been pursued energetically enough.[24]

The other concern of the biochemical community in this particular instance is that, as crystallographer Thomas L. Blundell of the University of London was quoted as saying, ''One must be careful that one is not justifying a huge space program on the basis of marginal experiments.''[25] This criticism has some merit, as the payoff from the microgravity program has not been large so far. The Soviets have had considerably more experience with microgravity, and nothing spectacular has come out of their program either. (But as Soviet microprocessor technology is considerably behind ours, and as the Soviet Union is a closed society, we have to be a bit cautious in drawing conclusions on the basis of their experience.)

Although I share the cautious approach to microgravity, it seems to me that the skeptics may be overstating their concern. It's hard for someone who knows just one disciplinary aspect of the space program to understand the extremely broad range of space science and technology. Sure, you couldn't justify the whole shuttle program just on the basis of the microgravity program. But you also probably couldn't justify it solely on the basis of oceanography, or earth sciences, or communications satellites (you could launch those with expendable rockets), or planetary exploration, or space plasma physics, or astronomy, either. Many people inside or outside the space program are largely aware of the part of the program that is closest to their line of work. They have to consider the whole range of space activities in order to decide whether an investment in an infrastructure of shuttles and space stations is worthwhile.

*    *    *

So where does this leave microgravity? In the four major areas of investigation, crystal growth and separation of biological molecules are two areas where a good deal of progress has been made. Containerless processing has led to the first commercial product from space (the tiny latex spheres that can be used to calibrate microphotography), but the market for these may be limited. The study of fluid flow in microgravity has not led to any commercial products but has been the basis for a good deal of research that could lead to useful products.

There's a fair amount of interest in using the microgravity environment to do basic and applied research on problems related to processing materials on the ground, such as the John Deere effort to understand the physics of cast-iron solidification. The other aspect of the microgravity program, making materials in space and selling them on the ground, is playing a more restricted role at the moment, with only one product on the market and nothing firmly in the pipeline. A few companies are beginning to get their feet wet, and others in the biotechnology industry are either being downright skeptical or waiting to see what happens to these first few adventurers. Is the microgravity business just pie in the sky, or is it the next industrial revolution? We just don't know right now. Some carefully done work, on the ground and in space, is the next logical step. Claims of multibillion-dollar sales in the next ten years are simply premature.

# Summary of Part II

The orbital vantage point, where one satellite can command a view of a large fraction of the earth, is the basis for many of the practical uses of space. Communications satellites in geosynchronous orbit are the major ongoing space industry. The capabilities of communications satellites are such that virtually any conceivable communications need can be met by technical upgrading of the system in the near future. These satellites weave a global network over our heads that brings people on different parts of the earth together. They have changed our lives in many subtle but significant ways.

The military uses of space also take advantage of a satellite's ability to view large fractions of the globe. The space age began because the American military needed an invulnerable reconnaissance satellite. These satellites have been invaluable monitors of arms control agreements. The military has also used the orbital vantage point to establish satellite navigation systems. The military's navigation system will be used by civilians, too; a related system is the search-and-rescue satellite system, which can locate ships and airplanes in distress and send aid. A much more visionary use of space for military purposes is the global ballistic missile defense system envisioned by the Strategic Defense Initiative. Political leaders apparently have diverging views about exactly what the Star Wars system can do, in principle and in practice.

Satellites can monitor the condition of the world beneath them in many ways, measuring wind speeds, temperatures, and colors as well as taking pictures. These measurements are used for practical purposes like weather forecasting as well as for a study of how the globe works. In recent years, it has become clear that the earth is a complicated, interrelated system of land, oceans, atmosphere, living systems, and even the magnetosphere. Sometime in the future, a comprehensive study of the earth as a complete system will provide us with an integrated view of our planet. Such a study will depend critically on satellite data.

The prospects for using the satellite environment to do experiments and to make materials in the absence of gravity are rather uncertain. It does seem clear that this environment can be used for experiments in which turning gravity off can teach engineers how to make better products on the earth. Much more speculative is the idea of manufacturing or separating products in space and selling these products on the ground. Too little is known about the cost of making such products and about the likely market for them to allow us to determine whether the microgravity industry will be making billions of dollars worth of materials or just pie in the sky in the twenty-first century.

# PART III

# Exploring the Near and Distant Universe

Space has opened up many human frontiers. Thirty years ago, most of the objects in the Solar System were mere dots with names or cloud-shrouded enigmas. Automated probes have flown by many of the major planets and countless satellites. They have revealed giant volcanoes on Mars, globe-girdling grooves on the tiny icy satellites of the giant planets, the black peanut in the core of Halley's comet, black holes in space, and more.

The exploration of the Solar System has progressed rapidly for a period of twenty years. It has opened up new vistas to stimulate the human imagination and has provided scientists with a cosmic laboratory in which to study the earth in comparison with other planets. Each planetary encounter has taught scientists that surprise is the rule, not the exception.

Mars occupies a very special place in the exploration of the solar system. In the past, it has dominated human imagination because it was the only planet with a surface clearly visible from the earth. In the future, it may be the next major destination for humans to visit and possibly even to settle. We still know very little about this red planet, and it is apparently becoming a major focus for the planetary exploration programs of both major space powers.

Exploration of the distant universe from the space vantage point has revealed its share of surprises, too. Telescopes above the atmosphere can access a much broader range of the electromagnetic spectrum than telescopes on the ground. Space astronomy has discovered black holes, ultrahot stars, protoplanetary disks, powerful dusty galaxies, and a host of new kinds of objects, providing a much richer universe for us to live in, understand, and appreciate.

A small but significant part of NASA's total program is an attack on one of the oldest of human questions: Are we alone? Technology has advanced to the point where a meaningful scientific attack on this question makes sense. What we know about the development of life on earth in the broader context of cosmic evolutions suggests that the universe may contain abundant life. If there is life in the universe, radio astronomers are patiently awaiting the signs of its existence.

# CHAPTER 8

# The Golden Age of Planetary Exploration

In early 1986, human eyes cast their first close-up glance on Halley's comet and the planet Uranus, thanks to the power of automated space probes. In the wake of the *Challenger* accident, the dramatic revelations of these encounters received less attention than they might have otherwise. The coal-black, peanut-shaped nucleus at the heart of Halley's comet and the V-shaped grooves on Uranus's tiny moon Miranda are only two of the surprises of 1986. For the last two decades, many other unexpected discoveries have overwhelmed astronomers and the general public. You and I are fortunate enough to live in an era in which our knowledge of the Solar System has expanded as explosively as our knowledge of the earth did about five hundred years ago, in what is now referred to as the Golden Age of Discovery. Numerous authors have drawn parallels between these two dramatic human ventures.

What was the Golden Age of Discovery? During the two centuries between 1400 and 1600, our conception of the world we live in changed completely. In 1400, as far as western Europeans were concerned, the world consisted of Europe and Asia, with uncharted oceans and lands, "Terra Incognita" surrounding them. Prince Henry of Portugal, a younger brother in the Portuguese royal family, established and led an enterprise that sent intrepid captains out into the unmapped oceans in the fifteenth century. The Portuguese caravels gradually worked their way down the coast of Africa, mapping the shores of the

Dark Continent bit by bit. By the end of the century, Bartolomeu Dias had rounded the Cape of Good Hope and had proved that Africa was not the northern extension of the mythical Southern Continent which, according to early maps, covered much of the southern part of the globe. The existence of the Americas came as a complete surprise. Columbus's expeditions would have dominated the network news for weeks had there been such a medium. Soon after, Magellan and Drake led voyages which demonstrated the vast extent of the Pacific Ocean. Although much remained to be done, by 1600 western Europeans knew that the world was mostly water, not land; that there were at least five major continents, not three (Australia and Antarctica were not discovered until later); and that a rich variety of people and places existed far from the familiar terrain around the Mediterranean Sea.

The first two centuries of the Age of Discovery coincided with the Renaissance in western Europe. That movement freed humanity from the narrow, cramped medieval worldview, in which life had been regarded as an unpleasant interlude to be nobly endured if one were to be admitted to the Utopia of heaven. Botticelli's famous painting "The Birth of Venus" is often thought of as a symbol marking this new age. It depicts a nude woman standing in a scallop shell, obviously enjoying life. It could be viewed as symbolizing humanity's emergence from the shell-like confines of the regulated medieval existence. The Age of Discovery, when explorers transcended the limits of the apparently uncrossable Atlantic and Pacific oceans, was part of the Renaissance.

It must have been a tremendous time to be alive, provided that you were in a position to participate in this revolution. Imagine living in Barcelona in 1493 and hearing that Columbus had discovered America. His report of the discovery, already written when Columbus landed, became a best-seller of its time and was quickly translated into many languages.[1] Thirty years later, the eighteen survivors of Magellan's voyage around the world had an even more dramatic story to tell. Antonio Pigafetta, a gentleman volunteer who had sailed as a crew member on the voyage, went from one medieval court to another, recounting the fleet's adventures and finally publishing a book about the voyage in Paris in 1525.[2] The communications technology of the time probably made it difficult for a large number of people to share the elation of discovery, as there were no Walter Cronkites on the docks at Seville to interview the weary mariners when they returned from Magellan's voyage on 8 September 1522. Yes, indeed, it must have been an exciting time, fitting the description of a "golden age."

Now, jump ahead nearly five hundred years in history and consider what we knew about the planets in the Solar System in 1950, the dawn of the space age. Eighth-graders memorized the names of the planets and a bunch of facts, often incorrect, about their distances from the sun, their sizes, and their rotation rates. Astronomers, believe it or not, knew little more; the moons of Jupiter and Saturn were mere dots in telescopes, dots with names and orbits. Mars was the only planet whose surface could be seen at all. Debates about its surface dealt with such unproductive questions as whether the dusky markings were canals or not. Planetary science was a backwater of astronomy, which itself was in a sleepy state, being practiced by a handful of experts isolated in a few mountaintop observatories. It's only a slight exaggeration to say that, when Carl Sagan got his Ph.D. degree in 1960, the number of American planetary scientists doubled (increasing from one to two). Our knowledge of the Solar System in the 1950s was indeed medieval, especially by comparison with what we know now.

Space-age exploration of the Solar System got under way in 1959, when the Soviet probes *Luna 1, 2,* and *3* hit the moon and returned the first pictures of the moon's far side. In the three decades since then, our conception of the Solar System has expanded explosively, in a way that is quite reminiscent of the discovery of the American continents, the Pacific Ocean, and the Indian Ocean five hundred years earlier. The satellites of Jupiter and Saturn have confounded our preconceptions and have turned out to be fascinatingly different, and more interesting, than the earth's moon, though they are roughly the same size. We've discovered rings around Jupiter and Uranus and debris around Neptune. A dozen astronauts have left their footprints in the barren soil on the earth's moon. Automatic landers have touched down on the two nearest planets, Mars and Venus. These two planets, the most earthlike in the solar system, have familiar yet bizarre features on their landscapes: canyons that would swallow up a hundred Grand Canyons, and volcanic mountains that would make the island of Hawaii, or Mount Everest, look like a tiny pimple.

Because of this grand increase in our knowledge, some planetary scientists have gilded the present, dubbing the period from 1965 onward the Golden Age of Planetary Exploration. The analogy has been pressed a bit too far; there are those space enthusiasts who expect the mere invocation of the memory of Columbus and Magellan to open the doors immediately to an expensive and expansive program of settlement of the inner Solar System. There is reason for the excitement. Thanks to modern technology, anyone can see the latest pictures from deep space almost as soon as they are telemetered to the computers at the Jet Propulsion Laboratory in Pasadena, California.

To compile a complete account of the significant discoveries of the past twenty years would take up an entire book.[3] I'll start with three chapters on them with some highlights of scientists' investigation of small bodies, planets and comets smaller than the earth. In the 1950s, this group of objects included the moon, Mercury, the comets, and a bunch of dots: the satellites of Jupiter, Saturn, and Uranus. Mercury had scarcely been mapped, and we didn't even know how fast it spun on its axis until the early 1960s. The satellites of the outer planets and the comets had names, but not much else was known about them. Although we could map the surface of our own satellite, the moon, nearly half of the moon is permanently invisible from the earth, and it, too, was "terra incognita," unknown territory like much of the surface of the globe in the 1400s.

Now we know what the surfaces of these smaller bodies are really like. These worlds are smaller than, but not very much smaller than, the earth. The surprising diversity of these worlds is not only intellectually stimulating but also of some pragmatic value. We can try to understand our own earth by analyzing it in comparison to the other worlds. Why does Mars have volcanic mountains that are like, but much larger than, comparable features on the earth? Why does Venus have such a thick atmosphere, an insulating blanket that makes its surface stiflingly hot? Why is the lunar landscape pitted with craters, whereas the surface of Jupiter's satellite, Io, has few craters, but nearly a dozen active volcanoes spewing sulfurous gas onto the surface? We can't test our ideas of how the earth works by experimenting with our own planet; such experiments are too expensive, not to mention dangerous. But the Solar System is a cosmic laboratory in which these experiments can be performed, in a sense, by asking why the twenty (or so) planets and small planetary satellites differ from the earth, and from each other.

## SMALL WORLDS OF ROCK AND ICE

The worlds in the Solar System vary enormously in size. The very smallest are not really separate worlds; these are the meteorites, rocks that wander through the Solar System and that come to our attention when they hit the earth. Most identified planetary satellites are somewhat bigger, many a few hundred miles across, though some are larger than the earth's moon and the planets Mercury and Pluto. Mars and Venus are the most earthlike in size, and the four biggest planets in the solar system are the gas giants Jupiter, Saturn, Uranus, and Neptune.

Perhaps the easiest way to make sense of the nine planets and the nearly three dozen known satellites in the solar system is to group them by size. Were it not for the orbital paths of many small bodies, which take them quite close to large objects like the giant planets (or the sun itself), this hierarchy of size would also correspond, roughly, to an increasing hierarchy of complexity, with the smallest planets being the easiest to understand. The reason is that planetary mass determines an object's thermal history, and its thermal history determines the complexity of its surface. When first forming, all planets and satellites acquire some internal heat. The smaller ones have a thin insulating blanket surrounding the hot core, which lets the heat dissipate quickly (on an astronomical time scale, quickly means in 100 million years, a small fraction of the age of the Solar System). The larger ones not only have more heat to start with but have thicker envelopes surrounding the hot cores, so that the heat leaks out much more slowly. The core of a large planet like the earth is hot even now, and Jupiter is so hot that it emits more energy than it receives from the sun.

So let's start with the best known small body in the solar system, the earth's moon. Its importance to humans goes far beyond its role as an excellent example of a small, geologically inactive body. For centuries, if not millennia, people and animals have marveled at the moon. Wolves howled at it. Nocturnal creatures used its light to hunt by. In centuries past, before electricity, men and women could travel at night because the moon would light their way. The earliest cave dwellers and their descendants kept track of the changing phases of the moon, trying to decode the celestial cycles producing the seasons and the tides. Generations of lovers hoped that the softness of moonlight would transform a commonplace landscape into something truly magical.

The romantic moon began to acquire another dimension nearly four centuries ago, when Galileo first turned his telescope to get a closer look at the lunar surface. The figure following shows a modern telescopic picture of our only natural satellite. The pattern of light areas and dark areas (if you want technical names for these, they are

A telescopic photograph of the moon, showing the circular impact craters that cover its surface. The larger dark areas are also circular or are formed from overlapping circles and are giant impact basins that have been filled in with darker colored magma. Small, inactive bodies should resemble the earth's moon in having cratered surfaces. (Lick Observatory photograph.)

the *mare* and the *highlands*) is visible to the naked eye as the features of the "man in the moon." A closer look at the moon shows a surface that is pockmarked with circular craters. Even the large dark areas are large, circular, filled-in craters. The two dark features covering most of the bottom half of the moon in the illustration (which, incidentally, are called Mare Imbrium, or "Sea of Rains," and Mare Serenitatis, or the "Sea of Serenity") are definitely circular.

All of these circular features, from the tiniest ones visible in the picture to the large basins filled with dark material, have been produced by impact. Small or large rocks whizzing around through space strike the lunar surface traveling at speeds of roughly ten thousand miles per hour (tens of kilometers per second). These high-speed bullets must suddenly rid themselves of their energy when they bury themselves in the lunar interior. This energy is explosively released beneath the lunar surface in the form of heat. As a result, the rock beneath the lunar surface has to expand rapidly, exploding and making a hole. Because the explosion is symmetrical, the hole is round no matter what shape the incoming object was. Telescopic pictures showed us, well before *Apollo*, what sculpted the lunar landscape: impact. But we knew little else.

On 20 July 1969, Neil Armstrong and Buzz Aldrin climbed down the ladder of the *Apollo 11* lunar module and embedded the first human footprints in the lunar soil. Once the hoopla of historic statements, conversations with the President, and flag waving was over, Aldrin and Armstrong began to collect rocks and deploy other scientific experiments, beginning a series of scientific investigations that continued through five additional *Apollo* missions.

The most significant scientific legacy of the *Apollo* program is the development of a time scale that can be used throughout the inner, and perhaps even the outer, Solar System. The primary ingredient in the lunar landscape is the circular crater, and it is craters that form the basis of this chronology. The older a planetary surface is, the more objects have hit it over the billions of years. A newly created planetary surface will be as smooth as a newborn baby's skin. As billions of years pass, it is struck by more and more incoming meteorites and becomes scarred with age, acquiring more impact craters.

Consequently, the simple, if tedious, task of counting craters on

Astronaut Buzz Aldrin on the surface of the moon. Neil Armstrong and the lunar lander *Eagle* are reflected in Aldrin's faceplate. (NASA.)

the surface of a planet can tell us whether the surface is old or young. How old? How young? Assigning absolute ages to a surface with a particular crater density is impossible unless one has some samples of material from the surface of one heavily cratered body. The moon rocks allow scientists to determine how old various parts of the lunar surface are, as their ages can be determined by radioactive dating techniques that geologists know well. This work provides a cosmic yardstick relating crater number to age. The yardstick then allows us to estimate the age of any other planetary surface by counting the number of craters on it.

## LUNAR SURPRISES

One great surprise of the *Apollo* studies was that the ages of lunar materials were not quite what we had expected. The heavily cratered lunar highlands are not primordial but are the result of a tremendous final bombardment about 500 million years after the moon formed, or 4 billion years ago. Scientists hoped that we would be able to find some very primitive rocks—dubbed "genesis rocks" by some journalists of the time—that could be key clues to the early history of the Solar System. The oldest rocks that came back from the moon were somewhat younger than the Solar System as a whole. There were, disappointingly, no genesis rocks on the moon. Our quest for the origin of the Solar System would have to go elsewhere.

Another surprise was the presence of some magnetic fields on the moon. The earth's magnetic field is produced by the motion of fluid in the earth's hot core. Measurements by the *Apollo* astronauts and by a number of magnetometers and particle detectors ejected into orbit around the moon indicated that, surprisingly, there was a magnetic field associated with the moon. This is not a global magnetic field like the earth's; rather, there are localized magnetic fields, apparently constant over distances of several hundred kilometers. The origin of these fields is a complete mystery. The moon does not spin fast enough to drive an internal dynamo. The alternative hypothesis that a magnetic field generated somewhere else (where?) would magnetize the lunar rocks doesn't work either.

The mysterious, spotty lunar magnetic field became even more puzzling when *Mariner 10* flew by Mercury. In photographs, Mercury looks very much like the moon, and it spins on its axis once every two months. In 1974, *Mariner 10* discovered that Mercury has a surface field about 1% as strong as the earth's. This field does form a magnetosphere around the planet. Here, too, the field's origin is quite mysterious indeed. The most reasonable possibility is that the planet has a currently active dynamo, suggesting that it has a large fluid core with a radius 70% of the radius of the planet.[4]

## SATELLITES OF JUPITER, SATURN, AND URANUS

Scientists expected few surprises from the other small bodies in the Solar System, such as the satellites of the giant planets Jupiter, Saturn, and Uranus. Because heat leaks out of their interiors so fast, the lunar analogy suggested that they should be geologically bland objects with nothing more than a bunch of impact craters on their surfaces. Satellite exploration was an incidental goal of the missions to the giant planets.[5] The limited instrumental capabilities of the Pioneer missions,[6] the first to the outer Solar System, could not produce useful pictures of these satellites.

One of the most successful planetary missions in the space program was the *Voyager* program, a stripped-down "grand tour" of the outer Solar System. In the heady *Apollo* years of the late 1960s, NASA scientists noticed that the planets would be in a rare position in the 1980s that would allow for multiplanet missions. Suitably targeted spacecraft could fly by Jupiter and Saturn, and the encounters with each of these planets would redirect the craft to Uranus, Neptune, or Pluto. The four proposed launches shrunk to two, and Pluto was dropped from the itinerary. *Voyager 1* flew by Jupiter, by Saturn, and then out of the Solar System. *Voyager 2* was launched on a multiplanet trajectory to take it past Jupiter (1979), Saturn (1981), Uranus (1986), and Neptune (1989). Because no other planetary missions have been launched since 1978, the *Voyager* program has kept planetary scientists active through the lean years of the 1980s.

The *Voyager* missions revealed many surprises, but none more unexpected than the weird worlds in the satellite systems of the giant planets. Only a few were ordinary cratered disks like the earth's moon. A simple examination of the photographs of some of the stranger satellites of the giant planets, shown in the four accompanying illustrations, shows that their landscapes have far more than impact craters. The pictures show Enceladus (a satellite of Saturn), Ariel, and tiny Miranda (both satellites of Uranus). There are a few outer planet satellites that are ordinary cratered disks, but many show the same sort of strange surface features seen here.

These satellites do contain some impact craters, indicating that they are not totally unlike the earth's moon. But note the huge rift valleys, extending halfway around these bodies, cutting through the impact craters. The floors of these valleys show very few craters, indicating (in the case of Enceladus) that they are less than a billion years old. The exotic V-shaped pattern on Miranda, nicknamed the Chevron, has light and dark areas in it. The picture of Miranda's entire landscape shows several circular swirls, which would, if seen directly

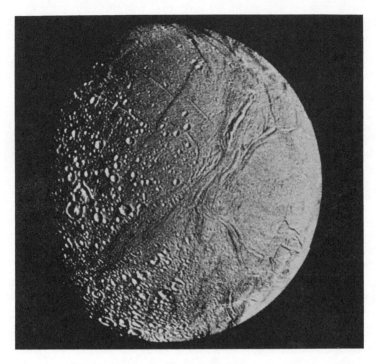

Enceladus, a satellite of Saturn, photographed from *Voyager 2* in August 1981. It is quite unlike the earth's moon, showing globe-girdling grooves and smooth, nearly craterless plains. (Jet Propulsion Laboratory, NASA.)

from above, be a giant bull's-eye of rocky cliffs. The shortage of craters on some of these circular features (called *ovoids*) indicates that they are young in geological terms.

What is going on? Ariel, the largest object illustrated, is only 1,200 kilometers in diameter. Enceladus and Miranda are even smaller, at 500 and 300 kilometers. These satellites are all much smaller than the earth's moon, which has a diameter of 3,476 kilometers. All of the primordial internal heat must have leaked out of these satellites eons ago, just as it did in the case of our own moon. But obviously, something is happening to modify their surfaces extensively. What is it?

The most common geological feature on these satellites, and on

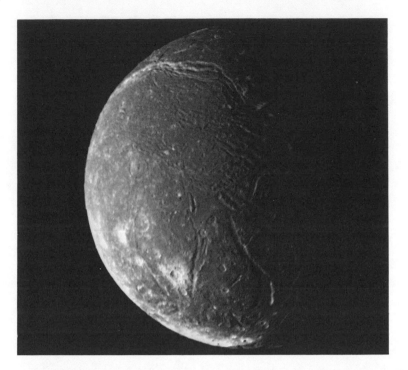

A mosaic of four high-resolution *Voyager 2* images of Ariel, a satellite of Uranus. The high crater density indicates an old surface, but this pitted terrain is crisscrossed by large fault valleys. (Jet Propulsion Laboratory, NASA.)

others like them, are huge rifts extending three-quarters of the way around the planet. Their nature is illustrated most clearly in the close-up photograph of Miranda, showing the ups and downs of the valleys silhouetted magnificently at the satellite's edge. The depths of the grooves are a few kilometers, a substantial fraction of the satellite's 300-kilometer diameter. Similar, globe-girdling grooves are found on virtually all of the other giant planetary satellites.

The satellites of Uranus present major puzzles, partly because they are unusual and partly because, at this writing, there has been little time for thoughtful interpretation of the pictures. Broadly speaking, scientists divide them into two major groups: the outer satellites (Oberon, Umbriel, and Titania, with relatively old surfaces) and the inner two (Ariel and Miranda), which are, in places, much younger. Astronomer David Stevenson has pointed out that, if there is some ammonia in the interior of an icy planet, that ammonia can act as antifreeze because ammonia–water mixtures (called *eutectics*) can remain liquid down to a temperature of 100°K, not much higher than the surface temperatures of these frozen, weird worlds. With such an interior, these small icy worlds could have fluid interiors even now. The difference between the two groups of Uranian satellites, then, presumably arises because Ariel and Miranda were spiked with more antifreeze when they formed than the outer satellites were. Another possibility is that they were originally endowed with a larger fraction

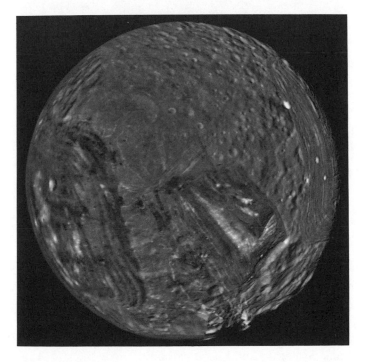

A computer-assembled mosaic of images of Miranda, the innermost and smallest of the five major satellites of Uranus, and just 480 kilometers (about 300 miles) in diameter. The surface is strikingly complex. Planetary scientists believe that this satellite may have been disrupted and reaccreted several times. (Jet Propulsion Laboratory, NASA.)

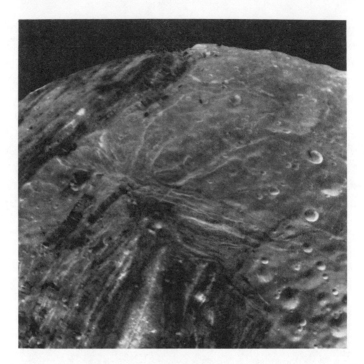

Close-up image of Miranda. The high resolution of 600 meters (about 2,000 feet) unveils many grooves and fractures, some of which are silhouetted against the planet's limb at the top. (Jet Propulsion Laboratory, NASA.)

of rocky material, so that radioactive decay of the uranium in the rock continues to warm their interiors.[7]

Another Uranian enigma is the high abundance of dark material on the satellite surfaces. Both the bright and dark areas on these satellites are quite colorless. The bright areas are probably water ice, as expected because water is quite common in the outer solar system. What is the dark matter? The smallest satellites and rings of Uranus are dark indeed, being darker than carbon black, dark as the darkest stuff in the Solar System. Could it be carbon?

Another possible influence on the peculiar surfaces of the satellites of Uranus is that planet's unusual tilt in the Solar System. Most planets spin generally in the same plane as they revolve around the sun. The earth's equator is tilted some twenty-three degrees relative to the earth's path around the sun, and other planets are tilted either about as much (Mars, Saturn, and Neptune) or considerably less (Mercury, Venus, and Jupiter). Satellites almost inevitably orbit in the same plane as the planet's equator. But some freak early accident tilted Uranus's rotation axis so that its equator is perpendicular to the ecliptic, and at the time of the *Voyager* encounter, its south pole pointed almost straight toward the sun. It's quite possible that some or all of the Uranian satellites were disrupted during this catastrophe and were reaccreted; this suggestion has been made specifically for Miranda, which looks very much like a conglomerate of different surface units.

Now, move inward in the Solar System and look at the satellites of Saturn, including the odd one Enceladus. Here, planetary scientists have had more time than a few months to consider the implications of the pictures. David Morrison of the University of Hawaii and his collaborators proposed a scenario to explain the rifts on these objects.[8] When a young icy satellite cools, the surface solidifies first, leaving an ice coating around a liquid water interior. When the liquid interior freezes, too, it expands; water is one of those few substances that expand when they solidify. This expansion will produce tension in the outer crust, which should then buckle and pull apart, producing rift valleys. Liquid water "magma" escaping from these rifts should resculpt the planetary surface, producing terrains of differing ages.

(I've used the word *scenario* deliberately. Usually scientific ideas are given more highfalutin' labels such as *hypothesis, theory, model,* or *explanation.* In science, at least, all of these words imply a degree of certainty that is lacking in planetary science, especially when it comes to the outer planets, where we have fragmentary information. These scenarios are stories that attempt to explain at least some of the data that we have. Their primary purpose is to suggest what types of future missions, measurements, and analyses would be useful. They are not written in stone.)

This idea of satellites expanding when they freeze seems intriguing, but we know it can't be the whole story. It can't explain the geological youth of these valley floors. Examination of the photo of Enceladus shows that there are parts of that satellite that have few if any impact craters. These valleys are certainly younger than a billion years old and may well be less than a hundred million years old, very young in astronomical terms. The Solar System as a whole, including these satellites, is 4.5 billion years old. These crater-free terrains were formed at least 3.5 billion years after these satellites formed. Even when we allow for the possibility of ammonia antifreeze spiking the watery interior of these small satellites, there is no way that an iceball 500 kilometers across, like Enceladus (or the even smaller Miranda, which also has terrain that is virtually crater-free), can stay melted for 3.5 billion years. What keeps the interiors of these satellites hot?

## VOLCANOES ON IO

A partial answer that works for the moons of Jupiter and Saturn was provided by another surprising discovery, this time involving the weirdest of all of these strange worlds, Jupiter's innermost large satellite, Io. (I've heard the name of this object pronounced as both *Eé-oh* and *Eyé-oh;* use whichever you like.) Incidentally, the story of this discovery should dispel the idea that automated, "unmanned" space probes work in ways that do not involve human beings. They are really extensions of our own eyes. This story also highlights the role of serendipity, of surprise, and demonstrates that even an "unmanned" probe can be quite flexible.

On 8 March 1979, *Voyager 1* took a historic picture, which initially was intended solely for the use of the spacecraft navigation team. The camera took a long-exposure photo of Io, a photo that washed out all of the details on the planetary surface but that recorded images of the background stars. Linda Morabito, a technician with the navigation team, was trying to use this picture to locate the position of the satellite. She noticed that the image of the satellite was not the partially lighted sphere that it should have been. Extending beyond the sunlit edge of the planet was a dim, crescent-shaped cloud rising hundreds of kilometers above the surface. A lesser person might have dismissed this apparently "impossible" cloud as an experimental artifact, but Morabito persisted, and on Thursday and Friday, 8 and 9 March, she and her colleagues dismissed all possibilities except the obvious and ridiculous one: that it was a cloud above the surface of Io.

One by one, other members of the imaging team realized that a tremendous volcano was the only correct explanation. The cameras of *Voyager 2* were thus redirected to photograph the volcanoes in more detail when it flew by Jupiter a few months later. Planetary scientists pored over all the *Voyager* pictures and found that Io had no less than eight erupting volcanoes, making it the most volcanically active object in the Solar System. (Most of the time, the earth has no erupting volcanoes, though several are "active" in the sense that they may blow their tops at any time.) A *Voyager* picture of a volcano is shown in the illustration.

Why does Io have volcanoes, and what do they have to do with the young surfaces of the satellites of Saturn and Uranus? Planetary scientists Stanton Peale, Ray Reynolds, and Patrick Cassen were ex-

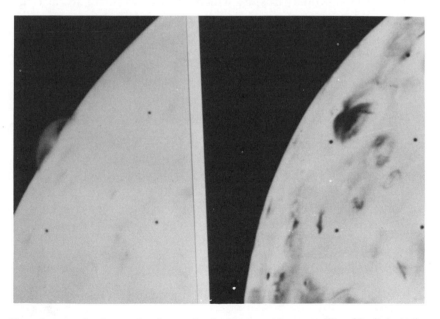

Two photographs show volcanism on Io, the innermost large satellite of Jupiter. At the left, gases escaping from a volcano are shown silhouetted against the satellite's edge. At the right, the black patch near the top center of the image of the planet are the sulfurous gases shadowing the satellite's surface. The surface is very young, containing no impact craters. (Jet Propulsion Laboratory, NASA.)

perts on lunar tides. They realized that Io, being so close to the giant planet Jupiter and also being caught in a curious orbital relationship with Europa and Ganymede, is subject to tremendous tidal stresses. Just before the volcanoes on Io were discovered, these scientists predicted that these tides would melt the satellite's interior, producing widespread volcanism. *Voyager* measurements indicate that thirty times as much heat flows out of a square mile of Io's surface area than flows out of a similar area on earth.[9] For Io, at least, it is the proximity of a big neighbor, the giant planet Jupiter, that makes it so unlike the passive environment of the earth's moon. (It may be that electrical interactions between Jupiter's magnetic field and Io's electrically conducting interior contribute to the warming.)

Hence, small bodies with big neighbors can have weird surface features. But can tidal stresses, which seem to work for Io, also account for the activity of Enceladus, Ariel, and Miranda, shown in the pictures? Can these ideas account for other oddities like smooth Europa, hamburger-shaped Hyperion, asymmetrical Iapetus (which has one dark hemisphere and one bright one), and the others? The best answer I can give is "Maybe; it's the best idea that anyone's had yet."

Media coverage of great occasions like planetary encounters may provide a misleading impression about how science is practiced. During one of these events, journalists ask the mission science team to produce "instant science," as if the explanations for new discoveries should leap out from the TV screens displaying the pictures. Scientists do the best they can in interpreting the data on the spot. It takes just one look at the pictures of Miranda to show that it's very strange. But why is it strange in the way that it is? The answers to these deeper questions take longer to emerge.

## COMETS: THE HALLEY ENCOUNTER

Comets are another spectacular type of small body that can put on a dramatic celestial show when they come close to a big neighbor. If you were to look at it closely when it was far out in the Solar System, Halley's comet would be a peanut-shaped, coal-black chunk of ice about nine miles long and perhaps five miles wide. (How can a chunk of ice be black? Don't expect an instant-science answer to that question, but it's an interesting one to ask.) Halley's comet is so small and so dark that it is barely visible to the largest telescopes until it comes really close to the earth.

The dimensions of this particular comet are known pretty accurately because the European *Giotto* probe and the Russian *Vega* probes flew quite close to the nucleus in March 1986. When this small.chunk of ice, called the comet's nucleus, comes close to the sun, the black outer layer heats up and bright jets of water vapor and other volatile substances come spewing out, producing a giant cloud of gas surrounding the nucleus. The action of the sun's radiation and the solar wind on these gases can push them away from the sun, forming the comet's famous tail. The result is a cometary display. A faint, fuzzy object with a tail, which can sometimes look like a tiny iridescent witch's broom in the sky, provides a celestial show for us.

For the average person, especially someone living in the northern United States or in Europe, the 1986 appearance of Halley's comet was a bust. The comet's path never took it very close to the earth this time around, producing the worst apparition of the comet in two thousand years. And when the comet was closest to the earth, in March and early April, it was well to the south of the earth's orbit, virtually invisible from north of latitude 35°N during this time period. Earlier in its 1986 appearance, Northern Hemisphere observers could see it—but (in my case, at least) only with binoculars. The brightness of the sky in light-polluted urban and suburban areas made spotting the comet doubly difficult. By the time it returned to the northern skies later in April, the bright moon overwhelmed it. Soon it lost its tail and faded into invisibility, not to reappear until 2061.

For the cometary scientist, the 1986 appearance of Halley's comet was a dream come true. It's normally quite a challenge to study comets. Most of them travel on orbits that take them way, way out into the Solar System, and when they dive in close to the sun, they are coming into the inner Solar System for the first time in human history. Because they are generally discovered only a few weeks before becoming bright, scientists can't plan a coordinated campaign. In the case of Halley's comet, in contrast, we've known for the last two hundred years that it would be coming in 1986. Scientists had time to prepare to send a flotilla of spacecraft by it in March of that year. The International Halley Watch, a coordinated campaign of careful observations from the ground, succeeded in making this the best-recorded comet in history.

Even so, Halley's comet was not the first to be visited by a spacecraft. The first encounter with a comet was made by the third *International Sun-Earth Explorer* (*ISEE-3*) satellite, later called the first *International Cometary Explorer* (*ICE*). When this satellite was launched in August 1978, it was part of the space plasma physics program. Its purpose was to remain on station 1.5 million kilometers sunward of the earth, monitoring disturbances in the solar wind. When it became clear that NASA was going to be conspicuously absent in the fleet of spacecraft greeting Halley's comet, Robert W. Farquhar began exploring possible ways in which existing spacecraft could be redirected toward a comet. He hit on a very complex trajectory, in which *ISEE-3* flew by the earth and moon four times, climaxing with a fifth pass that brought it 120 kilometers or only a bit more than 70 miles above the lunar surface. It then headed toward Comet Giacobini-Zinner, passing the comet in September 1985. Comet Giacobini-Zinner is a reasonably bright comet traveling on a compact orbit that brings it back to the sun every 6.5 years. It's named for its discoverers: M. Giacobini, who discovered it from Nice, France, in 1900, and E. Zinner who rediscovered it from Bamberg, Germany, in 1913. A byproduct of this complex trajectory was a deep exploration of the earth's magnetic tail well before the cometary encounter.[10]

The *ICE* mission was entirely complementary to the Halley intercept missions. The spacecraft carried instruments that could measure the density of ionized gas (plasma) in the comet's tail, and it flew through the tail of the comet, in contrast to the Halley missions, which all flew on the sunward side of the comet's head, opposite the tail. Like many other exploratory missions, *ICE* was full of surprises. For example, ions in the plasma surrounding the cometary nucleus are expected to be accelerated by the solar wind, but they shouldn't be going very fast.[11] However, the observed ions had ten times the expected energy, or three times the expected speed.[12] As another example, magnetic field measurements showed that the field plays a considerable role in shaping the plasma tail of this comet, being draped over the cometary nucleus like a bunch of limp spaghetti hanging off a fork, in nice (but unexpected) agreement with a theoretical prediction. This comet, tiny as it is, has a magnetosphere. Space plasma physicists can compare the comet's magnetosphere with similar structures around Earth, Jupiter, Saturn, and Mercury.

More cometary fanfare exploded when the comet fleet flew by Halley six months later. Two Japanese craft (*Suisei* and *Sakigake*), two Russian *Vegas,* and ESA's *Giotto* spacecraft all encountered the comet in early March. NASA was not along for the ride. A package of telescopes was to have gone aloft on the space shuttle to study the comet from earth orbit, but the *Challenger* disaster cut that mission short as well.

NASA did play a crucial role in organizing the International Halley Watch, a worldwide network of nearly a thousand astronomers, most of them amateurs, who monitored the comet on this particular passage. Amateur astronomers were trained, when necessary, in what to do to produce a scientifically valuable photograph (to start with, to record the exposure time and the time that the picture was taken). Investigators who later want to know what the comet looked like at, say, 3:46 A.M. Greenwich time on the morning of 6 March 1986 may well find that someone participating in the Halley Watch took just the picture that they need.[13]

International cooperation was vital for the Halley's comet missions. NASA was able to gather data from the Japanese *Suisei* mission from stations in Spain and California. *Giotto,* especially, was trying to get very close to a moving target; thus, precise information on the comet's position from the International Halley Watch, as well as from the Russian *Vega* mission, proved essential.

An estimated one billion viewers in fifty countries watched the data from the *Giotto* mission come in, relayed to mission headquarters in Darmstadt through Australian and American radio dishes located in Australia. It was a very tense moment, as the close encounter lasted only about ten minutes. A decade of work could go down the drain if a large chunk of cometary material hit the spacecraft at the wrong time. Something did hit *Giotto* just two seconds before its closest approach to the nucleus, and so it did miss getting some important pictures. Half an hour later the spacecraft stopped spinning wildly, and contact with it was restored.

The most dramatic results from these encounters were the first close-up pictures ever of a comet's nucleus, shown in the illustration.

This image of the nucleus of Halley's comet was taken by the Halley Multicolor Camera on board the *Giotto* spacecraft. The spacecraft was about 18,000 kilometers from the comet. The sun is to the left. The dark region on the right is the comet's nucleus; the very bright regions are two jets of dust. (Copyright 1986 by the Max Planck Institut für Aeronomie; image courtesy of the Max Planck Institut für Aeronomie, Lindau/Harz, Federal Republic of Germany.)

Both the *Vega* and the *Giotto* probes included cameras; *Giotto* flew closer to the comet's nucleus, but the difficulties of obtaining wide-angle views of the comet at close range and of accurately pointing the cameras at the nucleus mean that the information from both missions must be combined to get the best results. Analysis of the thousands of pictures that came back from the comet fleet has only begun. The illustration shown here shows the comet's nucleus as a black blob silhouetted against bright jets.

The pictures, as is often the case, produce the most readily comprehensible "instant science" from the cometary encounter—a reasonably good idea of what the nucleus looks like. The comet's nucleus is not spherical; it's 14 ± 1 kilometers long on its longest axis, and its short dimension is estimated variously as 7.5 ± 1 kilometers (by the *Vega* team) or 7–10 kilometers (by the *Giotto* team).[14] It's about as long as Manhattan Island and twice as broad. The surprising result is that it is quite dark; the comet reflects only 4% of the light striking its surface (this number, 0.04, is called the comet's *albedo*). Very few things are that dark; asphalt pavement has an albedo of 0.15. The darkest planet is Mercury, with an albedo of 0.12. It's not surprising that the comet turns out to be quite hot; at the time of the encounters in March, its temperature was 300–400°K, room temperature or warmer.

The dark surface can't be explained away as blackened ice, because at such high temperatures, ice would evaporate and gas would be produced all over the comet's nucleus rather than in places. It has to be a nonvolatile, hard crust of some sort. Evidently, what happens when the comet nears the sun is that the icy material below the crust evaporates and a jet of steam breaks through the crust, forming the one-kilometer-sized hole through which most of the cometary material escapes. The action of sunlight and the solar wind subsequently pushes this stuff away from the sun, forming the comet's spectacular tail.

A few of the probes, most notably the Japanese *Sakigake* probe, measured the plasma and dust environment of the comet. The findings are consistent with what was found at Comet Giacobini-Zinner, in that Halley's comet does disturb the flow of the solar wind and has a magnetosphere with field lines that are probably draped around the comet's nucleus. Many of the dust particles are quite small, smaller than a micron in size; this was not expected. The composition of the dust will be determined and compared to the dust of other primitive, small, unaltered objects in the Solar System.

As always, the analysis of the data will continue, and some questions will be addressed. What is the black stuff that coats the comet? Can we, by putting together information from various spacecraft and from the ground, figure out how the jets from the nucleus change while the comet rotates? Why does the dust have a peculiar composition? Why are such small dust particles associated with the comet? What is the role of the magnetic field around the comet in shaping the tail structure, and where does that magnetic field come from? The combination of four robot probes and a well-coordinated set of ground-based observations will produce a unique set of data that astronomers can sink their teeth into for decades to come.

## ASTEROIDS: ROSETTA STONES OF THE
## SOLAR SYSTEM?

Humankind has always been interested in origins. Some of the first scientists to join NASA pushed for an expedition like *Apollo* because of the hope of finding primitive material on the moon. Robert Jastrow, one of the first scientists to join NASA, recalled that Harold Urey marveled, "It could help us solve the mysteries of the origin of the solar system and the origin of life. What a heady thought. And all this from that cold, dead, miserable piece of real estate."[15]

The *Apollo* astronauts found that the moon was geologically dead, but not dead enough. The final bombardment four billion years ago and the flooding of the lunar basins erased the early record, burying the "genesis rocks" beneath miles of molten lava or smashing them to rubble. The moon is just a little bit too big. Perhaps it even deserves the name of being a "miserable piece of real estate," though that remains to be seen.

The search for genesis rocks may well lead humanity to the asteroid belt. There are a bunch of small planets orbiting the sun between the orbits of Mars and Jupiter. Variously called *minor planets* or *asteroids,* the biggest of these, Ceres, is 1,000 kilometers across, six times smaller in diameter than the moon. Countless kilometer-sized boulders remain undetected unless they come very close to the earth. For a time, they were occasionally called the *vermin of the skies.* The need to catalog them supposedly took time away from much more important tasks.

But there may be nuggets of gold in this particular astronomical trash pile out there in the asteroid belt. Although we've not sent any space missions to asteroids, a number of fragments from the belt have come to us for free. Giant, massive Jupiter is quite close to the asteroid belt, and its gravitational tug occasionally nudges an asteroid, sending it plunging into the inner Solar System on a trajectory that ultimately brings it close to one of the inner planets. Many of these fragments from the asteroid belt hit the earth, and we can find them as meteorites. Except for the moon rocks, these are the only sources of extraterrestrial material that we can analyze in the lab. The recent discovery that Antarctic ice fields contain many meteorites that hit the earth in the past and were then plunged into a deep freeze has given us a new, greatly enlarged sample of these extraterrestrial escapees from the asteroid belt.

The scientific value of the asteroids comes from their likely primitive nature. One group of asteroids is as black as the surprisingly black nucleus of Halley's comet. We have not visited an asteroid to measure the composition of its surface, but the colors of the dark asteroids are quite similar to the colors of a class of meteorites that have a very primitive chemical composition, similar to the composition of the sun.


## THE FUTURE

The small objects in the Solar System have revealed many surprises, and more investigations to more destinations could reveal a lot more. Except for the moon, every one of the objects mentioned in this chapter has been seen, up close, by only one or, in a very few cases, two missions in which the spacecraft zipped by and got a passing glance. A spacecraft flying by a planetary satellite sees only half of it, at one instant of time. When one probe goes through a satellite system containing a dozen objects, mission planners can usually pick out one to study really well and several others to study moderately well. The remaining satellites are glanced at from long range. Thus, many of the satellites of the giant planets have been looked at with only very low resolution, so that most of the interesting features are tantalizing blurs.

In the early 1980s, planetary scientists spent many hours sitting around long tables, arguing with each other, and hammering out a comprehensive program of Solar System exploration.[16] The *Challenger* accident and the ensuing disruption in the launch schedule has probably derailed this plan. The presumption of the program was that the existence of the space shuttle and powerful upper stages like the liquid-fueled Centaur rocket made launch opportunities "routine." The scientists proposed to send a series of inexpensive, small spacecraft to various destinations. The list of missions includes many hoped-for explorations of small bodies: a comet rendezvous and asteroid flyby, a similar rendezvous with an earth-approaching asteroid, a mission to the main asteroid belt that would fly by several asteroids, and a mission that would fly through the gaseous atmosphere, or coma, of a comet and collect material, returning it to earth for analysis. In addition, the list included a Saturn orbiter, a return visit to Uranus, and possibly a mission to Neptune. The idea of sending many inexpensive missions to many destinations was a great contrast to the approach of the 1960s and 1970s, when establishing a launch opportunity and a budget line for a mission was the major hurdle. Because it was apparently as hard to sell a $50-million mission as it was to sell a $500-million, planetary scientists had chosen to plan a few major missions, each of which was loaded with every conceivable instrument, making the spacecraft very expensive.

Before *Challenger,* only the inexpensive, low-cost Venus and Mars missions had obtained firm budgetary approval, but the rest of the package seemed likely, starting with the Comet Rendezvous/Asteroid Flyby mission (CRAF). In the CRAF mission, a spacecraft would head out to the asteroid belt and match velocities with a comet. Instruments would take pictures, measure the comet's surface composition, sample the dust around it, and determine the characteristics of its plasma environment. While the comet was passing through the asteroid belt, the craft could make some small detours to make the same measurements on one or more asteroids in the belt. The spacecraft would continue to follow the comet through one complete passage around the sun. Plans for CRAF are well developed, and it had seemed clear that it would start in the fall of 1986, with a launch in 1992 and a rendezvous with Comet Tempel 2 in 1996.[17] But CRAF was to have been launched with the Centaur upper stage, which cannot fly on the shuttle. It is not listed in the space shuttle manifest released in late 1986, which lists all major launches through 1994. (The space shuttle manifest is discussed in detail in Chapter 15.)

*Galileo* was to have been launched in May 1986. This mission, planned by scientists in the mid-1970s, was to be a combination of a long-lived orbiter (to study Jupiter's satellites extensively) and a probe that would descend into the whirlwind of Jupiter's atmosphere. On the way to Jupiter, it would fly by an asteroid, taking pictures and obtaining plasma data. The launch had originally been slated for January 1982 but was delayed because launching planetary probes with the space shuttle is rather complex. A powerful upper-stage rocket must be carried up in the cargo bay in order to propel a craft like *Galileo* into the outer Solar System. NASA managers considered and then scrapped various upper stages, and the spacecraft was repeatedly redesigned. One design called for separate rockets to launch the orbiter and the probe, requiring the design of a probe carrier. *Galileo* project managers may have to reconsider this possibility because NASA finally decided that the Centaur upper-stage rocket, which was to have launched *Galileo,* could not fly on the space shuttle for safety reasons.[18] The other upper-stage rockets that NASA is willing to fly on the shuttle are not as powerful as the Centaur and couldn't boost *Galileo* toward Jupiter.

The near future of the *Galileo* mission, like the future of the entire planetary exploration program, is unclear. NASA is now considering various options for launching it, including the use of unmanned rockets instead of the shuttle. The space shuttle manifest mentions launch opportunities for *Galileo* and the *Ulysses* missions (which both head toward Jupiter first) for November 1989 and October 1990. However, there are other concerns that could push the mission yet further into the future. The power source for *Galileo* is a radioisotope thermoelectric generator (RTG) in which the heat from decaying plutonium produces electricity. RTGs are much smaller and simpler than nuclear reactors. Nevertheless, various critics are asking whether the RTGs could survive the kind of explosion that blew up the *Challenger.* If they didn't, the environment around the launch site could be contaminated by radiation.[19] Another issue is that many of the scientific instruments on board were designed long ago, and that more modern instruments could make a better mission. Using new instruments, however, would require a redesigned spacecraft because their weight, power requirements, and so on would be different.

The European Space Agency plans an ambitious follow-up to the

*Giotto* mission to Halley's comet. One of the four cornerstone missions in the European Horizon 2000 plan for space science in ESA is a sample-return mission from a primitive body, either an asteroid or a comet. On such a mission, a spacecraft would rendezvous with the target, pick up material from its surface, and return to earth. Such a mission would require ion propulsion technology, which has not yet been tested and developed. The European plan discusses the possibilities of international cooperation; the initial studies are being done by France's Matra Espace.[20]

This sample-return mission may not happen until the twenty-first century, but NASA's foreign competitors have plans for other missions to small bodies that are closer at hand. In March 1985, Soviet scientists announced a series of ambitious planetary projects, including the Vesta mission, which includes an asteroid flyby. The original Soviet plan was to send one spacecraft to Venus in 1991. The craft would then split in two, with one portion matching velocities with the bright, peculiar asteroid Vesta. The other portion would drop a lander on Venus and then continue toward another asteroid.[21] A year later, revised plans appeared. France was brought in as a partner, the initial target now seems to be Mars, and the asteroid part of the mission will include encounters with at least five asteroids and comets, including the emplacement of a Soviet penetrator on one of them.[22] The Soviet Union has also announced plans for a lunar polar orbiter mission, scheduled for launch in 1989 or 1990. The purpose of this mission is to provide a comprehensive geochemical map of the lunar surface.

It remains to be seen how many of these missions will actually occur, and on what time scale, but the announced plans and the success of the missions to Halley's comet have firmly established the European and Soviet presence in the field of exploration of the small bodies. The ESA Horizon 2000 report confidently notes that "within planetary exploration, this [the study of primordial bodies] is an area where Europe could take the lead, following on the Giotto mission. The return of primordial material from primitive bodies, namely from asteroids and comets, constitutes a major theme in future planetary science."[23] The United States is still the only country investigating the strange, icy satellites of the giant planets; *Voyager* will send pictures of Neptune's two satellites in 1989, and *Galileo* will do an extended investigation of Jupiter's satellite system sometime in the mid-1990s or possibly later.

## MINING THE MOON AND THE ASTEROIDS

The use of "extraterrestrial resources" emerged as a new theme in the plans of planetary scientists in the early 1980s, at the same time that the concepts of space stations and space colonies were kicked around. Much of the discussion is in the preliminary, speculative stage, but some broad ideas are beginning to emerge. Consideration of this idea has rekindled some interest in the moon, not as a primarily scientific target but as a source of exploitable construction material.

The reason that it costs several thousand dollars per pound to launch material from the earth into space is that it has to be accelerated to an orbital velocity of nine kilometers per second. If you get your raw material for a space station or colony from the moon, you need only lift it out of the weak lunar gravitational field, saving bundles of rocket fuel because you need only a velocity of two kilometers per second. Of course, you have to have some kind of lunar base to do the launching. If you can be patient and clever, it turns out that an asteroid is better yet as a source of extraterrestrial material because the asteroid's gravity is even less than the moon's, and if you choose your orbit cleverly, you don't have to do much accelerating at all.

Is there anything on the moon worth mining? It depends on what you're looking for. It could be a very good source of materials for a large-scale space structure. Lunar soils contain common, useful elements like aluminum, silicon, oxygen, possibly some water (in the shaded polar regions), and ferrous metals like iron, nickel, and titanium. A NASA contractor reported that concrete made from lunar material is slightly stronger than terrestrial concrete.[24] Although astronauts have sampled a few localities on the moon, sophisticated remote sensing techniques allow us to determine the global mineralogical composition of the moon from orbit, as could be done by the Soviet lunar orbiter if its instrumentation was appropriate. NASA may launch such an orbiter, too. Such a mission would also answer some of the outstanding puzzles of lunar history, such as the origin of the local magnetic fields on the moon and some details of the thermal history of our satellite. Mining the moon was one of the central elements in a space colony design promoted in the late 1970s.[25]

Another aspect of "space mining" is bringing extraterrestrial materials back to earth for industrial purposes. Because it costs a great deal to bring material down to earth, initial discussions of this business focus on lightweight, valuable materials, like the spices that stimulated the exploration of the earth. Even when (if?) the cost of getting material up or down can be reduced to a hundred dollars per pound from the current figure of several thousand dollars, it will still be cheaper to dig ordinary material like iron and aluminum out of the earth. Some kinds of asteroids do contain metallic grains with platinum group metals in them that could be rather valuable. The extraction process required to refine these materials, similar to that used at mines like the Sudbury, Ontario, nickel mine, is simple enough to suggest the possibility of remote operation. The ore is mixed with carbon monoxide gas at a relatively low, but well-controlled, temperature and pressure so that the nickel and iron can be taken out, leaving behind a residue of cobalt, platinum metals, and even possibly some silver and gold. Prices for this residue are quoted at as high as $10,000 per kilogram. Next to this figure, the transportation costs to or from orbit may even look reasonable. Planetary scientist John Lewis concludes that these expensive, strategically valuable materials are the most attractive candidates for space mines.[26] Some kind of aerobraking scheme might make it very cheap to come back to earth, as the atmosphere rather than precious rocket fuel would be used to slow the spacecraft down.

The economics of asteroid mining are very uncertain indeed. But the first step—finding out how many earth-approaching asteroids there are and what they are made of—is the same step that is being suggested for purely scientific reasons. The NASA planning group specifically mentioned exploitation of extraterrestrial materials in their 1983 report, and so it seems to be at least respectable to talk about the idea.[27] This group specifically added the "provision of a scientific basis for the future utilization of resources available in near-Earth space" as a secondary goal for the planetary exploration program.[28] In this way, planetary scientists are following the tradition of Lewis and Clark in proposing a mission that will fulfill a number of different goals, not insisting on an artificial distinction between pure science and application.

*    *    *

The venerable, durable *Voyager* spacecraft keeps sending back more intriguing images of peculiar satellites of the outer planets, giant snowballs with surprisingly young surfaces. These small objects, previously expected to be bland cratered surfaces like the earth's moon, contain active volcanoes, globe-girdling crevasses, and young, mysteriously dark surfaces. An international comet fleet greeted Halley's comet and Comet Giacobini-Zinner on their recent forays into the inner Solar System and sent back the first-ever pictures of a cometary nucleus. The black, peanut-shaped nucleus of Halley's comet turned out to be a real surprise. Even the earth's moon, when visited by astronauts and remotely operated landers, proves to contain such surprises as the localized magnetic fields. Mercury looks very moonlike superficially but, surprisingly, has a rather strong global magnetic field much like the earth's.

The time scales for future exploration of the planets are uncertain. But the frontiers beckon. Even the small bodies of the Solar System contain surprising surface features like volcanoes, cracks, valleys, and oceans of water (possibly spiked with ammonia) underlying thin icy crusts. The idea of mining the moon or the asteroids may seem far-fetched and may really be pie in the sky, but it is a reminder that economic benefits may be derived from some of the most unlikely pieces of real estate in the Solar System.

# CHAPTER 9

# Earthlike Planets and the Magnet That Is Mars

Before *Sputnik,* scientists and science fiction writers both expected the two nearest planets, Venus and Mars, to have many similarities to Earth. The opaque clouds of Venus obscured its surface from ground-based telescopes, and science fiction authors of the 1930s wrote novels set in a steamy tropical paradise. Scientists knew a bit more about Mars because its atmosphere is relatively cloud-free. Still, the smallest features we could resolve were the size of Rhode Island. We could not measure the temperature of its surface remotely, using telescopes that existed before 1957. Thus again, scientists and science fiction writers alike persisted in believing that the surface conditions were approximately terrestrial. A series of novels by Edgar Rice Burroughs (who also created Tarzan) had the earthman John Carter gallivanting around Mars like a medieval knight, rescuing humanoid princesses who fell in love with him. The prevailing assumption was that humans could live on Mars or Venus without any kind of special equipment at all, or in some books, with the aid of modest oxygen masks. Like many preconceptions, these ideas that Venus, Earth, and Mars were a set of heavenly triplets turned out to be completely wrong.

Once the exploration of the moon was well under way, our attention naturally focused on these two worlds, two terrestrial siblings that also happen to be reasonably accessible. The combined efforts of the United States and the Soviet Union have accounted for twenty successful missions to Venus and eight to Mars.[1] But Mars and Venus, although still the most earthlike planets in the solar system, contrast sharply with our home planet. Venus is tropical in the extreme, with surface temperatures of 730–750°K (850–900°F). Mars is quite cold, with temperatures below freezing virtually all of the time, and an unbreathably thin atmosphere to boot. Oxygen masks and heavy jackets won't do the trick on Mars.

Although it might seem that the differences among Earth, Venus, and Mars put the latter two planets in the category of worthless rock balls, they are more valuable than they look. Exploration of these planets has put our earth into perspective, prompting questions like "Why is Venus so hot?" and "Why is Mars so cold?" Our growing understanding of our own planet makes it possible to begin to answer these questions and, in answering them, to gain a greater understanding of what makes our own planet work. This approach, called *comparative planetology*, has become increasingly powerful as more and more of the spacecraft data on these planets are analyzed and digested. And although humans on Mars will need sophisticated spacesuits to survive, the idea of sending humans there is still a powerful magnet attracting many space enthusiasts. It remains to be seen when such a mission will fly; the space program does not (and in my view should not) aim toward a single goal in space, as we did in the 1960s toward the Moon. There are many valuable goals to attain.

## SURFACES OF THE EARTHLIKE PLANETS: A GEOPHYSICAL LABORATORY

We expected the moon to be cratered and primitive because it's smaller than the earth. It turned out to be that way, though it is not a completely primordial remnant. Other low-mass objects in the Solar System also turn out to be cratered, unless a big neighboring planet provides energy by tidal heating and alters their surfaces by volcanism and cracking.

Mars and Venus are larger than these primordial satellites. Mars has about one-tenth the mass of the earth, and Venus has a mass equal to that of the earth. What do we expect these worlds to look like? Imagine the earth with the oceans stripped away, as neither Venus nor

Mars has oceans, and think about what's there. Huge, irregularly shaped continental land masses stand many kilometers higher than the lower, largely flat-bottomed ocean basins. Other features of the terrestrial landscape are folded mountain ranges like the Rocky Mountains; volcanic mountains like Mauna Kea, Hawaii; chains of volcanic mountains like the Cascades (which includes Mount Saint Helens); island arcs, which are also volcanic in origin; and huge rift valleys like the Red Sea and the East African rift valley. Would we expect to find such features on Venus and Mars?

To answer this question, scientists need to know the origin of the earth's topographic features, and this knowledge appeared only in the 1960s with the acceptance of the theory of plate tectonics. Even as recently as 1959, when I took a seventh-grade science course, the term *mountain building* was simply a verbal smoke screen that concealed our scientific ignorance from the unwary student. But in the early 1960s, Alfred Wegener's apparently absurd idea of continental drift, which had reposed on the scientific trash pile for decades, was accepted. This scheme transformed the massive, apparently immutable and changeless geological landscape into a slow but still turbulent scene of continents drifting apart and crashing into each other, a truly restless earth. Measurements of magnetic fields in the middle of the Atlantic Ocean demonstrated that this huge low-lying basin opened up in the last hundred million years, in the last 2% of the earth's history. If the entire earth's history were to represent one year, the Atlantic Ocean would start opening on December 21, in the final pre-Christmas rush.

This idea of continental drift, generalized to a more complete theory called *plate tectonics,* accounts for most big highlights of the terrestrial landscape. The earth's surface is broken up into a dozen or so plates that are moving around in different directions. Many of these plates are the size of continents, like the North American plate, or include most of an ocean (e.g., the Pacific plate). A few are smaller. When two continental plates bash into each other, in the way that the Indian subcontinent plows into Asia, land is pushed up and folded to create a mountain range like the Himalayas. Oceanic plates meet continental plates, and the ocean bottom plunges beneath the continent. This collision digs a deep ocean trench and produces a chain of volcanoes because of the scrunching and scraping of the two plates against each other deep beneath the earth's surface. Rift valleys occur when a continental plate splits in two, or when two continental plates pull apart from each other. And mid-ocean ridges form when a rising plume of molten rock pierces the low-lying ocean plates, creating a new ocean bottom and causing this plate to expand.

The driving force for all these tectonic motions is the flow of heat through the outer surface of the earth. The heat itself comes from the decay of radioactive atoms and the heat left over from the earth's formation. It flows by convection through a thick layer of the earth's crust. Convective heat flow involves mass motions, and yes, the earth's crust is moving, but slowly. This heat flow proceeds at different rates on different planets. More massive planets have larger heat sources, and the insulating mantle that throttles the heat flow is thicker than it is on smaller worlds.

In short, more massive planets can be expected to be tectonically more active. A scientist with unlimited powers (like the fictional, evil characters on Saturday morning cartoons) could test this expectation by actually making a world with composition identical to the earth's, but with twice as much mass, or half as much mass, and waiting a few billion years for tectonics to develop to see the results. Real scientists can't conduct such a global experiment even in this era of ''big science.'' But the triad of planets in the inner solar system presents us with a natural way of conducting just such an experiment. The plate tectonics model, in the context of comparative planetology, predicts that the earth-sized Venus would have similar surface features from well-developed plate tectonic motions. Mars, a bit smaller, should be intermediate between the geologically active earth and the immobile, dead surface of a cratered world like the still smaller moon. Spacecraft data bear out these expectations to some extent, but they also present some puzzles.

Let's take Venus first. Because its atmosphere contains some very thick white clouds completely obscuring the surface, its surface was mysterious until the 1960s. At that time, scientists used radar to bounce radio waves off its surface and were able to map our sister planet with varying degrees of comprehensiveness and resolution. The first images of terrain near the Venusian equator showed surprisingly moonlike, depressingly dull, cratered plains.

Venus and Mars taught us that it can be a mistake to base a global impression on a small part of the planet. Later experiments, including a number of radar maps made from spacecraft that had orbited Venus, demonstrated that there were several irregularly shaped areas looming above the smooth plains. The highest of these, Maxwell Montes, turns out to have a number of parallel grooves in it, as well as a circular feature, probably the caldera at the summit of a volcano.[2] Two other continental areas have been roughly mapped and show irregular edges. High plateaus with irregular edges are not impact features and are clear indications of some kind of tectonic activity on Venus.

Mars, too, provided some misleading first impressions. It is the only planet whose surface can be seen at all clearly from the earth. Even then, it appears as a tiny reddish disk in most telescopes, with a fascinating pattern of light and dark. Seventy years ago, *the* big question put to most planetary astronomers was whether the dark areas were linear features called *canals*. The straightness of the dark areas turned out to be an optical illusion, where the human eye and brain conspire to interpret a collection of fuzzy dots in the simplest possible terms: a line connecting them. The first Martian flyby, in 1965, zipped over the most "interesting" light and dark areas. It sent back pictures of a sandy moonscape dotted with craters.

*Mariner 9,* beginning to orbit Mars in 1971, revealed a very different world. Approaching Mars during a giant dust storm, it returned images to us showing four dark spots sticking up above the clouds. When the dust cleared, scientists learned that these were the summits of giant volcanoes. The largest of these, appropriately named Olympus Mons (Mount Olympus), has a peak 20 kilometers above the surrounding plain, and a base 700 kilometers in diameter. This giant volcano is illustrated here. Were such a feature to exist on earth, it would cover the entire state of Pennsylvania, and the highest clouds surrounding it would come only halfway up this huge volcanic mountain. The closest terrestrial counterpart would be the island of Hawaii, about 10 kilometers higher than its 100-kilometer-diameter base at the bottom of the Pacific Ocean. Hawaii is smaller than Olympus Mons, half as high and one-seventh as wide, and has active volcanoes at its summit. The giant Martian volcano may be extinct and probably last erupted a hundred million years ago.

Mars has a few other features that are generally similar to earth's,

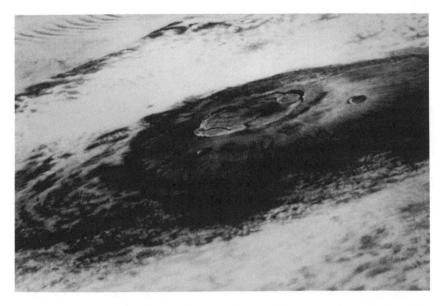

Oblique view of the summit of Olympus Mons, the largest volcano in the Solar System. The circular caldera at the summit is 90 kilometers (54 miles) across. White clouds in the Martian atmosphere obscure the rest of the mountain; the entire mountain is 700 kilometers (420 miles) across at its base, and its summit is 25 kilometers (15 miles) higher than the surrounding plains. (NASA.)

but differing from their terrestrial counterparts in important ways. Valles Marineris is a huge canyon system that extends almost one quarter of the way around the planet for 4,000 kilometers. In places, it is more than 4 kilometers deep. Next to it, the Grand Canyon looks like a mere scratch. Mars has polar caps, somewhat like the earth's, but made of both water ice and dry ice (frozen carbon dioxide) instead of just water ice. Most enigmatic are ancient riverbeds, channels on Mars that must have been cut by flowing water (more about these later).

The naive predictions based on the idea of plate tectonics, that Mars should be less active than Earth and that Venus should be similar, at first don't seem to fit the data at all. The tectonic structures on Mars are bigger, not smaller, than Earth's. Venus seems quite different from Earth, in stark contrast to our expectation that it should be similar.

Scientists reflected on these naive expectations and soon realized that the Martian landscape, at least, is consistent with the theory that tectonic activity increases with increasing planetary mass. Consider Olympus Mons and Hawaii, for instance. It turns out that smaller, not bigger, volcanoes are to be expected on a tectonically more active planet. The Hawaiian islands were formed by a hot spot underneath the floor of the Pacific Ocean breaking through the ocean floor and depositing magma on the surface. Because the Pacific plate is drifting, this hot spot has broken through the ocean floor in various places and has not had time to build up a huge volcano the size of Olympus Mons. The Hawaiian islands are a chain of volcanoes, and the smaller islands located northwest of Hawaii are the leftovers from the time that the hot spot was somewhere else. But horizontal tectonic movements are less extensive on Mars. The hot spot that produced Olympus Mons has stayed in one place and has had time to build up this gargantuan volcanic mountain.[3]

Venus is less well understood, largely because its entire surface cannot be mapped by radar at resolution comparable to the photographs of Mars taken from orbiting spacecraft. The abundance of smooth, cratered low-lying plains led scientists to speculate about such ideas as *scum tectonics,* where a small number of smaller, thinner continental blocks floated around and occasionally crashed into each other.[4] Recently, the results from all the radar experiments—American and Russian orbiters and terrestrial radars—were compared. It became clear that horizontal tectonic motions vaguely similar to the large-scale tectonics on earth were evident in the folds on top of Maxwell Montes and in other places.[5] The problem with Venus is that we know just enough to know that it's interesting (in other words, something more than a cratered, boring moonscape), but that we do not have the information needed to understand how Venus works.

Planetary scientists in both the United States and the Soviet Union wanted a better map of Venus. A high-priority planetary mission in the early 1980s had been a satellite to obtain a radar map at high resolution. Considerably descoped and renamed *Magellan,* it is slated for a space shuttle launch in 1989. In the meantime, the Soviet Union sent *Veneras 15* and *16* to Venus in 1983 and 1984, and the radar data from these missions (which are only now being published and made available to the West) cover the northern part of Venus at a resolution of 1 kilometer, apparently the same as that of *Magellan*. Although *Magellan* will survey more of the planet than the *Veneras* did, there is considerable overlap between the capabilities and purposes of the two missions, and the *Veneras* did get to Venus first.[6]

## ATMOSPHERES OF THE TERRESTRIAL PLANETS

Take a deep breath. Even if it's a hot and humid summer day, or if you're in the middle of a freezing winter storm, the temperature and makeup of the earth's atmosphere fall within a relatively narrow range, so that the air is always breathable for earth creatures. We don't usually think of something so natural as the breathability of the earth's air as being remarkable, but in comparing other planets with ours, we realize how lucky we are.

Think about Venus. Venus is the same size as the earth, and only a little closer to the sun. But its atmosphere is quite different, fatally different for us. Imagine yourself on the surface of Venus, taking a deep breath. Your lungs would be burned by the 850-degree air, if your skin wasn't roasted first. (This temperature is in Fahrenheit degrees). Your lungs would be overwhelmed by the high air pressure. Your body couldn't extract oxygen from the air because there's hardly any there. If your body had withstood all of these insults, you'd dry out very quickly because the humidity on Venus is so low. Venus is not the tropical paradise that some science fiction writers have imagined. It's a noxious oven.

And think about Mars. Mars is a little farther from the sun than the earth is, a bit smaller, and its atmosphere is much thinner. It will take more than a tiny oxygen mask to make Martian air fit for human consumption. At its best, in the summertime in the Martian tropics, the temperature is that of Antarctica. The atmosphere, mostly carbon dioxide with only faint traces of oxygen and water vapor, is far thinner than the air on top of Mount Everest.[7]

In fact, if you ignore a couple of things like temperature and oxygen content, the most earthlike atmosphere in the Solar System is found on Saturn's moon Titan. It's rather unconventional for me to include this Saturnian moon that comes out of the deep freeze in a chapter on "terrestrial planets," but it's a freak that does not fit anywhere else and, from the point of view of atmospheres, is surprisingly earthlike. Titan's atmosphere is mostly nitrogen, like earth's; its surface pressure is in fact even a bit greater than earth's. There are definitely puddles, possibly lakes, and maybe even oceans on its surface. But count Titan out as a vacation spot, unless you like swimming in liquid methane and ethane, with temperatures around −300°F. Table 6 summarizes the temperatures of these four objects.[8]

## BALLOONING ON VENUS

A most ingenious expedition to Venus was undertaken in June 1985. For centuries, scientists and explorers have used balloons to take themselves and their equipment aloft in the earth's atmosphere. Jacques Blamont of France thought it might be fun and useful to use the same technique elsewhere, on another planet. The French and Soviet equivalents of NASA studied the possibilities in the 1970s, and

Table 6.    Planetary Atmospheres

| Object | Surface Gravity ($m/s^2$) | Surface Temperature (K) | Surface Pressure (Millibars) | Major Gases | Minor Gases (Parts per Thousand) |
|---|---|---|---|---|---|
| Earth | 9.78 | 288 | 1013 | Nitrogen (0.78) Oxygen (0.21) Water vapor (0.01) | Carbon dioxide (3) Argon (9) Neon (0.018) |
| Venus | 8.88 | 730 | 90,000 | Carbon dioxide (0.95) Nitrogen (0.04) | Water vapor (0.1) Argon (0.07) Neon (0.004) |
| Mars | 2.18 | 218 | 7 | Carbon dioxide (0.95) Nitrogen (0.03) | Argon (16) Water vapor (0.3) Oxygen (1.3) |
| Titan | 1.44 | 93 | 1600 | Nitrogen (0.8?) Methane (0.05) Argon (0.15?) | Hydrogen (2) Carbon monoxide (0.1) Ethane ($C_2H_6$) (0.02) Propane ($C_3H_8$) (0.004) |

Blamont's scheme eventually became a part of the Russian *Vega* mission that went first to Venus and then to Halley's comet. Two balloons were deposited in the atmosphere of Venus, and radio telescopes on the earth tracked these balloons as the winds of Venus carried them to and fro.

As usual, ideas that sound elegantly simple turn out to be a bit harder to put into practice. Scientists needed to use two radio telescopes located at opposite ends of the earth, in tandem, to determine the precise position of the balloon. Keeping track of two balloons over a period of two days each required a network of twenty radio telescopes in ten different countries, including the United States and the Soviet Union. The balloons floated at an altitude of fifty kilometers, in the cloud layer.

Because Venus has a thick atmosphere and rotates slowly, you might expect the weather to be comparatively bland. But each balloon was blown 11,500 kilometers (6,900 miles) around the planet by 150-mile-per-hour winds. We had known of these winds from photographs of the cloud patterns in Venus, but we had not expected them to be so turbulent. The balloons were buffeted upward and downward by intense downdrafts, up to three meters per second and lasting for four hours.

One of the most interesting aspects of this particular mission was its international flavor, involving close collaboration between Russian, French, and American scientists, along with the cooperation of scientists in seven other countries. The crucial stage of mission planning occurred in the late 1970s and early 1980s, when diplomatic relations between America and the Soviet Union were at a twenty-year low. The USSR had invaded Afghanistan, had crushed the Solidarity movement in Poland, and had shot down a civilian Korean airliner that had wandered off course over Siberia. The Reagan administration had put arms control negotiations into the deep freeze and talked about the Soviet "evil empire." Yet, scientists were able to put together a very complex mission that required an unprecedented degree of international cooperation even in this rather chilly era.

The Soviet Union seems to be so excited about ballooning that it has announced plans to use balloons for the Martian part of the *Vesta* mission, which is planned for the 1990s. The *Viking* spacecraft obtained magnificent photos of the Martian desert from one location on the Martian surface. Scientists would like to have broader coverage of the surface of the red planet. One way to get such a panoramic view is to put mobile landers on the surface, with caterpillar treads instead of landing pads. These craft would be complex (and therefore expensive). The Soviet scheme is to put balloons carrying cameras on them on the surface of Mars. During the daytime, sunlight would heat the balloons, causing them to lift up a few thousand feet and drift over the red sands of Mars, taking pictures of the terrain beneath. (This mission is still in "definition," and plans may change.)[9]

## THE MARTIAN ENIGMA

As seen now, Mars is a frozen polar desert. But its surface provides a very puzzling clue to its past history, shown in the illustration. The streaming tentacles that reach toward the top of the picture blend together as they work their way downward. These tentacles are the channels through which Martian streams and rivers flowed in more temperate times. Toward the top of the picture, the rivulets are small; they merge together and become large streamflows as they flow downward toward their ultimate watery destination. On the extreme left-hand side, there are the signs of multiple flow patterns as the massive cascade of water cut various channels through the low point in the surrounding terrain.

Why do planetary scientists insist that the illustration shows riverbeds? They look for key signs of structure that are found in comparable terrain on earth. There are tributary structures, streamflow lines, and teardrop-shaped features that would-be islands left over from when the river was flowing. All of the features that are part of the natural landscape of terrestrial riverbeds in arid environments are present in the channels of Mars. Crater counts tell us the age of these features. Between three and four billion years ago, rivers flowed on Mars.

But there is no way that these rivers could form under present Martian climatic conditions. The Martian atmosphere goes above freezing at the equator only in the summertime, for a brief interval at noon, and on the average has a temperature of 218°K ($-55$°C, or

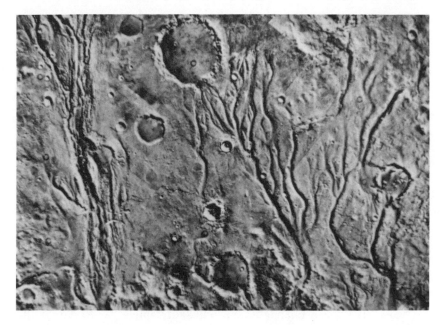

Two systems of Martian channels, Maja Vallis (to the left) and Vedra Vallis (to the right). Tributary structure is evident at the right; multiple flow lines are evident in Maja Vallis on the left. These channels flow toward the *Viking 1* landing site in Chryse Planitia. (Jet Propulsion Laboratory, NASA.)

−67°F). Even more of a problem is the low pressure on the surface of Mars, only 0.6% of the surface pressure on the earth, equal to the pressure 35 kilometers (20 miles) up in our atmosphere. Even if we could somehow produce a large volume of water on the surface of Mars, it would evaporate almost instantaneously.

The channels tell us very clearly that Mars once had a thicker atmosphere, and that the planet was probably warmer. How come? There is a naive scenario that provides an explanation. In the beginning, Mars had an atmosphere not unlike the earth's, thick enough to warm the surface by the so-called greenhouse effect, and thick enough so that liquid water wouldn't evaporate very fast. Under these conditions, Martian rivers could exist. Rains fell, or perhaps snow fell and then melted. Icy dams may have held water in glacial lakes. One way or another, rivers were made. If you had been on Mars at that time, perhaps you could have sat on the red sand and listened to the burble of a little brook flowing downhill. Or perhaps you would have been wary of flash floods, where a sudden event sends a wall of water crashing through the desert, devastating everything in its path.

If you were to sit on Mars now, neither the refreshing sounds of a little brook nor the frightening rumble of a flash flood would disrupt the cold, frozen landscape. The remotely operated *Viking* spacecraft photographed the surface of Mars repeatedly; one such photograph is shown in the illustration. Only a patina of frost disturbs the surface during the Martian year. There's very little water in the Martian atmosphere, which is too thin for river flow anyway.

What happened to change things? Where did the thick Martian atmosphere go? There are a couple of possibilities. The carbon dioxide atmosphere could have combined with the Martian soil to form carbonates, as happened on the earth. The earth is different because it's geologically active, so some of the carbonates are mashed and heated in the terrestrial interior and are eventually recycled to the atmosphere. On Mars, that doesn't happen, and so the atmosphere is eaten up by the planet itself and is locked away in surface carbonate and oxide compounds. The second idea is that the surface gravity of Mars is some-

The rock-strewn landscape northeast of the *Viking 2* lander. A small channel winds from the upper left to the lower right, a remnant of days when water could flow on Mars. The horizon is tilted because one of the lander's three legs is perched on a rock and the spacecraft is tilted eight degrees. (NASA.)

what less than the earth's, allowing atmospheric gases to escape more easily. Carbon dioxide is heavy enough so that even the weak Martian gravity can hold onto it, but if it is split into single carbon and oxygen atoms high in the Martian atmosphere, these could get away over time.

This picture of Mars suggests that the red planet is a planet in a permanent deep freeze. One reason that planetary scientists and others are interested in the Martian climate is that they want to use it as a test bed for investigating similar climatic changes on earth. The earth has not always been as habitable as it is now. About eighteen thousand years ago, huge ice sheets covered all of Canada, New England, New York, Michigan, Wisconsin, Iowa, and Minnesota and buried substantial pieces of Ohio, Indiana, Illinois, Missouri, Nebraska, and the Dakotas. Human life under such conditions would certainly be possible—precivilized humans did, after all, live at that time—but life would be quite different. The occurrence of several ice ages in the recent past suggests the possibility that, in another second of geological time, we would be plunged again into the global refrigerator. Understanding the causes of ice ages on Mars can help us understand why terrestrial ice ages happen as well.

## TITAN

Titan, a curious object to consider in a chapter on earthlike planets, is quite unearthly but, like Mars, may represent a permanently frozen, primordial earth. Titan, the largest satellite of Saturn, is the only satellite in the Solar System that is known to have a substantial atmosphere. Nitrogen is the dominant constituent of the atmosphere, just as it is for the earth. Atmospheric composition is the primary, but not the only, reason to consider Titan an "earthlike planet." Its atmosphere contains hydrocarbon compounds, possibly including complex organic polymers similar to the chemicals that are the building blocks of life. No one is suggesting that there is life on Titan, but it's quite possible that what we find there is similar in some respects to the chemicals that existed on the early earth, just before the remarkable transition that separated nonlife from life and chemical evolution from biological evolution.

We've known that Titan was unusual since 1945, when Gerard Kuiper, at the time the only American astronomer seriously interested in planets, detected the presence of methane gas around Titan. *Voyager 1*'s spectrometers detected nitrogen in the Titanian atmosphere, and it and subsequent ground-based investigations detected a variety of complex hydrocarbons in the atmosphere. The dark orange clouds, the only features that are seen in the rather bland photographs of this planet,[10] are probably methane droplets, contaminated with complex compounds produced by the interaction of light with simpler chemicals in the Titanian atmosphere. Many of these reactions are similar to the reactions that produce the Los Angeles-type smog, where sunlight plus nitrogen plus hydrocarbons produces a veritable witches' brew of irritating chemicals.

What we know about Titan suggests the intriguing possibility that it may be the only Solar System body other than the earth with rain, oceans, and rivers. Planetary scientists who model Titan's atmosphere suggest that the surface pressure is high enough so that methane ($CH_4$), or possibly ethane ($C_2H_6$), or both, could exist as a liquid on its surface. This hypothetical "ocean" could also contain ammonia, water, and nitrogen. At the surface, some of these compounds could be solid. Water-ice icebergs may float in an ocean of ammonia, methane, and ethane. Exotic trace compounds in its atmosphere—ethane, propane, acetylene, ethlyene, hydrogen cyanide, methyl acetylene, and more—are detected in Titan's upper atmosphere, but they condense at some level in the atmosphere and may fall as rain to the surface. This rain would collect and form rivers. *Voyager* provided little information about Titan's weather but did measure some brightness differences around the planet that could be a sign of winds.

Very little is known for certain about Titan's interior, but the possibilities are intriguing. It's likely that the interior contains a core of rocky material. In this context, *rock* means some form of silicates like terrestrial rocks, iron, sulfur, solid carbon-rich compounds, and other stuff that is solid at terrestrial temperatures. But much of the interior is composed of a mixture of water ice and ammonia. The mixture of the two will partially melt into a eutectic fluid at a temperature of 173°K. (A eutectic is a mixture of two substances that melts at a lower temperature than either of the two substances alone.) This eutectic is lighter than water ice and would rise in Titan's interior, producing water–ammonia volcanoes. It's possible that large-scale fluid motion in the interior could produce tectonic processes and even water–ice continents on the surface. These volcanic processes would vent water to the surface, and the water could float in the ocean as icebergs.

The last two paragraphs sound like science fiction, but they aren't. We don't know how many of these exotic possibilities are actually there, but everything just mentioned comes from a sober scientific summary of what we know about Titan.[11] A dedicated mission to Titan could be modeled on the *Galileo* probe that (some day) may orbit Jupiter. It could map the surface with radar, analyze the composition of the atmosphere, and send in an entry probe that could take pictures of the clouds. Current plans call for a modest mission with a hard lander that would be destroyed in impact.[12] However, a soft lander could be produced quite simply: a parachute is deployed, and a floating probe or boat is landed in the Titanian ocean. The pictures coming back might evoke Samuel Taylor Coleridge's "Rime of the Ancient Mariner":

> All in a hot and copper sky,
> The bloody Sun, at noon,
> Right up above the mast did stand,
> No bigger than the Moon.
>
> Day after day, day after day,
> We struck, nor breath nor motion;
> As idle as a painted ship
> Upon a painted ocean.
>
> Water, water, every where,
> And all the boards did shrink;
> Water, water, every where
> Nor any drop to drink.[13]

The poem's magnificent imagery conveys a picture of what might be there, when we go and sail the seas of Titan. I doubt that human beings will physically land on the surface, but we could use an automatic probe to serve as an extension of our eyes and enjoy much of the visual experience without the dangers. It would be worth sending a soft lander to Titan. It could float in the oceans, take pictures, and show us a real world even stranger than some of the colorful, surreal, psyche-delic images associated with the film *2001: A Space Odyssey*.

In a more serious vein, Titan is fascinating to compare with the earth because it's the only other solar system body with an atmosphere similar in composition and total pressure to the earth's. The two bodies are similar in that each has or may have rain, oceans, continents (?), and volcanic or tectonic processes. But the similarity may extend to processes of biological interest as well.

The orange stuff visible in the bland pictures of Titan is a complex photochemical smog that forms far above the Titanian surface. Similar chemical reactions may have taken place on the primitive earth. In its early days, the earth had, broadly speaking, an atmosphere like Titan's present one, containing nitrogen (possibly molecular, and probably in the form of ammonia), hydrocarbons, and no free oxygen. All the oxygen was locked up in carbon dioxide and hydrogen dioxide (water). This unearthly atmosphere interacted chemically to produce the precursors of life: amino acids, nucleic acids, and other complicated chemicals that, perhaps, eventually came together to form the first living creatures. Although the idea is speculative, Titan could be a global laboratory that can be used to investigate the chemistry of the primitive earth. Titan is too cold to harbor living organisms similar to the ones we find on earth; we're not looking for slimy Titanian creatures. Rather, Titan could show us what chemical reactions preceded origin of life on earth.

## THE SEARCH FOR LIFE ON MARS

Although comparative planetology, exotic landscapes, and extraterrestrial resources can serve as a magnet to draw us to the planets, the search for extraterrestrial life has been a Holy Grail for a considerable number of scientists and laypeople. A number of pioneering space scientists thought of another planet as a global exobiological laboratory: Mars.[14] Much of the space on the *Viking* probe to Mars was devoted to a set of biological experiments seeking life on Mars. It seems to me that the inconclusive, generally negative results from this search should serve to guide future mission planners, alerting them to the dangers of focusing too narrowly on one particular way of asking one particular question. Leonard David of the National Space Institute neatly summarized the results of the *Viking* mission: "Is there life on

Mars? After years of work, *Viking* simply answered: 'Would you repeat the question?' "[15]

How would you look for alien life with a robot spacecraft if you didn't know what to look for? My own answer today would be quite simple: Send a microscope with a TV camera hooked up to the eye-piece. You could look at a variety of soil samples and, if the lander was mobile, search in different environments. (I don't know what the pundits' answer would be. To my knowledge, the unsatisfactory results from *Viking* have discouraged scientists from similar searches for life.) High school biology students look at a drop of pond water to examine all of the microscopic creatures in it, and a similar experiment done with the Martian soil would probably be a key part of any future remotely operated life-detection equipment.

If scientists had tried to send a microscope to Mars using 1960s technology, the lander would have become much too big to afford. None of the science instruments on *Viking* occupied much more than a cubic foot of space. Microscopes could fit in such small spaces, but the TV cameras of that time could not. The cameras that photographed the landscape from *Viking* were special ones that scanned one vertical slice of the picture at a time, taking twenty minutes to complete a panorama. Such cameras would obviously be unsuitable for microscopy. The only cameras available in the 1960s that would take real two-dimensional pictures quickly were big klutzy things, bigger and heavier than foot-ball players. Cameras like these are still found in TV studios, where weight isn't a problem, but the lightweight "minicams" and home video cameras were simply not available then. So a microscope was out.

Given the weight limitations, *Viking* scientists and mission planners devised a miniature chemistry lab that fitted into the allowable cubic foot of space. They knew that one characteristic of living organisms is that they eat and grow, exchanging chemicals with the surrounding environment. The three "biology" experiments on *Viking* exposed Martian soil to water, nutrients, sunlight, and warmth—stimuli that might cause any Martian microorganisms to start growing. The growth might be detected in one of three ways, by different experiments. The labeled-release (LR) experiment fed the soil with a veritable witches' brew of likely foods, all of which contained a small amount of radioactive carbon ($C^{14}$). Putative organisms in the soil would eat this stuff and then breathe out slightly radioactive carbon dioxide ($CO_2$). The pyrolytic release experiment was similar but operated in the reverse direction, exposing the soil to radioactive $CO_2$ and then heating the soil to see if any of the atmosphere was incorporated into living organisms. The third experiment was a bit more complex, monitoring the composition of the air above an unlabeled soil sample as the sample was exposed to nutrients and other stimuli.

The puzzling surprise of *Viking* was that all three of these experiments produced results that were, at first glance, indicative of life. CO and $CO_2$ were incorporated into Martian soil. Nutrients did react chemically with something in the soil and released labeled $CO_2$ and oxygen, though the time pattern was wrong. We would expect bacteria to grow and reproduce, producing a stronger signal with time, and the actual release of gas was initially rapid and then declined. However, in the initial interpretation of the biology experiments, this difference between a biological signal and what was actually observed was not regarded as being particularly important.

Did these findings mean that there were creatures on Mars? One last crucial experiment indicated that the *Viking* biology experiments were probably detecting exotic soil chemistry rather than Martian bacteria. A fourth instrument searched for organic matter in the Martian soil and found none, though it would have been sensitive to one organic molecule per billion molecules of soil material. Terrestrial soils, even from remote places like the dry valleys of Antarctica, contain organic matter in concentrations thousands or millions of times higher than *Viking* could have detected on Mars. This organic matter (which gardeners call *humus*) is found in all terrestrial soils and is either the excreta or the decomposed remains of living creatures. There is no humus on Mars.

Virtually all of the *Viking* scientists believe that the sterility of Martian soil deals a fatal blow to any biological interpretation of the pseudopositive results of the other three biological experiments. And so it was back to the drawing board. In the late 1970s, a number of teams worked over the *Viking* data and showed that if there were a "superoxide" in the Martian soil, perhaps a combination of hydrogen peroxide and a similar iron compound, the *Viking* life-detection experiments would be fooled. This is a chemical rather than a biological explanation of the data from the three experiments. Most scientists would agree with the conclusion cautiously expressed by Harold Klein, a life scientist at NASA's Ames Research Center:

> The results of the biological experiments on Viking, taken in isolation, allow the possibility that at least some of the data (i.e., from the LR experiment) might be of biological origin. However, when considered together with all of the other Viking results and when the subsequent ground-based investigations are also considered, it would appear more reasonable to ascribe all of the biology "signals" to nonbiological causation. Given the inherent limitation of the biology instrumentation, . . . one can only speculate whether *other* tests or *other* approaches to "life detection" would have provided credible evidence for biology in the samples that were tested. Nor is it possible to rule out that the Viking metabolic tests, applied to samples from other martian sites, . . . would have given data unequivocally indicative of living systems. Finally, it is not possible, on the basis of the Viking data, to exclude the presence, in the Viking samples, of exotic forms of life not based on carbon chemistry and, therefore, not related to current notions about the origins of life.[16]

In other words, *Viking* didn't find life on Mars, but maybe it didn't look hard enough. Klein and I agree that if you are to look harder for life on Mars, a much more comprehensive approach is necessary. Perhaps the best way to search for life on Mars would be through a sample-return mission, where a spacecraft would land on Mars, collect samples, and return them to earth. This mission could, of course, do much more than search for life, and it is under active discussion as the next major initiative in the planetary sciences.[17]

## LESSONS FROM *VIKING*

As I write about the results of the biology experiments, I feel a sense of acute frustration. These experiments occupied nearly half the space available for science on the *Viking* lander, and we really learned relatively little from them. Those who have scientific, theological, or personal reasons for fervently wanting to believe in life on Mars can always point to the loopholes in the *Viking* experiments. In retrospect, much of the space devoted to the biology experiments would have produced more science if it had contained some of the geology or meteorology instruments that had been sacrificed in order to save weight (and costs). The search for life was so appealing to mission planners and to influential scientists that the final experiment package turned out to be quite unbalanced and quite unsatisfying in every way.

My point in making such arguments is not to point fingers of blame but to suggest how we can learn from the failure of the *Viking* biology experiments. The mission as a whole was not a failure. Much of our knowledge about the geology and meteorology of the red planet came directly from *Viking*. I know of no other part of the space program ever in which so much effort has been put into answering such a narrowly specialized question. Norman Ness, one of those whose instruments were sacrificed to make room for the life detection experiments, pointed out to me that, at the time *Viking* was launched, one-third of all of the money NASA had spent on science up to that time was put into *Viking*. As half of the effort of *Viking* was allocated to biology, this means that one-sixth of all of NASA's science dollars went into this ultimately unsatisfying search for life on Mars.

As I see it, the problem is that the *Viking* biology experiments were premature, a departure from the usual sequence of planetary exploration. Scientists knew next to nothing about Mars when the concept of the *Viking* mission matured in the mid-1960s, and little more when the final design solidified around 1970. The *Viking* results suggest that, before planning an expensive, specialized space mission, scientists should do a more general investigation.

The exploration of a particular body can be separated into three phases. The *reconnaissance* phase comes first and usually involves flyby missions. These change a blur in the telescope or a mere dot with a name to a world whose gross properties we understand in such terms as the relative youth of its surface, the thickness of its atmosphere, and the importance of various geological or atmospheric processes. The *exploration* phase follows, involving orbiters and soft landers, and our coverage of the planet becomes truly global and can last for an extensive period of time. Next comes *intensive study,* such as has been done with the *Apollo* lunar samples, using soft landers and possibly with sample-return missions to make detailed measurements. NASA occasionally adds a fourth phase, *exploitation* or utilization, which has not occurred so far but which has been discussed in connection with the moon and the earth-approaching asteroids.[18]

The *Viking* biology experiments were characteristic of the intensive-study phase of planetary exploration and therefore departed from this normal sequence of missions, as they were planned during the reconnaissance phase and skipped over the exploration phase. The question probed by the various biology instruments was very specific: Is there something in the soil that is involved in a very specific set of chemical reactions? Had scientists known how globally bleak the planet was, mission planners might well have realized that the chances of finding carbon- and water-based life were quite slim and might have redesigned the lander accordingly.

*Viking* suggests that we should approach any aspect of the space program in an orderly, evolutionary manner. We should not invest thousands of millions of dollars in any one particular area until we have done the spadework, which can often be done for tens of millions or even less. As a specific example, consider the microgravity program. We just don't know whether microgravity will be the next major space industry. It would be a mistake to build a space station that could do only microgravity research. The space station currently under discussion is a very flexible one and should not suffer the ignominious fate of the *Viking* biology experiments.

Another lesson to be learned from *Viking* is that we should pay attention to future technological change when considering what to include in a particular orbiter or lander. I don't know if anyone in the mid to late 1960s could have anticipated the microelectronics revolution, in particular its ability to make very compact imaging devices. NASA, perhaps because of the *Viking* experience, has designed a number of new missions in a modular way that allows for upgrading. For example, technologically obsolete first-generation instruments on the Hubble Space Telescope can and will be replaced by second-generation, updated instruments.

## THE FUTURE

There's a place for Venus and possibly Titan in NASA's future. The *Magellan* mission has begun, having reached the magical "new start'' status in 1984. When a mission becomes a new start, money is spent actually to build hardware rather than to come up with designs of varying degrees of sophistication. *Magellan* can map Venus at high resolution and provide crucial information on its geological history. However, the USSR's *Venera* mission may have skimmed some of the cream in this area. A mission to Titan, in which many of the instruments on *Galileo* would be used to survey Titan, was the fourth priority on the SSEC's list of "planetary observer" missions. An additional proposed mission to Venus is the Venus Atmospheric Probe, which could clear up some of the ambiguities in the data on the abundance of rare gases on Venus, providing key clues to its history. Only *Magellan* is in the space shuttle manifest, scheduled for launch in April 1989.

Yet, Mars remains a powerful beacon, luring us onward as it has for fifty years. Even from a strictly scientific viewpoint, it is as enigmatic as Venus is, if not more so. Because it's smaller than the earth, but not so small that it lacks an atmosphere or geological activity, it sits in a key spot in any comparative planetological scheme. However, the attraction of Mars is much more than strictly scientific. We can send people to Mars, and we can't send people to the surface of Venus. Venus is just too hot. Humans have worked in the vacuum of outer space, and so walking around on Mars should be comparatively easy (though it will take more than an oxygen mask). The idea of Martian colonies is not at all far-fetched from a strictly technological standpoint. If humans are to colonize the Solar System, Mars is the only logical starting point.

A human expedition to Mars would be a tremendously ambitious undertaking. Were it to be done in the same way that the *Apollo* mission was done, with every last drop of rocket fuel, radiation shielding, water, and so on lifted laboriously out of the earth's deep gravitational hole, the cost would be prohibitive. But a group of people calling themselves the Mars Underground, mostly graduate students in various science and engineering fields at the University of Colorado, began to think about alternative technologies in the late 1970s. They have held a couple of conferences exploring both the reasons for going to Mars and some technological advances that will make it cheaper. Ingenious rocket engineers have devised transportation schemes to save rocket fuel by, for example, using Mars's atmosphere to apply the brakes when a spacecraft arrives at Mars. The primary obstacle seems to be the psychological and physiological difficulties of surviving the trip to Mars and back, which (in most mission scenarios) is a nine-month journey each way. Some mission profiles require a year's stay on Mars, making the total mission two and a half years long.

Whether we should send people to Mars or not is probably a decision for the future. The National Commission on Space has proposed such an expedition for the twenty-first century; a commitment to such a mission does not need to be made now. The scientific exploration of Mars does not require people; if the only objective is science, the nation would be better off investing the billions of dollars in sophisticated remotely operated landers rather than in life support systems. However, there are many reasons beyond space science for exploring Mars, because the space program is multifaceted. A remote lander cannot serve as an ambassador or as a symbol of humanity in the same way that astronauts do.

When we ask what should be done about Mars right now, we need to keep the possibility of going to Mars firmly in mind. Some of the younger people who are reading this book may well be alive when the first human expedition touches down on the wind-swept sands of Mars. The exploratory missions that are absolutely necessary as long-range precursors to a human expedition are of great interest in their own right. The prospect of human expeditions to Mars can serve as a long-range focus for a program of exploration that would be valuable even if we ultimately decide that it isn't worth the trouble to send people there.

What missions to Mars await us? Two further missions seem established to the point where reasonable people expect them to happen: the American *Mars Observer* (originally the *Mars Geoscience/Climatology* orbiter) and the Russian *Phobos-Deimos* mission. The American mission was the second of a series of low-cost "planetary observers" proposed by NASA's Solar System Exploration Committee (NASA-SSEC) in the early 1980s. This spacecraft will go into polar orbit around Mars, allowing for global coverage, and will map the mineralogical and atmospheric composition around the whole planet over at least one Martian year (687 Earth days). A magnetometer on the spacecraft will finally determine whether Mars has a magnetic field or not. This mission was the top-priority one for Mars and the second overall in a proposed American planetary science plan.[19] An indication of the parallel goals of the scientists and the Mars settlement movement is that the Mars Underground[20] independently (as far as I know) picked this very same mission as their first choice. The two groups had completely different motives, with NASA-SSEC being motivated by science and the Mars Underground chasing a dream of people on Mars. The *Mars Observer* has been approved for new-start status, with possible launch dates being 1990 or 1992.

In an uncharacteristic display of openness regarding their future plans, the Soviet Union has discussed its ambitious mission to Mars's satellites Phobos and Deimos in some detail.[21] (The acronym *Ph-D* is occasionally used to refer to this mission.) A photograph of Phobos is shown in the illustration. The Phobos mission, slated for launch in 1988, involves two identical spacecraft. One will attempt to match orbits with Phobos. Phobos is one of Mars's two tiny satellites, with a mean diameter of 21 kilometers and gravity so weak that an average human being would weigh a couple of ounces and could easily jump off the moon using legpower alone. The spacecraft will hover some 50 to 100 meters above the surface of this moon, taking pictures and doing remote chemical sampling. The most dramatic part of the mission involves the planned use of a laser beam to vaporize material from the surface of this planet. (Incidentally, these lasers will probably be much less powerful than those that might be used in a Star Wars missile-defense program.) Because Phobos's gravity is so weak, the vaporized material will float up toward the spacecraft, where its composition can be analyzed. A lander will then descend to the surface and hop around from one point to another, taking pictures and analyzing rocks. If all, or most, of these ambitious analytical schemes work, the second spacecraft will be sent to repeat the process at Deimos, Mars's other satellite.

A number of missions beyond these two are in the talking stage. Additional missions to explore Mars's upper atmosphere have been discussed by both Americans (who call the mission the *Mars Aeronomy Orbiter*) and Europeans (*Kepler*).[22] Other small-scale missions to the surface involve sending a probe with an instrument (called a

The *Viking 2* orbiter produced this photograph of Phobos, the most detailed available. The craters are produced by impact; the origin of the grooves is unclear. (NASA.)

*gamma-ray spectrometer*) to do a detailed chemical analysis of the Martian surface. A similar instrument was flown on the *Surveyor* lunar missions, but it was left off *Viking* in order to make room for the biology experiments.

Another way to survey the Martian surface is to use penetrators. An orbiting craft would hurl a dozen or so lances at the Martian surface. Heat shields and parachutes would slow them down as they entered the atmosphere. The long, missile-shaped objects would penetrate the soil and come to a stop. A small, simple box attached to these lances could then land softly on the Martian surface and provide information on the Martian weather, on the chemistry of the surface at that point, and on any Marsquakes. A half dozen or so such stations could provide us with an inexpensive, global view of Mars to supplement the detailed but narrow perspective of the desert surface provided by the *Viking* lander.

But closest to the hearts of those scientists involved in the exploration of Mars is the most ambitious mission of all: a mobile lander.[23] Look at the illustration of the Martian surface, and envision yourself standing there, protected from the frigid winds by your spacesuit. You would want to go beyond the next hill, to see what's on the other side of that sand dune. The difference between the appearance of Mars at the two *Viking* landing sites suggests how varied the surface may be. Although it may be some time before people can go to Mars, a mobile lander could allow humans to do almost the same thing. If caterpillar treads are put on the legs of a lander like *Viking,* it could roam about the Martian surface, taking pictures. Human controllers on the earth, viewing the pictures, could decide where to go next. The command

View of the surface of Mars from *Viking 1*. The Martian landscape is quite different at the two widely separated sites, with sand and large boulders much more prevalent at the *Viking 1* site. (NASA.)

would go up to the lander, and off it would rumble. It would be almost like being there and walking around, using the television cameras as our "eyes."

A mobile lander mission could open some of the most interesting Martian features to close examination. The *Viking* landers had to land on some of the geologically blandest terrain on Mars because we didn't want them to land on the side of a hill and fall over. A mobile lander could travel up some of the long, sinuous channels that once carried water over the Martian surface. They might even be able to climb up the steep cliff that flanks the huge volcano Olympus Mons and climb the highest mountain in the Solar System. Because the lander wouldn't run out of food or water, or need to keep a date with a rendezvous vehicle to carry it back to earth, it could explore Mars for a long time. It would be almost as good as being there ourselves.

The most sophisticated lander mission would have provisions for returning samples to terrestrial laboratories. The roving craft would separate from the lander, which would serve as a base. It would wander about Mars, selecting samples from various geological terrains. Automated shovels and picks would pick up fragments of larger rocks as well as fine soil. The samples would not be sterilized, and an attempt would be made to keep them in an environment resembling that of Mars on their return to earth, so that life detection experiments could be done in terrestrial laboratories (or perhaps on board the space station) rather than remotely. This mission has been studied fairly intensively and could be launched in the late 1990s after the completion of the space station.[24] However, it would cost a lot and may not be started very soon.

Because both the Soviet Union and the United States are interested in Mars, the two nations have resumed discussions on international cooperation.[25] In 1982, as a way of protesting the declaration of martial law in Poland, the United States failed to renew a previous agreement between the two countries calling for international cooperation. A ten-member team headed by General Lew Allen, director of the Jet Propulsion Laboratory, visited Moscow in September 1986 to discuss thirteen separate proposals for cooperation. Many of these proposals specifically involve Mars, including coordination of the Soviet *Phobos-Deimos* mission with the American *Mars Observer* mission.

The Soviet Union proposed to coordinate studies of the *Venera 15* and *16* radar data in connection with the planning and operation of the *Magellan* radar mapper. Informally, the Soviet Union has invited three American planetary scientists to participate in the *Ph-D* mission. No final agreements have been signed yet; some U.S. State Department and Defense Department officials have raised questions about technology transfer. Others on Capitol Hill, most particularly Senator Spark Matsunaga of Hawaii, are quite anxious to promote international cooperation for its own sake. It is likely that the informal cooperation, in which each country's scientists talk to the other's about mission plans, will continue. More extensive cooperation involving, for example, the use of the American Deep Space Network to communicate with Soviet spacecraft, has been done already in connection with the Venus ballooning experiment.

*    *    *

Mars is a natural first target for space exploration beyond the earth's neighborhood. Its surface, once thought to be extremely earthlike, is cold, wind-swept, and relatively hostile. The planet still may contain some resources that will benefit future missions. Its landscape contains huge volcanic mountains, globe-girdling canyons, and enigmatic channels, the remnants of a much more temperate period in its climatic history.

Mars, Venus, and Titan have great scientific value as a global scientific laboratory that we can use to test our understanding of how the earth works. The different tectonic styles on Mars, Venus, and Earth provide us with landscapes that are fascinatingly different, yet similar enough so that each can help us understand the others. Titan, with its earthlike atmosphere and hydrocarbon smog, may well be the primitive earth in a deep freeze, a global laboratory that will allow us to study what the earth was like before life began.

We'll come back to Mars, somehow, some day. At some time, automatic rovers will roam over the surface, climbing the next Martian hill under the guidance of their human operators. Perhaps, within the next fifty years, a team of astronauts will set foot on the frigid Martian surface, to gaze over the sands of the red planet.

# CHAPTER 10

# Voyage to the Land of the Giants

Beyond Mars, the Solar System takes on a very different cast. The most earthlike objects in the outer Solar System are not the planets themselves, but the planetary satellites. Four of the five planets in the outer solar system—Jupiter, Saturn, Uranus, and Neptune—are far larger than Earth. These giant planets have a most unearthly composition—mainly hydrogen and helium. We couldn't land on any one of them because none has a solid surface (except at the very center, where there are cores of rock or ice). All except Uranus emit more energy than they receive from the sun, being more like stars than like planets, real giants when compared to Earth.[1]

Until very recently, we knew very little of these giant planets. The largest and nearest, Jupiter, is a blurry disk as seen from ground-based telescopes. Weather patterns that are smaller than North America are too small for us to see from the earth. Even the crude cameras of the *Pioneer* mission, which flew by Jupiter and Saturn in the early 1970s, demonstrated how complex these worlds are. Each *Voyager* encounter has resulted in the discovery of a whole new group of planetary satellites.

In some ways, the biggest surprise among the giant planets was the discovery of ring systems. Before 1977, we knew only that Saturn had rings. Elementary schoolchildren would send me letters asking why; I responded to those letters with a few words explaining our ignorance. Now we know that all four giant planets have rings or ringlike arcs around them. These ubiquitous rings turn out to be fascinatingly complex.

## SATURN

The rings of Saturn, shown from *Voyager 1* in the illustration, have been known for nearly four centuries. Galileo, in a sense, discovered them in 1610 because he noticed that Saturn had a strange appearance, seeming to be a "triple planet." In the 1650s, the Dutch astronomer Christiaan Huygens used better telescopes to reveal that these "companions" to Saturn were in fact a ring system girdling the planet. Ground-based observations indicated that there are three components to the ring system: an outer ring, an inner ring, and an innermost, filmy "crepe" ring.

In following centuries, the mathematicians Pierre Laplace and James Clerk Maxwell deduced the fundamental nature of the rings. They realized that the inner part of a planetary ring, being closer to the planet and in a stronger gravitational grip, had to rotate around the planet faster than the outer part if the centrifugal force produced by the ring's rotation were to balance gravity. If the rings were a solid sheet, this differential rotation would rip them apart. Consequently, they had

The planet Saturn, seen from *Voyager 1* several days after the encounter. The sun is to the right of the picture. The shadow of the rings on the planet shows up as a thin line, and the shadow of the planet falls on the rings. (Jet Propulsion Laboratory, NASA.)

to be something else: a swarm of small, icy particles, each in its own orbit around the giant planet.

As happens so often in astronomy, serendipity demonstrated that the rings of Saturn were not a unique feature in the Solar System. The disk of Uranus is so blurry in the telescope that it is hard to determine just where the edge of the planet is. Consequently, the precise numerical value of the planet's radius was known rather inaccurately in the mid-1970s. A number of teams of astronomers sought to clean up this problem by observing an eclipse (technically called an *occultation*) of a star by Uranus in March 1977. The orbital motion of Uranus around the sun carries it across the sky, and very occasionally, the planetary disk passes in front of a star. If one times just when the starlight cuts off, and when it reappears again, one can measure the planetary radius quite precisely. Many scientists felt that this project was worthwhile, though it was not easy to convince NASA officials to send their airborne observatory on a long trip just to pin down a number more precisely. No one had any inkling that the project would lead to a fundamental discovery.

## URANUS

A number of teams of astronomers trained their telescopes on the star, called SAO 158687, on 10 March 1977. Because the occultation was predicted to occur at far southern latitudes, Jim Elliot of MIT had successfully persuaded NASA to send its airborne infrared telescope to the Southern Hemisphere, where it would make a number of other measurements of the infrared sky. This telescope sticks out of the belly of a C-141 airplane. The usual reason for using the airplane is to take the telescope well above the lower layers of the earth's atmosphere, which interfere with the measurement of infrared radiation. But in this case, the idea was to take the telescope into the southern part of the Indian Ocean to see the planet occult the star.

Occultation astronomy is quite exciting because the instruments and equipment have to be ready at the same time as Nature is; one can't come back the next day and repeat the measurement. In film terms, there's only one "take." On 10 March, the C-141 and its equipment were ready. A team of astronomers boarded it and ventured southwest from Perth, over the turbulent Southern Ocean. They aimed the telescope precisely at the star, expecting to see a steady beam of light from it until Uranus would block it.

A major surprise was in store for the observing team. As they were checking the light levels reaching the telescope from the star and from the background sky, a discovery occurred, one of the kind that is a highlight of any scientific career. The reactions of the participants were tape-recorded, and below is an excerpt of the conversation. Jim Elliot led the team, bringing graduate student Ted Dunham and programmer Doug Mink aboard the aircraft. Of the others who were tape-recorded, Al Meyer operated the star tracker, Pete Kuhn ran the infrared radiometer monitoring the water vapor in theatmosphere above the plane, and Jim McClenahan was one of two mission directors:

DUNHAM: (*referring to the sky brightness*): One seventy.
ELLIOT: One seventy?
DUNHAM: Yeah.
?: Well, subtract two hundred to get one thousand six hundred fifty.
DUNHAM: What was that? What was that? (*The signal from the star had dipped unexpectedly, and Dunham noticed the dip on the chart recorder.*)
ELLIOT: What?
DUNHAM: This!
ELLIOT: I dunno. Was there a tracker glitch?
MEYER: Nothing here.
MINK: Uh-oh.
DUNHAM: No, I don't think it's anything here, it's clearly duplicated in both of those. (*It's not clear what's being referred to here; I suspect that the team was making sure that the dip appeared in all of the light sensors and was not just an electronic glitch.*)
ELLIOT: Yeah, I mean clouds or . . . ?
MEYER: Ask Pete.
DUNHAM: Pete, what's your water vapor?
KUHN: Eight point nine.
DUNHAM: Well, that's pretty low.
McCLENAHAN: What happened?
ELLIOT: Well, we got a dip in the signal here which was either due to a loss or a momentary glitch in the tracker, or a cloud whipping through.[2]

What had happened was not tracker error, or clouds, or electronics. The thin outer ring of Uranus had passed between the star and the

Kuiper Observatory, cutting off the light from the star. This interpretation became clear only after several other dips were observed, and after the rings were observed to occult the star again, after the star passed behind Uranus. The conversation continues:

ELLIOT: We got blips again.
MINK: What was the time about?
DUNHAM: Maybe about a half-minute before that.
. . . (*intervening dialogue omitted*)
MINK: That was one at 21:47.
ELLIOT: They're real.
?: Yeah.[3]

The conversation conflicts with some popular images of how science is done. Science is often viewed as a solitary enterprise, but many discoveries are made by teams like this one. It's also clear from the tape that it takes some time before one recognizes something as a major discovery, especially if it's unexpected. The team's first reaction was the typical scientist's reaction to something unexpected: What piece of equipment fouled up this time? Or was it clouds? These possibilities were finally excluded when the team learned that everyone who was observing the occultation saw the dips, a discovery removing any possibility that some sort of error in a single observation was the cause.

Even after determining that the dips were real, no one was sure that rings had caused them. The only ring system known at the time was Saturn's, in which the rings are very broad. Narrow rings seemed outrageous, and for a few days, Elliot's team thought that they had detected a belt of satellites around Uranus. But eventually, when all the observations of this occultation were put together, the scientists interpreted the results as showing rings.

But why were the rings so narrow? And why were they so dark? Once they were discovered, Keith Matthews and colleagues used the Palomar 200-inch telescope to obtain an infrared picture of them. (The infrared was used because the planet is much dimmer in those wavelengths.) The rings were visible, but just barely. They absorb 98% of the light that hits them, reflecting a feeble 2% for Matthews's instrument to detect. The only terrestrial substance that is this black is carbon black, the pigment used in the ink that makes up these words.

## JUPITER

Jupiter joined the collection of planets with known rings in 1979. The story behind the discovery of Jupiter's rings has its own sense of drama, too, but in a different way. I have heard it a few times from friends; Jim Elliot and Richard Kerr delightfully recount the story in their book *Rings*.[4] The hero is real-life planetary scientist Toby Owen of the State University of New York at Stony Brook. Curiously, Owen was not a ring specialist; he studies planetary atmospheres and was one of many members of the imaging team that was to point *Voyager*'s cameras at interesting targets as it flew by Jupiter.

In retrospect, the rings had been seen in 1974 by the *Pioneer 11* spacecraft. (Earlier claims of optical observations of a ring are not demonstrably related to the ring that is actually there.[5]) When *Pioneer* passed through Jupiter's equatorial plane, there were unexplained dips in the number of charged particles striking the *Pioneer* detectors. Mario Acuna and Norman Ness, writing an article describing Jupiter's magnetic field, mentioned a ring as a possible explanation. They wrote, "Finally, although we consider it remote, the possibility exists that the two minima are due to sweeping effects by an unknown satellite or ring of particles at approximately 1.83 $R_J$ not yet visually observed."[6] Ness and Acuna wrote several letters trying to interest people in a search for a satellite or ring.

Toby Owen also wondered whether Jupiter had rings. I asked Ness what role the *Pioneer* results had had on Owen's thinking, and neither he nor I know the answer for certain. Owen took it on himself to propose a search for Jovian rings, not a simple proposal to bring to fruition. The difficulty was not that *Voyager*'s cameras weren't capable of seeing rings. Rather, the precious moments of a close encounter with a planet have to be budgeted very carefully. The team of scientists, who had paid their dues patiently sitting through interminable briefings by NASA engineers as *Voyager* was being built, finally cashed in their chips by planning just where on the giant planet the

*Voyager* cameras would point during the close encounter. If *Voyager* was to discover a gossamer ring, the cameras would have to take a long exposure just when the spacecraft crossed Jupiter's equator, so that it could look through the rings and pick up something, no matter how thin. Most of the team members believed that it would be a waste of time to move the cameras away from the turbulent mix of clouds on the giant planet simply to take a picture of what they expected to be empty space.

In October 1977, shortly after *Voyager 1* was launched, Owen made his final case. He and Edward Danielson of Caltech started by asking for four exposures to search for rings; in the final bargaining, they were cut back to one eleven-minute exposure and were told they could take it only if the instructions to the on-board computer did not take up much memory space. Hardly anyone really expected to find anything. David Morrison, a member of the imaging team, wrote later that this picture was planned "not really with any great expectation of a positive result, but more for the purpose of providing a degree of completeness to Voyager's survey of the entire Jupiter system."[7]

Where were they to look for rings? *Voyager* had only two cameras. One could cover the entire region of space from the planet to Amalthea, the innermost satellite of Jupiter. The other took a picture of a postage stamp right in the middle of that range, where the middle of a scaled-up Saturn ring would be. Candice Hansen, a technician with *Voyager,* was able to keep the other *Voyager* scientists happy by devising a set of computer instructions that would use only eight words of computer memory to control the cameras.

Owen expected that the best results would come from the wide field camera that covered every spot where a ring could possibly be. But this photograph was overexposed because the radiation and the scattered light were higher than expected. If Owen had obtained four or even two shots, he could have tried a number of different exposure times to get around this problem, but there was only one chance. The *Voyager* team's efforts were not in vain. The rings were shown by the narrow field camera, which was pointed exactly in the right place. The discovery picture is shown here.

Owen and the *Voyager* team had a little trouble, at first, figuring out just what was in this picture. The wiggly lines resembling broken

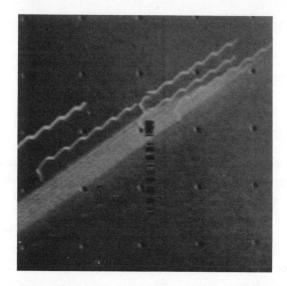

The discovery image of Jupiter's ring. The spacecraft's gentle rocking motion creates the bizarre hairpin patterns, which are the images of stars. The broad band is the ring. The dark lines are places in the picture where data were lost at the tracking station. (Jet Propulsion Laboratory, NASA.)

hairpins are the images of stars, trailed out by camera and spacecraft motions. The diagonal white streek, however, is not a star. Three days of groundwork followed, during which calculations showed that the streak (which actually is a composite of six images of the ring) was exactly in the plane of Jupiter's equator, where a ring should be. And so Jupiter, too, joined the club of ringed planets. Now that Owen had found Jupiter's ring, the other members of the imaging team were willing to give up some of their pictures of clouds, and *Voyager 2* was reprogrammed to obtain many more ring pictures. Jupiter's ring is a thin, tenuous sheet of matter, going completely around the planet and having dimensions of 6,000 by 30 kilometers in cross section.

By 1980, the question was not "Why does Saturn have rings?" but "Why doesn't, say, Neptune have rings?" It turns out that Neptune does indeed have a ring system, but of a different sort. Several observations made at various times in the 1980s—and one made many years earlier—showed occultations of the same kind seen in the case of

Uranus. But the observations were inconsistent with each other. If there were rings, scientists would expect to see a pattern, where one occultation on one side of the planet would be matched on the other side of the planet, or in a later observation. Finally, scientists observed enough of these occultations to dismiss the possibility of clouds or equipment failure. Scientists now believe that Neptune is surrounded by arcs of material rather than a complete ring.[8]

Rings are complicated but fascinating. Planetary scientist J. O. Burns has reportedly said that "with rings, as with perfume, a small amount of substance can be very revealing."[9] The illustration shows the many ringlets in the ring system of Uranus. The sun, ring, and

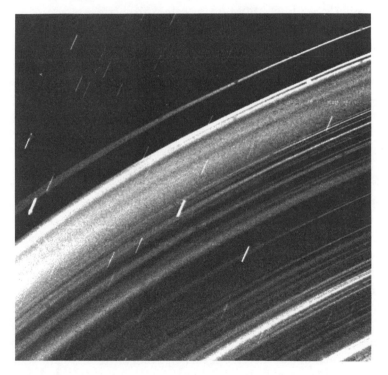

This *Voyager* 2 photo of the rings of Uranus shows the complexity of the Uranian ring system, which has countless ringlets. *Voyager* took this picture when it was in the shadow of Uranus. This geometry of sun, rings, and camera makes very fine dust particles visible. (Jet Propulsion Laboratory, NASA.)

camera were almost perfectly aligned when this picture was taken. In this geometry, sunlight has to be deflected only a little bit in order to bounce from the sun, off the rings, to the waiting cameras of *Voyager*. You can see very tiny particles, only one micron (forty-millionths of an inch) across in a picture like this. Such particles are much harder to see in pictures taken face-on. Most of the main nine rings discovered from the ground can (just barely) be picked out in this picture, but an observant viewer can count many more than nine rings. The fuzzy ring at the top of the picture is made up entirely of tiny particles, a micron across or smaller, and it doesn't show up in the other ring pictures.

Jupiter's ring, and parts of Saturn's rings, is also made up of very small particles, about the size of the particles in smoke. Smoky rings like Jupiter's are quite different in origin from rings made of larger, snowball-sized particles. Small ring particles don't last very long in a ring system. They are small enough so that a photon of light hitting one of them can deflect its path significantly. The rings are close enough to the planet so that, in the case of Jupiter, Uranus, and the inner part of the ring of Saturn, the plasma environment of the planet itself will slow down the orbiting ring particles, causing them to spiral into the planet in a few thousand years. The ring material must be continuously replenished. One possible scheme to account for the continued existence of these rings is that the smoky ring material is chipped off larger, kilometer-sized particles. This hypothesis may explain the ring structure of Uranus as well.[10]

One ingenious way to detect structure in the rings of the outer planets is to play the stellar occultation game again. If the brightness of a star as *Voyager* whizzes by the rings is monitored, the starlight will appear and disappear as the motion of the probe takes it past successive ringlets. Some very small features indeed can be probed. The smallest feature resolved by this technique, at Uranus, is 40 meters (44 yards) across, about the size of an Olympic swimming pool. The planet Uranus is never less than 1.5 billion miles from earth, and scientists are performing a neat trick by resolving something so small when it's billions of miles away.

Why do the rings have the structure they have? Detailed examination of ring pictures (which at this writing has not really been possible for Uranus) shows several basic types of structure. Let's consider as a starting point a feature which is evident from the pictures of the Uranian rings shown here. The rings are very thin in radial extent, being only tens of kilometers at most from inner edge to outer edge. Jupiter's ring is also very thin, as are some of the ringlets of Saturn.

Astronomers Peter Goldreich and Scott Tremaine of Caltech predicted, in the late 1970s, that "shepherd satellites" on either side of a thin ring could keep the particles in line and could keep the ring thin. *Voyager* gave their theory a real boost by discovering just such shepherding satellites in the ring systems of Saturn and, most recently, near the outer ring of Uranus, as shown in the illustration. But *Voyager,*

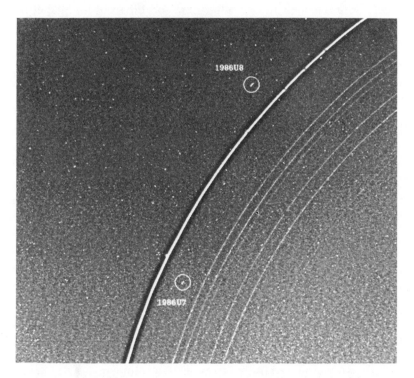

Two "shepherd" satellites, indicated in circles, are visible on the inner and outer edge of Uranus's bright "epsilon" ring in this image from *Voyager 2*. Image-processing techniques designed to make narrow features more apparent produce many of the blips as well as the dark halo surrounding the outer ring. The two satellites keep the ring material from spreading out. (Jet Propulsion Laboratory, NASA.)

searching for such satellites around some other thin rings, was unable to find them. Perhaps the shepherd satellites are too small and dark, or perhaps satellites can regulate planetary rings in another way. A different model (proposed by Stanley Dermott, Thomas Gold, and Carl Murray of Cornell) suggests that small satellites are hidden within each ring. These moonlets shape the ring both by their gravitational force on the ring material and by colliding with it.[11]

Where did the rings originally come from? This, the first natural question to ask about rings, turns out (as always) to be one of the hardest to answer. The classic answer—that a satellite came too close to a planet and was ripped apart by tidal forces—may be only partly correct. Some of the rings of Saturn are too far from the planet to be explained in such a simplistic way. Another possibility is that the rings condensed out of material that was forming the planet itself. Neither of these scenarios can account for the rings of small, dusty particles seen around Jupiter or Uranus; those rings must be replenished every thousand years.

## WEATHER ON THE GIANT PLANETS

As with the rings, telescopic observations of atmospheric features on these giant planets provided a mere glimpse of what *Voyager* subsequently discovered. For three centuries, astronomers have seen a red oval on Jupiter's surface. They dubbed it with a descriptive if unromantic name: the Giant Red Spot. Curiously, the region of red color spans a space twice the size of the earth and has persisted at least since the invention of the telescope. The color faded a bit in the nineteenth century but eventually came back.

Although the basic nature of the red spot could be deduced from ground-based observations, detailed measurements of its motion could be made from *Voyager*. A still picture, like the one here, shows much detail suggestive of turbulent eddies in the atmosphere. The smaller eddies circle around the red spot, last for a week or two, and then merge with it. The circulation direction shows that the red spot is a giant, semipermanent high-pressure area (or anticyclone) on Jupiter. Infrared measurements indicate directly that the atmospheric pressure

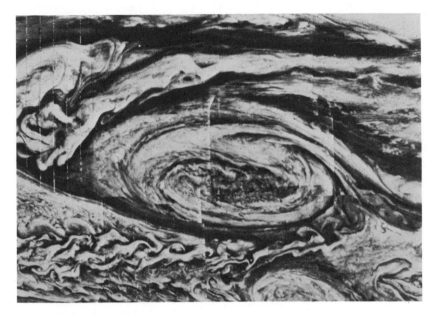

Close-up of Jupiter's Giant Red Spot shows a variety of complex circulation patterns. The spot itself is about twice the size of the earth; many smaller eddies are visible in the picture. (NASA.)

in the interior of the red spot is higher. Other persistent spots exist on Jupiter, such as a white oval known as BC, right next to the red spot, that appeared about forty years ago and has lasted ever since.[12] The circulation pattern shows that these, too, are high-pressure areas.[13] *Voyager* discovered a similar feature on Saturn, which was nicknamed Anne's Spot.[14]

The existence of the red spot and the related white ovals prompts two questions: Why is the red spot red? Why has it lasted for so long? Let's take the first question first. The chemical content of the Jovian atmosphere, combined with its strong surface gravity and its distance from the sun, produces an enormous variety of chemical compounds at levels of the atmosphere containing the visible clouds. Complex organic substances like ethane ($C_2H_6$), acetylene ($C_2H_2$), and phosphine ($PH_4$), have been detected in its atmosphere.[15] We don't know exactly what kind of exotic substance is responsible for the red color.

Some have proposed that the phosphine reacts in some way to form elemental phosphorous, which is red. Sulfur compounds like those found on the surface of Io are another possibility. There's no shortage of reasonable explanations; we just don't know which one is correct.

Understanding the longevity of the Giant Red Spot is more of a problem. Most weather systems on earth last for a week or so; there are a few high-pressure areas like the Bermuda High, a high-pressure area that sits in the Atlantic over Bermuda and gives the East Coast hot summer weather, which last for, at most, a few months. But the Giant Red Spot has been seen on Jupiter for three hundred years. Two possible explanations have been proposed. One has it that the weather on Jupiter and Saturn is really quite deep-seated. These planets have no solid surface like the earth, and it's possible that, despite the vaguely similar appearance of the red spot to a terrestrial weather system, its roots lie buried deep within the planet, where circulation currents change slowly. A second idea is that the tremendous shear flows on these planets—where their rapid rotation stretches out planetwide jet streams into uniform bands—stabilize the long-lived circulation patterns like the red spot.

The surfaces of Saturn and Uranus are quite bland when compared with that of Jupiter. In fact, Uranus looks like a featureless bluish-green disk when seen from the earth. It is only by pushing the image-processing capabilities of the *Voyager* system to extremes that we see any detail at all. The difference between colorful Jupiter and the others seems to be that the temperatures and pressures in the upper layers of Jupiter's atmosphere are those in which complex chemistry will occur, producing the colors. If these reactions occur at all on Saturn or Uranus, they occur deeper down, in layers of the atmosphere that we cannot see.

The deep-seated nature of the weather patterns on the giant planets may have another surprising manifestation: no seasons on Saturn or Uranus. Seasons occur on the earth because the earth's axis is not perpendicular to the plane of its orbit around the sun. We don't expect seasons on Jupiter anyway because its spin axis is perpendicular to the plane of its orbit. We do expect them on Saturn, where the axial tilt is rather like the earth's. Uranus should have the most extreme seasons of all because of the bizarre tilt of its spin axis, in the orbital plane, at right angles from where it should be. Each pole of Uranus has forty-two years of daylight and forty-two years of darkness. Scientists expected that tremendous winds would flow from the heated pole to the darker pole.

There are some very small differences between the summer and winter hemispheres of Saturn, but nothing terribly dramatic. As it turned out, one of the poles of Uranus was pointing almost straight at the sun when *Voyager* flew by, and so we were looking at the planet when one hemisphere was in midsummer and the other was in midwinter. *Voyager*'s infrared instrument measured precisely the same temperature in both hemispheres.[16] The winds did not blow from one pole to the other; *Voyager* followed a wispy white cloud for ten hours and saw it whizzing around the planet, following a circle of constant latitude. Why are there no seasons on these planets? The explanation may be quite simple, in that the atmosphere of Uranus is simply so thick and cold that it takes hundreds of years to cool off. Saturn and Uranus rotate very rapidly, spinning once every 10.7 and 17.3 hours, respectively, and their high rotational speed may stretch out any seasonal weather patterns into winds that just spin around the planet.

## COMPARATIVE PLANETARY MAGNETISM

The usual approach of comparative planetology involves comparing pictures of different planets and using geological techniques to determine the origin of their surface features and the different tectonic patterns on them. Recently, planetary scientists in the space plasma physics area have proposed a similar approach to the analysis of magnetospheres. A number of objects in the distant universe have magnetospheres, too, and scientists hope that this comparative approach can let us use our understanding of planetary magnetism to help interpret observations of pulsars and giant magnetized galaxies.

Scientists expected an unusual situation at Uranus because of this planet's strange axial tilt. In most planets, the magnetic field axis points in more-or-less the same direction as the rotation axis. This is the case on the earth, and as a result compasses work. A magnetic compass points toward the magnetic North Pole, which is located pretty close to the North Pole, in northern Canada. Jupiter and Saturn also have magnetic and rotation poles which are not too far apart, though in both cases the north magnetic pole is near the south geographic pole. If Uranus's magnetic axis was similarly aligned, it would point towards the sun, creating an unusual magnetosphere.

It turned out that the magnetic field of Uranus, like everything else about this planet, is sideways. The magnetic axis is tilted by about sixty degrees relative to the rotation axis. It would be like having compasses point to the Caribbean Sea. Furthermore, the magnetic axis passes not through the middle of the planet but one-third of a Uranian radius away from the center. The field is not at all like the field produced by a simple bar magnet, in contrast to the fields of the other planets.

Once again, why? Norman Ness and his colleagues wrote cautiously that "the possibility that we may be observing a polarity reversal, well known for the case of the terrestrial magnetic field or the more general case of a nonsteady dynamo, cannot be ignored."[17] The earth's magnetic field flips on occasion, reversing the earth's North and South Poles. The cause is not understood in full detail, but it is expected that the turbulent currents of conducting fluid in the earth's field switch on occasion from one flow pattern to another, causing the magnetic field that they generate to reverse direction. We may be seeing this happen at Uranus right now.

How can one use the comparative approach to understand magnetospheres and planetary magnetism? The first step is to construct a table something like Table 7, which lists a number of significant planetary properties.[18] Magnetospheres are produced by the interaction of the solar wind with the planet's atmosphere (if any), its intrinsic magnetic field (if any), and its satellites, which can be either sources or sinks of plasma (if any). The planets in our Solar System provide a wide variety of configurations, as shown in the last two columns of the table; virtually any combination of these ingredients is available.

Consideration of the origin of planetary magnetic fields and the information in this table can raise a number of interesting questions about planetary magnetism. One explanation for the origin of the mag-

Table 7.   Comparative Planetary Magnetism

| Mercury | Core? | Rapid Rotation? ($P < 1$ day) | Intrinsic Magnetic Field[a] (Gauss) | Plasma Sources |
|---|---|---|---|---|
| Moon | Small or none | No | 0.000 000 15 | Wind |
| Mercury | Must have | No | 0.003 50 | Wind |
| Venus | Yes | No | <0.000 3 | Atmospheres |
| Earth | Yes | Yes | 0.31 | Wind plus atmosphere |
| Mars | Yes | Yes | 0.000 65? | ? |
| Jupiter | Yes | Yes | 4.2 | Wind, atmosphere, satellites |
| Saturn | Yes | Yes | 0.2 | Wind, atmosphere, satellites |
| Uranus | Yes | Yes | 0.23 | Wind, atmosphere |
| Comets | No | No | None | Plasma atmosphere |

[a]At the planet's surface, at the equator.

netic field on most planets is that a dynamo exists in a fluid core, and that planetary rotation may be a significant force in driving this dynamo. The most puzzling phenomenon is the existence of a magnetic field on Mercury. Because Mercury is so small, thermal history calculations suggest that it does not have a substantial fluid core. Because it rotates so slowly, a substantial core, extending 70% of the way from the planet's center to its surface, is necessary to generate a field. Planetary magnetism is telling us that Mercury must have an active fluid core, despite its geologically inactive appearance and small size.

Additional questions are suggested by comparisons between worlds. Jupiter and Saturn, similar in size and rotation period, were expected to have similar magnetic fields. Why is Saturn's field so much weaker? Let's compare Earth and Mars next. The volcanism on Mars indicates that there was fluid material beneath the surface of Mars relatively recently. Mars and Earth, then, spin at the same rate and have fluid cores and should have similar magnetic fields. Why, then, is the field on Mars so weak, if it exists at all? Because magnetometers were not carried on any of the American spacecraft that orbited Mars, it is not even known whether the indications of a magnetic field obtained from other missions indicate a global Martian field. Other comparisons, and other questions, will no doubt arise in the future.

## THE FUTURE OF PLANETARY EXPLORATION

The *Voyager* mission to the outer Solar System may have been one of the most productive single planetary missions of all time. It has popped up in all three chapters dealing with planetary exploration, for it has turned the tiny satellites of the outer planets into worlds of their own, has shown us how weird yet potentially earthlike Titan is, and has produced images of the giant planets themselves as well as their rings. Many people, scientists as well as the general public, enjoy the fruits of *Voyager* in the 1980s and tend to forget that the efforts that went into *Voyager* date from a considerably earlier time. *Voyager* was developed during the 1970s and launched in 1977. What's next?

A perspective on the future of planetary exploration can be gained from a glance at Table 8, which is a list of all the planetary missions that have been undertaken. I include this table for a couple of reasons. First, many books, like this one, mention various missions out of order, which may confuse readers. Even space aficionados get to the point where they forget just what mission went where. Second, an examination of this table reveals some long-term trends for the planetary exploration program.

The long list of missions shows how active the planetary exploration program has been as well as indicates some of the changing trends over the years. The list for the 1960s illustrates an overt emphasis on the moon. Moreover, the milestones show clear competition between the two major superpowers' space race. The Soviet Union was ahead in the beginning with the 1959 *Luna* missions and was the first to soft-land on the moon in 1966, with the Americans following a few months later. The United States eventually caught up, and at some time in the very late 1960s, the Russians abandoned their attempt to send people to the moon, sending remotely operated landers instead.

In the race to the other planets, the United States started out ahead, but then American preeminence faded. Provided *Magellan* is launched on schedule, more than a decade will have passed with no American planetary mission being launched. NASA's Solar System Exploration Committee laid out an ambitious list of inexpensive planetary missions that, they hoped, would be launched before the year 2000. Only the first two were in place in 1986.

Table 8.  Spacecraft Missions to the Solar System[a]

| Year[b] | Destination | Name | Science Highlights |
|---|---|---|---|
| 1959 | Moon | *Luna 1–3* (USSR) | First lunar flyby (4 Jan.), impact (12 Sept.), and pictures of lunar farside |
| 1960 | Deep space | *Pioneer 5* (USA) | First mission beyond lunar orbit |
| 1962 | Venus | *Mariner 2* (USA) | Thick atmosphere, hot surface |
| 1964 | Mars | *Mariner 4* (USA) | Twenty-one pictures of cratered surface |
| 1964 | Moon | *Ranger 7* (USA) | Impacted on the moon, returned high-resolution pictures |
| 1965 | Moon | *Ranger 8, 9* (USA) | Lunar impact probe |
| 1965 | Moon | *Luna 5–8* (USSR) | Soft landing attempts; all crashed or missed the moon |
| 1966 | Moon | *Luna 9, 13* (USSR) | First successful soft landing (31 Jan.) |
| 1966 | Moon | *Luna 10–12* (USSR) | First successful lunar orbiters |
| 1966 | Moon | *Surveyor 1–2* (USA) | Soft landers; *Surveyor 1* successful (2 June); *Surveyor 2* crash-landed |
| 1966 | Moon | *Lunar Orbiters 1, 2* (USA) | Lunar photography of potential *Apollo* landing sites |
| 1967 | Moon | *Orbiters 3–5* (USA) | Follow up to *Lunar Orbiter 1, 2* |
| 1967 | Moon | *Surveyor 3, 5* (USA) | Soft landers |
| 1967 | Moon | *Surveyor 6* (USA) | Soft lander |
| 1967 | Venus | *Venera 4* (USSR) | First successful entry probe; failed 25 kilometers above surface |
| 1967 | Venus | *Mariner 5* (USA) | Flyby |
| 1967 | Moon | *Explorer 35* (USA) | Lunar magnetic field, plasma, radiation |
| 1968 | Moon | *Surveyor 7* (USA) | Emphasized science; determined composition of lunar surface |

[a]Data from NASA's Solar System Exploration Exploration Committee, *Planetary Exploration through the Year 2000* (Washington, DC: Government Printing Office, 1983); *Aeronautics and Space Report of the President: 1984 Activities* (Washington, DC: Government Printing Office, 1984); K. Garland *et al.*, *The Illustrated Encyclopedia of Space Technology* (New York: Harmony Books, 1981); J. Kelly Beatty, "The High Road to Halley," *Sky and Telescope* 71 (March 1986): 244–245.
[b]Year of launch.

(*continued*)

Table 8.    (*Continued*)

| Year[b] | Destination | Name | Science Highlights |
|---|---|---|---|
| 1968 | Moon | *Zond 4–6* (USSR) | Unmanned flights of Soyuz lunar cabin; probably precursors to manned missions |
| 1968 | Moon | *Apollo 8* (USA) | Manned flight around moon |
| 1969 | Venus | *Venera 5, 6* (USSR) | Atmospheric entry probes |
| 1969 | Moon | *Apollo 9, 10* (USA) | Rehearsals of lunar landing |
| 1969 | Moon | *Apollo 11* (USA) | First landing of humans on moon (20 July 1969) |
| 1969 | Moon | *Zond 7* (USSR) | Flight around moon |
| 1969 | Moon | *Apollo 12* (USA) | Landed near *Surveyor 3* craft |
| 1969 | Mars | *Mariner 6, 7* (USA) | Flyby, 201 pictures |
| 1970 | Venus | *Venera 7* (USSR) | First soft landing on Venus |
| 1970 | Moon | *Luna 16, 17* (USSR) | Unmanned soft landing; sample return to earth; automated lunar rover |
| 1971 | Moon | *Apollo 14, 15* (USA) | More varied rock samples; used mobile lunar rover |
| 1971 | Mars | *Mariner 9* (USA) | Orbiter; first global view of Mars |
| 1971 | Mars | *Mars 2, 3* (USSR) | Failed on landing |
| 1971 | Moon | *Luna 18, 19* (USSR) | Orbiter (*Luna 18*); soft lander (*Luna 19*) |
| 1972 | Moon | *Apollo 16, 17* (USA) | First scientist-astronaut on moon (Harrison Schmitt, *Apollo 17*) |
| 1972 | Moon | *Luna 20* (USSR) | Orbiter, soft lander; sample return |
| 1972 | Venus | *Venera 8* (USSR) | Soft lander; successful data return |
| 1972 | Jupiter | *Pioneer 10* (USA) | Flyby; plasma measurements in deep space |
| 1973 | Moon | *Luna 21* (USSR) | Lunokhod 2 roving laboratory |
| 1973 | Mars | *Mars 5, 6* (USSR) | Failed on landing |
| 1973 | Jupiter, Saturn | *Pioneer 11* (USA) | Flyby of Saturn |
| 1973 | Venus, Mercury | *Mariner 10* (USA) | First high-resolution photos of Mercury |
| 1974 | Moon | *Luna 22, 23* (USSR) | Orbiter; lander; partial failure |
| 1975 | Mars | *Viking 1, 2* (USA) | Soft lander; pictures; search for life |

Table 8.   (*Continued*)

| Year[b] | Destination | Name | Science Highlights |
|---|---|---|---|
| 1975 | Venus | *Venera 9, 10* (USSR) | Soft landers; returned pictures |
| 1977 | Four planets | *Voyager 1, 2* (USA) | Grand tour mission to Jupiter (1979); Saturn (1980–1981); Uranus (1986); Neptune (1989) |
| 1978 | Venus | *Venera 11, 12* (USSR) | Orbiter; soft lander; pictures |
| 1978 | Comet G/Z | *ICE* (USA) | Probe diverted to Comet Giacobini-Zinner |
| 1978 | Venus | *Pioneer 12, 13* (USA) | Orbiter; probe; radar map |
| 1981 | Venus | *Venera 13, 14* (USSR) | Lander |
| 1983 | Venus | *Venera 15, 16* (USSR) | Good radar map |
| 1984 | Venus and Halley | *Vega 1, 2* (USSR) | Balloons; Halley pictures; extensive foreign collaboration |
| 1985 | Comet Halley | *Giotto* (ESA) | European Space Agency; pictures |
| 1985 | Comet Halley | *Planet A* (Japan) | Two spacecraft (*Sakigake* and *Suisei*) |

Planned Missions[c]

| | | | |
|---|---|---|---|
| 1988 | Phobos (Martian satellite) | *Ph-D* (USSR) | Hovering over Phobos, chipping off surface, surface analysis and pictures |
| 1989 | Venus | *Magellan* (USA) | High-resolution radar map of surface |
| 1989 | Moon | *Orbiter* (USSR) | Polar orbiter; launch in 1989 or 1990 |
| 1990 | Jupiter | *Galileo* (USA) | Orbiter plus probe; originally scheduled for 1982 launch; possible 1989 launch |
| 1992 | Mars | *Mars Observer* (USA) | Surface and atmospheric composition; possible launch in 1990 |

[c]The distinction between ''planned'' and ''contemplated'' missions reflects the likelihood that they will happen. ''Planned'' missions are those with new-start status, where money has been appropriated to build hardware, in the case of American missions. The other missions listed as ''planned'' have been discussed in public meetings (see M. M. Waldrop, ''A Soviet Plan for Exploring the Planets,'' *Science* 228, 10 May 1985: 698, 703). I have selected several missions that I call ''contemplated'' from a much larger list of those under discussion.

(*continued*)

Table 8.    (*Continued*)

| Year[b] | Destination | Name | Science Highlights |
|---|---|---|---|
| 1995 | Mars, then asteroid belt | *Vesta* (USSR-France) | Mars lander; asteroid and comet encounters |

| | | Contemplated Missions | |
|---|---|---|---|
| | Comet and Asteroid Belt | CR/AF (USA) | Comet rendezvous and asteroid flyby; third in the SSEC list of moderate-cost planetary missions |
| | Titan | USA | Titan probe and radar mapper; fourth in the SSEC list |
| | Comet | USA or ESA | Comet or asteroid sample return |
| | Mars | USA | Mobile lander; sample return? |
| | Mars | *Kepler* (ESA); *Mars Aeronomy* (USA) | Upper atmosphere investigation |
| | Mars | USA, USSR | Manned landing (if it occurs, it will be in the twenty-first century) |

Morale in the American planetary science community is exceedingly low in the wake of *Challenger*. I hear words like *disarray* and *shambles* used to describe the situation. The planetary program, and particularly exploration of the outer solar system, will probably suffer more in the short run than any other part of NASA's space science and applications program. Sending vehicles into the outer Solar System challenges the capacity of launch vehicles in a way that few other space activities do. The constraints on our launch capability presented by *Challenger* and by NASA's past refusal to use expendable rockets make the short-term outlook bleak indeed.

In addition, the planetary exploration program is probably the only area of space science that will benefit little from the space station.[19] One of the station's greatest benefits for space science will be its ability to repair and refurbish instruments, and planetary probes are well out of reach. The only planetary mission that makes significant use of the space station's capabilities is the Mars sample-return mission.

Six missions are sufficiently far advanced so that I feel confident that they will happen before the end of this century. Three of them are American, but two of those have been delayed enough to discourage planetary scientists from hailing their likely completion. When the *Galileo* mission reaches Jupiter, its instruments will be based on twenty-year-old technology (if they are not rebuilt). The *Magellan* radar mapper will produce maps of roughly the same quality as the maps from the Soviet *Venera* mission, which flew in 1983. Only the *Mars Observer* mission is left as a representative of cutting-edge technology in the American planetary program. The other three missions all take the Soviet Union to Mars, indicating that the USSR has perhaps shifted its focus from Venus to Mars.

Looking further into the future is difficult, particularly for the American program. Many words have been spoken about the need to launch space science missions, and particularly planetary ones, on expendable rockets. This way of launching planetary probes was politically impossible in the early 1980s because NASA was committed to the shuttle as the exclusive space transportation system, the only way of reaching orbit. After *Challenger,* NASA officials admitted that there was a need for a balanced fleet, but it is not clear how NASA will make expendable rockets available for use on scientific missions.

The only missions to the outer Solar System that seem as if they will happen before the year 2000 are the encounter of *Voyager 2* with Neptune in 1989 and the *Galileo* mission. It is remotely possible that the Titan probe, the fourth on the SSEC list of ''moderate'' planetary missions, will also occur before the century's end; in some mission scenarios, this mission would include a Saturn orbiter. The SSEC plan included several other missions to the outer Solar System, such as flyby and entry probe missions to Saturn, Uranus, and Neptune, and an initial reconnaissance of Pluto. The Neptune and Pluto missions must swing by Jupiter in order to reach the outer Solar System in less than ten years. Because Jupiter takes twelve years to orbit the sun, the right alignment for a trip beyond Jupiter will occur only every twelve years. The next set of opportunities will be in the early 1990s and will clearly be missed. The next realistic chance to go to Neptune or Pluto will occur in the early part of the twenty-first century.

Other considerations in the planetary program will probably result in a greater emphasis on the inner Solar System, particularly the planet Mars. The report of the National Commission on Space extensively discussed the possible settlement of the inner Solar System, focusing attention on Mars at the expense of uninhabitable, distant places like Jupiter and Saturn. Although these giant outer planets are intriguing, a long time may pass between our first reconnaissance and further, in-depth explorations.

\*    \*    \*

The outer Solar System contains a number of very strange objects. Each of the giant planets is surrounded by a ring or by ringlike arcs. The planets themselves have deep, turbulent atmospheres and strong, occasionally strange magnetic fields. They are very different from the planet Earth.

In the early days of the space program, most of space science focused on the moon and the planets, the near universe. However, by the early 1970s, it became clear that space offered a unique vantage point for studying the distant universe as well. Astronomers who stay on the ground are limited to a narrow view of the universe because visible light is only a tiny piece of the broad reach of the electromagnetic spectrum. The view from space revealed black holes, planets in the process of forming, and other new objects in a remarkably varied and violent universe.

# CHAPTER 11

# Exploration of the Distant Universe

## PROLOGUE

The telescope looped along its path high above the Atlantic Ocean, circling the earth once every twenty-four hours. It did not follow the precise circle that the communications satellites did. The rocket that boosted it to orbit wasn't quite powerful enough to put it in the circular geostationary orbit. Other satellites in similar high orbits were mostly communications satellites, the highways that carried billions of dollars (in the form of electronic impulses) between the financial markets of New York, London, and Frankfurt. This telescope's mission, as well as its orbit, was different, lying far beyond the confines of tomorrow's stock prices of such giant enterprises as IBM, Comsat, Philips NV, Avions Marcel Dassault/Breguet Aviation, Aerospatiale, Messerschmitt-Boelkow-Blohm, and other high-tech companies around the world.

The telescope—the International Ultraviolet Explorer, or IUE for short—looks outward to the distant universe. Ultraviolet radiation such as that emitted by the hottest stars simply cannot penetrate the earth's atmosphere. Astronomers seeking to use this window on the universe must use telescopes orbiting in space, high above the atmospheric blanket. The IUE, a small, eighteen-inch telescope that worked with this high-energy radiation, collected the feeble trickle of ultraviolet photons, allowing scientists to fathom the deep mysteries of cosmic evolution.

On 26 February 1986, Ron Pitts, Richard Wasatonic, Nancy

Evans, and I used this telescope in orbit to obtain a spectrum of one of the hottest stars in the universe. We were not in orbit or anywhere near outer space. Rather, we were in a rather dingy federal office building at NASA's Goddard Space Flight Center in Greenbelt, Maryland. I could transport my eyes twenty thousand miles upward, above the earth's atmosphere. Wasatonic sat in front of a keyboard, issuing commands to the telescope, and moving it around the sky as though it were in a dome right next to us. At long distance, we did many of the things that we would do if we were to become scientists-astronauts and use telescopes in space. Of course, when we looked out the window, all we saw were magnificently ordinary parking lots and office buildings rather than a gleaming earth below. To be sure, I'd like to go into space and see the earth from above, for fun. However, I can do my work from the ground by operating these instruments remotely, and American taxpayers don't have to spend billions of dollars putting me and the hundreds of other astronomers who use this telescope into orbit.

## ASTRONOMY FROM SPACE

Even though the stars in the night sky look like identical dots of differing brightness, the astronomical universe contains a magnificent variety of fascinating objects. The smallest objects we can study directly are neutron stars and black holes, end points of stellar life cycles. A neutron star is a tiny ball, 20 kilometers (12 miles) in diameter, in which a billion tons of starstuff are mashed into a volume no bigger than a peanut. Black holes are even more extreme; their gravity is so strong that light cannot escape from them, and in the center of a black hole, an entire star is squashed to become smaller than a pinhead. Fuzzy pinwheels in the nighttime sky are galaxies, huge island universes of hundreds of billions of stars, ten million billion miles ($10^{16}$ miles) across.

Bigger yet are the huge superclusters of galaxies, wispy chains of thousands of galaxies extending $10^{24}$ meters from end to end. (I'm hard put to translate $10^{24}$ meters into human terms. I can tell you that $10^{24}$ meters is 1 million billion kilometers, or roughly 2 million billion miles.) We live at one end of one of these huge superclusters, and the light currently reaching us from the other end started on its journey about the time that ancestral organisms started crawling out of the primeval ocean onto the land, 300 million years ago. But if either of these statements gives you a mental picture of just how big a super-cluster is (apart from it's being bigger than you can possibly imagine), you're better than I. I have been doing calculations with just such astronomical numbers for over fifteen years, and I have no mental picture of how big a cluster of galaxies is.

Size is far from the only respect in which the astronomical universe presents us with some extremes. The temperature in the interior of some astronomical objects reaches the billion-degree mark. The quasars, galaxies that contain exploding nuclei, produce $10^{40}$ watts of power, 100 billion billion billion billion times brighter ($10^{38}$ times brighter) than the hundred-watt light bulb illuminating the page you're reading. The gas between the stars, in much of interstellar space, has a density of 0.01 atoms per cubic centimeter. The air you are breathing is a mere trillion billion ($10^{21}$) times as dense as this gas. At the other extreme are the white dwarf stars and the neutron stars. White dwarfs have a density of $10^7$ grams per cubic centimeter, ten million times as dense as ordinary matter, which has a density of 1 gram per cubic centimeter. Neutron stars have a density of $10^{15}$ grams per cubic centimeter, a million billion times the density of water.

Why is it that space offers us a uniquely valuable perspective on this very diverse universe? In comparison with all other scientists, we astronomers are handicapped, even the astronomers who study the near universe of planets rather than the distant universe of stars and galaxies. We can't touch anything we study, nor can we get even close to examine it. All we can do is measure the intensity, the time variability, the position in the sky, the wavelength, and the polarization state of the light and other forms of radiation as they reach us from celestial bodies. Before the space age, only visible light and radio waves could be used to probe the universe because only these forms of electromagnetic radiation could penetrate the earth's atmosphere.

Visible light is a tiny slice in the middle of the electromagnetic spectrum. Trying to understand the universe by using visible light alone is somewhat like watching a baseball game through a knothole in the outfield fence. There's much that we can understand, but even more evades our grasp. The true drama unfolds before us when we can get the entire picture. For the electromagnetic spectrum, this includes (in order of decreasing wavelength) radio waves, infrared radiation, visible light, ultraviolet photons (a photon is a particle that carries the energy in electromagnetic radiation), X-rays, and gamma rays.

A reason for probing the entire electromagnetic spectrum is that the wavelength of a photon, or its place in the spectrum, is related to its energy and to the temperature of the objects that emit it. Low-energy, long-wavelength photons are emitted in general by low-temperature objects. High-energy, short-wavelength photons are emitted by hot objects. Visible light tends to be emitted by objects that have temperatures similar to the temperature of the sun, and so a view of the universe in visible light is basically limited to a view consisting of stars. And even here, the limitations of visible light alone are quite apparent.

Take a cluster of galaxies as an example of the limitations of the view of the universe provided by visible light. Galaxies are huge collections of millions or billions of stars; we live in one such galaxy, the Milky Way, containing 100 billion stars much like the sun. The first illustration shows one such galaxy. If you look at a cluster of galaxies, such as the one in the second illustration, in visible light, you will see a number of galaxies, some spirals like our Milky Way, containing stars. Generations of astronomers thought that galaxies basically contained stars; in the twentieth century, we realized that there were some dust and gas between the stars.

In the years following World War II, radio astronomers opened another window on the universe that can be reached from the ground. They recorded radio waves from energetic electrons, which were the first indication that stars and gas were not all that was there. None of the other forms of electromagnetic radiation can penetrate the earth's atmosphere readily, and the remaining windows on the universe are best looked through by going into space. (Specific wavelengths of infrared radiation can penetrate fairly far down into the atmosphere, and high-flying aircraft or observatories on high mountains can make useful measurements. Even in this wavelength region, observations from space made a great deal of difference.)

Photograph of NGC 1300, a barred spiral galaxy. This galaxy is unlike our Milky Way in that it has a bar passing through the nucleus. (Mount Wilson and Las Campanas Observatories.)

A view of the universe through only one or two windows misses a number of important constituents of particular astronomical objects. Consider a cluster of galaxies as an example. The diagram on page 253 shows what we're missing if we look at the universe through only one or two windows. In the 1970s, the UHURU X-ray satellite discovered that a hot ionized gas or plasma permeated clusters of galaxies. (The names of astronomical space missions are listed in Table 9.) The IRAS infrared survey discovered that many galaxies emit almost all of their energy in the infrared part of the spectrum; an infrared view of this cluster would show some of the same galaxies, but some of the faintest in visible light would be the brightest in the infrared. We believe that this infrared radiation comes from heated dust, and the infrared galaxies are exceptionally dusty ones. Why are they dusty? We don't know. We don't yet know what the gamma-ray picture of this cluster of galaxies would look like. Gamma rays are the shortest wavelengths of electromagnetic radiation, and this window on the universe has been explored relatively little. Based on other gamma-ray work, we would

A cluster of galaxies showing a number of unusual types. The spiral galaxies in this picture are similar to our own galaxy, the Milky Way. If you were to place an ordinary pin on top of one of the galaxies in this picture, the head of that pin would cover as many stars as can be seen in the nighttime sky. (Palomar Observatories.)

expect that quasars, explosive energy sources in the nuclei of galaxies, would show up as powerful gamma-ray sources. Yet, there may be equally powerful, currently unknown sources of gamma-ray emission.

Excluding the study of our own sun, space astronomy is relatively young compared to other branches of space science like space plasma physics or planetary exploration. A reasonably complete, but probably not exhaustive, list of astronomical space missions, excluding the forty missions that could observe only the sun, is given in Table 9. It is only half the length of the similar list of planetary probes given in Table 8. A similar number of missions were launched to observe the sun, and well over a hundred missions have been devoted to space plasma physics, including studies of the magnetosphere. A further contrast between astronomy and planetary science is in the timing of missions, with few done before 1970, and in the countries participating, with a greater role for Europe and a comparatively smaller role for the Soviet Union.

Why so many telescopes? In any particular part of the electromagnetic spectrum, missions tend to follow a sequence of increasing complexity (and hence cost). The first missions are generally survey missions, designed to gain a view of the entire sky with relatively broad wavelength coverage. Such a mission can tell us whether there's anything exciting there. Surveys cannot, however, do a complete job. If the survey does discover new objects, a fairly common difficulty with a survey mission is that the location of sources of new forms of radiation is not known very precisely. Unless there is an unusual object near the new source, astronomers have to work hard to identify the source with an optical counterpart. Further, many surveys (including some of the early ones) could identify only a few of the brightest sources.

Follow-up work requires greater sensitivity. In addition, much of

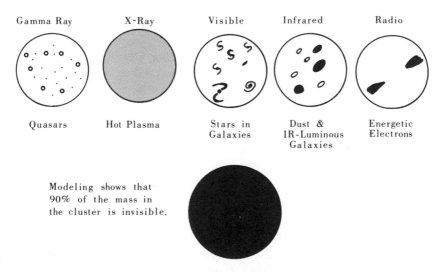

The circles show the types of objects we see if we look at a cluster of galaxies in different wavelengths of the electromagnetic spectrum. The motions of galaxies in most large clusters indicate, through modeling, that most of the matter in the clusters is invisible.

Table 9.    Space Astronomy Missions[a]

| Date | Name and Country | Remarks |
|------|------------------|---------|
| 1961 | *Explorer 11* (US) | Gamma-ray background |
| 1964 | *Cosmos 51* (USSR) | UV, gamma radiation; stellar background |
| 1968 | *Explorer 38* (US) | (RAE 1): Radio radiation |
| 1968 | *OAO-2* (US) | Ultraviolet survey and spectra |
| 1968 | *Cosmos 215* (USSR) | Visible, UV, and X-rays |
| 1970 | *Explorer 42* (US) | (SAS-1) = *Uhuru*: X-ray survey |
| 1972 | *Explorer 48* (US) | (SAS-2): Gamma-ray survey |
| 1972 | *Copernicus* (US) | OAO-3: Ultraviolet spectroscopy |
| 1972 | *TD 1A* (ESA) | All-sky ultraviolet survey |
| 1974 | *Ariel 5* (UK) | X-ray astronomy |
| 1975 | *Explorer 53* (US) | (SAS-3): X-rays |
| 1975 | *ANS* (Holland) | X-ray and ultraviolet survey |
| 1975 | *COS B* (ESA) | Gamma-ray survey |
| 1977 | *Prognoz 6* (USSR) | Ultraviolet, X, and gamma rays |
| 1977 | *HEAO-1* (US) | X-, gamma, and cosmic-ray all-sky survey |
| 1978 | *HEAO-2* (US) | (EINSTEIN) Deep X-ray images |
| 1978 | *Prognoz 7* (USSR) | UV, gamma rays |
| 1978 | *IUE* (US, UK, ESA) | Ultraviolet spectrometer |

[a]Acronyms: RAE = Radio Astronomy Explorer; OAO = Orbiting Astronomical Observatory; SAS = Small Astronomy Satellite; ANS = Astronomical Netherlands Satellite; COS = Celestial Observation Satellite; HEAO = High Energy Astronomical Observatory; IUE = International Ultraviolet Explorer; IRAS = Infrared Astronomy Satellite; GRO = Gamma Ray Observatory; ROSAT = Roentgensatellit; COBE = Cosmic Background Explorer; XTE = X-ray

the most valuable information about an astronomical source of radiation comes only when we split the radiation up into different wavelengths and obtain a spectrum of the source, or a graph of the intensity of radiation at different wavelengths.

Before *Challenger,* 1986 was to have been a banner year for space astronomy, as the comparatively long list of planned missions in Table 9 indicates. The Astro mission, a cluster of ultraviolet telescopes on the European-built spacelab pallet, was scheduled for launch in March and a second flight in November. The Hubble Space Telescope and the Ulysses mission to study the sun's polar regions were to go in May. These missions will eventually fly, as the hardware is completely built, but they will have to wait their turn on the shuttle manifest. One

Table 9. (*Continued*)

| Date | Name and Country | Remarks |
|---|---|---|
| 1979 | *HEAO-3* (US) | Cosmic ray survey, gamma-ray spectra |
| 1979 | *Hakucho* (Japan) | X-ray astronomy |
| 1983 | *IRAS* (US, UK, Neth.) | Infrared Sky Survey |
| 1983 | *EXOSAT* (ESA) | Pointed X-ray observations |
| 1983 | *TENMA* (Japan) | Medium-resolution X-ray spectroscopy |
| 1980s | *Spacelab* | Some X-ray, ultraviolet, and infrared astronomy on *Spacelab 1, 2* |

Planned missions

| | | |
|---|---|---|
| 1988 | *Hubble Space Telescope* (US, ESA) | Ninety-inch general-purpose telescope for ultraviolet, visible, infrared work |
| 1989 | *ASTRO* (US) | Ultraviolet spectroscopy on *Spacelab* |
| 1990 | *GRO* (US) | Sensitive gamma-ray survey/spectroscopy |
| 1990+ | *ROSAT* (Germany, US) | X-ray spectroscopy |
| 1990+ | *HIPPARCOS* (ESA) | Stellar positions and distances |
| 1990+ | *COBE* (US) | Background radiation from the Big Bang |
| 1992+ | *EUVE* (US) | Extreme ultraviolet survey/spectroscopy |
| 1990s | *XTE* (US) | X-ray variability |
| 1990s | *ISO* (ESA) | Deep infrared observations |

Timing Explorer; ISO = Infrared Scientific Observatory; EUVE = Extreme Ultraviolet Explorer. ESA = European Space Agency. Note: This list excludes missions on which astronomy was done as an adjunct to other activities, as well as instruments that exclusively measure radiation from the sun. As in Table 8, in Chapter 10, the planned missions are those that have been given a "new start."

of the missions, the Astro payload, was scheduled for three flights in two years and has now been cut back to one flight before 1994.

## STARDEATH

What have we learned from space astronomy, and what's in the immediate future? In most chapters, I have tried to take a reasonably balanced, comprehensive approach to these questions. But I'm not even going to try to write a one-chapter survey of space astronomy. We've flown only two dozen major space-astronomy missions, but the impact of space observations has been so pervasive that a summary of space astronomy would quickly turn into a summary of astronomy itself. Because this is my own field of work, I will take a rather personal approach to space astronomy and deal primarily with its impact on the late stages of stellar life cycles, fields of astronomy that I am quite close to. 1 will briefly mention some highlights of other astronomical fields to give you an idea of the impact of the view from space on the rest of astronomy.[1]

I often refer to my specialty in astronomy as *stellar geriatrics*. Through the usual combination of chance decisions that leads someone to a particular career niche, I have ended up studying the late stages of stellar evolution. Stars like the sun are giant balls of gas with nuclear furnaces inside. These nuclear furnaces consume fuel—the sun fuses 500 million tons of hydrogen each second in order to produce the energy we see as sunlight. Eventually, this fuel supply will run out, and the sun will pass through a later stage in its life cycle when it will be completely different from what it is now. First, it will turn to alternative nuclear fuels, swell up in size, and become a red giant star, boiling the earth's oceans away and swallowing the innermost planet Mercury, and possibly Venus. It will then shed its outer envelope and become a hot dying cinder, a white dwarf star.

White dwarf stars are weird. They are dying, glowing cinders, no larger than the earth; one-third of the mass of the star that they once were is packed into this tiny volume. A cupful of white dwarf stuff would outweigh two dozen elephants. If you were to stand on the surface of a white dwarf star, you would weigh about twenty tons. If you dropped a cupful of matter onto the surface of a white dwarf, its powerful gravity would accelerate it so that when it hit the surface of the star, it would release the energy of a twenty-kiloton nuclear bomb (the size of the one that flattened Hiroshima).

The hottest of these white dwarf stars are among the hottest stars in the universe. But before the space age, we could only guess at their temperatures. The first observations of hot white dwarf stars came from two satellites launched in the 1970s, the *ANS* (*A*stronomical *N*etherlands Satellite) and the *Apollo-Soyuz* Test Project. This latter project was basically intended to celebrate detente and to ensure that American astronauts would rescue Russians (and vice versa); the astronomical telescope was added at the last minute. In these cases, the telescopes were pointed at previously identified hot white dwarfs to see what kind of radiation they emitted.

The star I was looking at in my IUE observing run, mentioned at the start of this chapter, turned out to be one of the hottest stars known (not quite the hottest of all). Although this discovery is not as important as the discovery of the first real black hole (which we'll turn to soon), I know what happened in detail. Recounting our steps to understanding may help to explain how astronomy is practiced and why we astronomers seem to have an insatiable appetite for big telescopes.

John Nousek, a young astronomer at Penn State University, had identified my IUE target as the seventh brightest source of low-energy X-rays in the sky seen in a sky survey he was working on. It had a name: H1504+65.[2] He took an optical photograph and a rather poor spectrum of that part of the sky and surmised that the X-rays were coming from a faint blue star, tentatively identified as a particular type of star: a hydrogen-rich white dwarf star. John and I met at the summer 1984 meeting of the American Astronomical Society in Minneapolis, where he gave a short talk on his results. John suggested to me that we work together on the interpretation of this star. I have a number of computer programs that predict the type of radiation that a star of a given temperature and composition should emit, and the idea was to combine the predictions of my programs with John's observations in order to figure out just what this star was.

I tentatively interpreted the data I had on this star as indicating a moderately hot white dwarf star of dominantly hydrogen composition. However, my friends Jay Holberg and Jim Liebert at the University of Arizona, who knew of my work, then obtained much better spectra of this star, which showed something very strange: *no hydrogen,* at least in the visible part of the spectrum. Jay also obtained some spectra with the *Voyager* ultraviolet spectrograph, an instrument that was intended primarily for work on planetary atmospheres but that has done a good job on stars, too. Jay and Jim were convinced that the star had no hydrogen at all. Jay suggested to me that I get a small-aperture IUE spectrum in order to settle the issue of whether there was hydrogen or not; a small-aperture IUE spectrum would be much more sensitive to the presence of hydrogen than a ground-based spectrum would be. As for the analysis, it was back to the drawing board.

It was several months before we arrived at a satisfactory, if still tentative, solution to the mystery of H1504+65. At the same time that Jay and Jim were obtaining some of their data, I had my IUE observing run, in which I made a small-aperture observation of this freakish star. The result was the same: *no hydrogen*. During that spring, the observations by the European X-ray satellite, *EXOSAT* of this star became available to us, and Nick White and Paolo Giommi of the *EXOSAT* staff joined the collaboration. Four of the seven of us who were to be co-authors on our final paper met for lunch at the American Astronomical Society's meeting in Charlottesville, Virginia, in June 1985. Over sandwiches and beer, we pooled our thoughts and came to the tentative conclusion that this was an ultrahot white dwarf star that contained neither hydrogen nor helium. I spent the next month or so of the summer working more-or-less full time with the results of my computer programs, and I confirmed that the star was very hot (we estimate 160,000°K, thirty times the absolute temperature of the sun) and hydrogen-poor. I'm reasonably certain, but not positive, that it doesn't have any helium either.

This longish story illustrates some important points about how space astronomy is practiced. Once again, the use of "unmanned" satellites involves a great deal of human activity. Although satellite operators are not up in space, we scientists and satellite controllers were able to use the IUE satellite in a hands-on mode, making it seem as though we were there to touch it. Observations at many different wavelengths were needed to solve the puzzle. One telescope or even two could cover only a narrow wavelength range, and successful interpretation of this star required observations at many widely separated wavelengths. (In the end, we used four telescopes in space and three on the ground to do the job.) Moreover, "instant science," where a scientist is asked to interpret some data in a few days, does not provide all the answers. Weeks and months of analysis, discussion, and argument go by before the final solution is reached.

White dwarf stars are not the strangest things that a star can turn into. A star somewhat larger than the sun will end its life with a mass greater than 1.4 solar masses, the maximum possible mass for a white dwarf, and will form an object called a *neutron star,* with densities $10^9$ times the density of white dwarf matter. A cupful of neutron star stuff would outweigh ten billion elephants, not just twenty-four. But neutron stars, too, have a limit. We don't know exactly what it is, but it's somewhere between three and seven solar masses. Try to make a neutron star bigger than this, and it, too, will collapse under its own weight, becoming the strangest final state of stellar evolution: the black hole. One of the greatest triumphs of space astronomy in its early decades has been the discovery of black holes in space.

## BLACK HOLES

A black hole forms when an object becomes so compressed that its gravity will not let anything escape from its grasp, not even light. If you were to go too close to a black hole, inside the "event horizon" or point of no return, you could never come back out again. Scientists and science fiction writers alike have been fascinated with these one-way tickets to nowhere. Space astronomy took black holes from the fantasy world to the real world.

The hypothetical existence of these objects dates back to the eighteenth century, well before we really understood what light was. The mathematically correct description of real black holes was developed almost immediately after Albert Einstein developed his general theory of relativity in 1916. General relativity theory has withstood seventy years of scrutiny and is well established as the correct mathematical description of what makes things fall. In most ordinary situations, it is for all practical purposes identical to the laws of motion proposed by Isaac Newton nearly three hundred years ago. The major difference between Einstein's and Newton's descriptions of gravity arises when something is compressed to near black-hole dimensions.

For fifty years after the publication of the general relativity theory, black holes were to remain mathematical curiosities. Scientists could hypothesize about the curious way in which clocks would behave when they fell into black holes. More speculative papers described how black holes might be the gateway to another universe. (It turns out that these speculations are incorrect.) The sole connection made between black holes and reality were a few papers written in the 1930s that demonstrated that black holes could form when very massive stars die, leaving very massive cores. But saying black holes *could* exist is a very different thing from saying that they *do* exist.

In 1964, X-ray astronomy began with a few rocket flights. A few gutsy pioneers built some inexpensive telescopes, persuaded NASA to let them use some surplus rockets to launch them with, and hurled them above the atmosphere. On a rocket flight, there are only five minutes in which data can be gathered. A few places in the universe emitting X-rays, called *X-ray sources,* were found. They were named for constellations: Cygnus X-1, X-2, and X-3; Taurus X-1; Sagittarius X-1; and so on.

Astronomers rounded up the usual suspects and compared a list of important objects and astronomical freaks with the locations of celestial X-ray sources. Some matches were made, and X-radiation could help astronomers understand them. For example, Sagittarius X-1 was identified with the center of the Milky Way Galaxy, our local swarm of 100 billion stars. But the others remained mysterious because even their location in the sky was not known very well. One measurement with one telescope does not allow us to analyze the true nature of an object. Just as it took seven telescopes to unravel the true story of the ultrahot white dwarf star my colleagues and I found, it took many telescopes to solve the riddle of most of these X-ray sources, in particular Cygnus X-1, the black hole.

In the late 1960s, a group at MIT flew a telescope on a rocket to pinpoint the location of Cygnus X-1. Around the same time, radio astronomers detected a new radio source in the constellation of Cygnus, and the *Uhuru* X-ray observatory detected a marked change in the nature of the X-rays that came from this bright X-ray source. The name of this mission, more formally designated *Explorer 42,* or *SAS* (Small Astronomy Satellite) *1,* is the Swahili word for freedom. *Uhuru* was launched from a platform off the coast of Kenya on the fifth anniversary of Kenyan independence. These independent measurements identified this bright X-ray source with a rather ordinary blue star in the constellation of Cygnus. Annie Jump Cannon had classified this star at the turn of the century. Otherwise there was nothing to distinguish it from hundreds of thousands of other relatively dim stars.

Visible light still plays a large role in astronomy, and astronomers using ground-based telescopes were a key part of the story of Cygnus

X-1. Careful measurements of the spectrum of this moderately faint blue star proved that the star was moving back and forth along the line of sight, going through one orbital cycle every 5 ½ days. The visible star was orbiting around an invisible companion, and something in the system was emitting X-rays. The visible blue star seemed no different from other large blue stars in the galaxy, and these other blue stars are very massive ones. The simplest explanation of these disparate bits of information was that the blue star was circling around an invisible companion with a mass of at least eight solar masses. Gas flowing from it onto the invisible companion swirls around the companion, forming a hot disk of gas. This hot, turbulent disk of gas produces the X-rays that we see from the system.

This star, and the interpretation of the X-rays, would not have attracted a great deal of attention were it not for the large mass of the companion star. Many similar systems exist where matter swirls onto a neutron star or white dwarf and emits X-rays. The invisible companion in Cygnus X-1 is far too massive to be a white dwarf or neutron star. Its mass indicates that it must be a black hole.

Scientists must be the most skeptical people in the world. When this interpretation of Cygnus X-1 was first advanced in the early 1970s, other scientists did not react by saying (or feeling), "Wow! We've found a black hole! Give him a Nobel Prize!" but, rather, "Wow! They think they have found a black hole. I wonder what else it could be." (MIT astronomer Jim Elliot had the same reaction when his team unexpectedly found the rings of Uranus—he wanted to be sure that he wasn't overinterpreting instrument failure or clouds in the earth's atmosphere.) In the next several years, many ingenious devil's advocates proposed complex models for the Cygnus X-1 system that avoided the need for black holes. Other scientists shot down these models (or tried to). The problem was quite simple. Because there was only one solid black hole candidate around, we couldn't be sure that some complicated Rube Goldberg scheme explaining what we know about this system wasn't right.[3]

The only way we could eliminate complicated explanations of Cygnus X-1 was to somehow find a way of appealing to Occam's razor. In the fourteenth century, William of Occam (or Ockham) enunciated a principle that has been one of the guiding lights of science for a long time. I've seen it stated in many different ways, which range from highfalutin' obfuscatory pedagogy to elegant simplicity: "The simpler the explanation of a natural phenomenon is, the more likely is it to be correct''; ''Nature may be subtle, but she is not malicious''; or ''Keep it simple, stupid.'' The last phrase is not strictly a statement of Occam's principle, but it is a maxim that engineers follow in designing things, and it is known by the lovely acronym of *KISS*. This principle is sometimes called Occam's razor because it slashes away complications.

What does Occam's razor have to do with Cygnus X-1? The most popular way of avoiding a black hole in this system suggested that it was a triple star system, where a low-mass neutron star would be at the center of the accretion disk and a high-mass star would be the object that the blue star orbited around. Scientists had to invoke some very special conditions in order to avoid detecting the light from this massive star and to figure out a way of dumping enough mass onto the neutron star. However, interpreters could never absolutely rule out the possibility that these special conditions applied to Cygnus X-1 as an isolated object. It would be very unlikely that these special conditions would be found operating in the same way in two double stars, or three. Thus, what would clinch the black hole explanation of Cygnus X-1 would be the discovery of another double star system like it.

An X-ray source called LMC X-3 had been discovered quite early in the days of X-ray astronomy. Astronomers could not investigate it very easily because of its location in the sky and its faintness. The letters *LMC* stand for "Large Magellanic Cloud," a blotch of light visible from the Southern Hemisphere of the earth and first brought to the attention of the Western world by the Magellan expedition. This galaxy, and stars like LMC X-3 in it, never rise above the horizon from the perspective of someone living in the United States, even in Hawaii.

It was some time before astronomers followed up on LMC X-3. Until recently, there have been few large telescopes in the Southern Hemisphere, and so it was not easy to obtain the measurements of LMC X-3. In addition, the position of LMC X-3 was somewhat uncertain, and it was not clear just which star in the sky was the counterpart of the X-ray source. Yet another problem is that LMC X-3 is much further away than Cygnus X-1. The Large Magellanic Cloud is a small irregular galaxy, physically distinct from the Milky Way Galaxy, which contains both the solar system and Cygnus X-1. Because LMC X-3 is far away, it is comparatively faint, and lots of precious telescope time would have to be used to investigate it thoroughly. There seemed to be no reason to look at this particular X-ray source rather than some other one.

Nevertheless, Anne Cowley of Arizona State University and her colleagues David Crampton and John Hutchings of the Dominion Astrophysical Observatory in Victoria, British Columbia, decided that, despite the inconveniences, it would be worth investigating the X-ray sources in the Large Magellanic Cloud. The *EINSTEIN* X-ray Observatory, launched eight years after *Uhuru*, provided a good, precise position, identifying the X-ray source with a faint blue star, hundreds of times fainter than Cygnus X-1. They obtained spectra of this star in the early 1980s. Analyses published in 1984 showed that the faint star was orbiting something like every 1.7 days. The mass of the companion is at least seven, and most likely ten, solar masses.[4] A significant distinction between Cygnus X-1 and LMC X-3 is that the visible star in LMC X-3 has a mass that is less than that of the possible black hole. Consequently, it is much more difficult to appeal to a triple star solution for the system. Another distinction is that we know the distance to this system quite accurately and can therefore rule out other complicated models.

But to me, the real importance of LMC X-3 is that there were now two black-hole candidates rather than just one. With two systems, you can rule out *all* models for these types of massive X-ray binaries that require special circumstances in order to avoid having a black hole. A unique combination of circumstances is unlikely to occur twice, according to Occam's razor. Discovery of the second example of a particular type of object may not make newspaper headlines, but it's important for scientists.

If two black holes weren't enough for the most extreme skeptics, a third black hole was to emerge shortly. Our ability to appeal to Occam's razor advanced one more step in 1986 when Jeffrey McClintock of the Harvard-Smithsonian Center for Astrophysics and Ronald Remillard of MIT identified a third black hole. In 1975, the British *Ariel 5* X-ray satellite discovered an "X-ray nova," an X-ray source that flared up and became for a while the brightest X-ray source in the sky. Its nature remained obscure until it, too, was identified with a star that is quite faint in visible light. Detailed analysis of the faint star showed that it, too, was like Cygnus X-1 and LMC X-3 in that it had a massive companion, too massive to be a neutron star. This object, designated by the telephone-number nomenclature of A0600-00 (where *A* means *Ariel* and 0600-00 refers to position in the sky), joined the small group of real black-hole candidates, not just black-hole suspects.

## A NEW LOOK FOR OUR SUN

Black holes are perhaps the strangest, most exotic of the new objects to be discovered by space astronomy. But even as familiar and nearby a star as our own sun had some surprises in store for us. When you look at it through the narrow optical window on the universe, you see a placid, glowing ball. The only signs of activity are some funny black spots on the surface. These sunspots, first discovered by Galileo, come and go in an eleven-year cycle. Occasionally, the solar surface near a sunspot brightens abruptly in what is called a *solar flare*. There is no way to tell from visible light alone that sunspots and flares are anything more than minor pimples on the solar surface, unimportant disturbances in the smooth flow of light from our own star.

However, it turns out that the sunspots are one manifestation of solar activity, a very wide range of complicated phenomena making the outer layers of the sun's atmosphere a very complex and dynamic place. Thus, the outer part of the sun is far different from the quiescent, almost unchanging surface we see. The results of solar activity have some very practical effects on humans in space and on the earth itself.

NASA's long interest in the sun is shown by the dozens of missions that have been flown since the start of the space program. Eight OSOs (Orbiting Solar Observatories), ten IMPs (Interplanetary Monitoring Platforms), nine Pioneers, six OGOs (Orbiting Geophysical Observatories), three Solrad explorers (to measure solar radiation), and numerous particle detectors on board other missions have provided a great many data on the sun. Astronauts conducted major investigations of solar physics as part of the Skylab space-station program. The *Solar Maximum Mission,* launched in 1980, became the first satellite to be repaired in orbit in 1984. (The large number of these missions prohibits their listing in the table of missions in this chapter.)

A series of pictures of the sun taken by the Skylab astronauts in the early 1970s demonstrated how complex the sun's outer atmosphere is. The picture shown here, taken with an X-ray camera, shows how the sun would appear if you looked at it with X-ray vision. X-rays come from hot, million-degree gas in the outer solar atmosphere. This outer atmosphere is far from the smooth corona scientists expected on the basis of observations made during solar eclipses. Almost all of the coronal material is concentrated in giant loops of hot gas that channel themselves along magnetic fields. Parts of the outer solar atmosphere

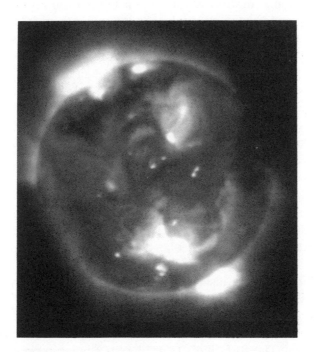

Soft X-ray images of the sun obtained from the Skylab mission. The looped structures and the bright points of X-ray emission are both evidence of complex gas and magnetic field structures in the corona. (NASA.)

contain very little gas at all; the solar wind flows from these coronal holes. Occasionally, giant bubbles of coronal gas erupt from the sun and race away from it. All of these coronal disturbances are associated with changes in the solar magnetic field.

Solar flares are another unpredictable manifestation of high-energy solar activity. The visible manifestation of a solar flare is a sudden increase in the brightness of a small area on the solar surface. This is, however, a relatively minor sideshow of the whole event. A flare occurs because the magnetic fields and ionized gases in the solar atmosphere coexist uneasily and unstably. Any microscopic disturbance of the precarious balance can be suddenly amplified, accelerating the electrons in the gas to very high speeds. The sudden increase in surface brightness, abrupt radio emissions, and other flare phenomena are all produced by this burst of high-speed particles. The flare can accelerate atomic nuclei and other solar particles to high speeds, and these particles can harm human beings if they hit them. These particles travel toward the earth as solar cosmic rays, and if they strike astronauts, they can be quite harmful. The earth's magnetosphere protects astronauts in low orbit from some of the danger. Because "manned" missions beyond low orbit, such as a mission to Mars, are under active consideration, solar flares are a concern to the "manned" space program.

The kinds of instabilities that produce solar flares are observed in laboratories on the earth, but that does not mean we can forecast the future development of the sun's "weather." There are difficulties in trying to understand plasma structures that are millions of miles long by experimenting with laboratory plasmas with dimensions of meters or less. Despite decades of observations, we are a long way from understanding the causes of solar activity, much less being able to predict it. The major problem with the observations is that our pictures of the solar surface are insufficiently sharp; what happens when a flare goes off is seen as a blur of activity.

Space astronomy has added another dimension to understanding solar activity because it has allowed us to detect the outer atmospheres of other stars. The surface of a star like the sun emits an insignificant quantity of high-energy ultraviolet and X-ray photons. However, the outer solar atmosphere (the so-called chromosphere and corona) is considerably hotter and does emit photons of higher energies, which have been detected from the coronas of a wide variety of stellar types.

The disturbances associated with solar activity can affect the earth's environment because the particle flows near the sun propagate outward into the inner Solar System. As a result, the study of solar activity has some practical importance as well as being an extremely difficult intellectual challenge. The dangers of solar activity to unprotected humans working in space have been recognized for some time. More recently, the speculation that solar activity affects terrestrial weather patterns has become scientifically respectable with the accumulation of supporting data.

Solar activity can be a danger to "manned" spaceflight. A solar flare is often associated with a great increase in the flow of high-energy charged particles from the surface of the sun. The most dramatic such increase in recent times occurred in August 1972, when the intensity of high-energy charged particles increased ten thousand times and lasted for about an hour.[5] Neither American nor Russian astronauts were in space at that time. So far, the radiation dose to astronauts has been relatively minimal because the astronauts have spent almost all of their time inside space vehicles, which provide a significant amount of shielding. The maximum amount of measured radiation exposure to American astronauts was 7.8 rads on the eighty-four-day *Skylab 4* mission. This figure is considerably greater than the dose of 0.2 rads per year that the average ground-living American is exposed to, but below the 25–50 rad dose that is needed for humans to show even limited immediate symptoms. One needs to get up to a dose of 200 rads before starting to worry about radiation killing somebody right away, though extra doses of radiation are associated at some level of risk with an increased long-term risk of cancer.[6]

Despite NASA's relatively happy experiences so far, solar flares and the radiation associated with them are a concern for the future. If an astronaut had been out in space, protected only by a spacesuit, when the big flare of August 1972 arrived, she or he could have been killed. In the future, astronauts may spend more extended periods of time out in spacesuits, fetching satellites that need fixing or constructing a space station. There is therefore some practical value as well as intellectual merit in the continued study of solar activity and its causes.

The practical dimensions of solar physics extend downward from earth orbit to include human activities. Solar activity and solar flares are associated with auroras, with disturbances in the earth's upper atmosphere, and possibly with weather in the lower atmosphere. Low-frequency radio waves bounce off the ionosphere. Because solar flares change the properties of the ionosphere drastically, they can produce unpredictable changes in communication. For example, there have been times when I could not tune in to a local AM station on my car radio because some station a thousand miles away was drowning it out, thanks to the ionosphere.

Solar flares have even been cited as the cause of electrical power failures. A geomagnetic storm on 24 March 1940 knocked out power in parts of New England, New York, Pennsylvania, Minnesota, Quebec, and Ontario. The August 1972 solar flare mentioned earlier caused a power transformer in British Columbia to fail. Solar flares have even produced corrosion in long oil pipelines (the long pipeline is a long electrical conductor), though this seems to be merely a nuisance rather than a serious problem.[7]

The most tentative aspect of a solar-terrestrial connection involves the earth's weather. For years, scientists have speculated about a relation between the eleven-year sunspot cycles and such meteorological phenomena as droughts. The recent discovery that the sunspots disappeared altogether for seventy years in the latter part of the seventeenth century, a time that coincided with unusually low temperatures in Europe, has added fuel to this fire. There has been a recent claim that droughts in the American West occur in phase with the steep decline in the number of sunspots in alternate sunspot cycles.[8] A stronger correlation appears to exist between the advance and retreat of alpine glaciers in northern Europe and the long-term existence or absence of solar activity. But how solar activity influences the weather is unknown, and some scientists even question whether this correlation has been established beyond a reasonable doubt.[9]

But leaving the possible sunspot–weather connection aside, it is clear that there are a number of other areas in which solar activity does have its effects on human affairs. Thirty years of studying the sun from space have demonstrated that our star, placid though it seems to the eye, is the center of a huge, turbulent plasma, containing hot gas and magnetic fields that act and react on each other, occasionally producing streams of high-speed particles. The very existence of these energetic interactions was unsuspected before the space age.

## PROTOPLANETARY CLOUDS, SUPERBUBBLES, INFRARED GALAXIES, AND MORE

Black holes and energetic solar activity are two examples of astronomical phenomena that were purely hypothetical—if they were suspected at all—before we had the ability to hurl telescopes above the atmosphere and use the entire sweep of the electromagnetic spectrum rather than the tiny sliver of an optical window to examine the entire universe. In the twenty years since we began to use an orbital vantage point to look beyond the Solar System, we have discovered a number of other new objects, ranging from relatively small protoplanetary clouds surrounding a few nearby stars to the huge clouds of gas in galaxy clusters, millions of light-years across, mentioned earlier. Let me give you some examples of how space astronomy has changed our view of the distant universe.

### Protoplanetary Dust Clouds

Astronomers have long suspected that planets outside our Solar System existed, but until the mid-1980s, there was no direct evidence to support this belief. In the current picture of the origin of the Solar System, the sun was surrounded by a rotating cloud of dust and gas as it was in the process of forming. Such clouds of dust and gas have now been found around other nearby stars, suggesting (although not proving definitively) that the formation of planets is a relatively common event. The Infrared Astronomy Satellite (IRAS), a joint project of the space agencies of the Netherlands and Britain, working with NASA, uncovered a number of new phenomena, but none of such widespread interest as the indication that some other stars were surrounded by solid material. IRAS observations of the bright star Vega were one of the first sets of observations analyzed because this bright star is often used as a calibration standard. Vega turned out to be much brighter in the infrared part of the spectrum than had been anticipated. The only reasonable explanation for the excess infrared brightness is that Vega is surrounded by a cloud of solid material that absorbs radiation from Vega, heats up, and reradiates it as infrared radiation. We have no real idea of what this material is—it could be dust, baseball-sized particles, or cometary nuclei. IRAS observations of other nearby stars have shown that nearly two dozen other stars also show an infrared excess. Once these stars were recognized as being unusual, follow-up observations of several of them have confirmed that they are surrounded by dusty or cometary disks.

## Ultrahot Interstellar Gas

The stars are far, far apart in interstellar space. What fills the space between them? For years, our answer to this question was prejudiced by pictures of interstellar gas surrounding hot stars, as shown in the illustration, where the radiation from the star itself illuminated the relatively dense gas. We thought that the space between the stars would be filled with relatively cool gas like the material shown in the picture, which has a temperature of about 10,000°K (18,000°F). However, observations by the *Copernicus* satellite in the early 1970s showed that most of space is filled with much hotter gas. A high-resolution spectrograph on this telescope split the radiation from distant stars into various wavelengths, and it turned out that the interstellar gas absorbed some radiation. No real surprise at that, but what was unexpected was that some radiation was absorbed at a wavelength that can be absorbed only by oxygen atoms with five electrons missing (spectroscopists call this O VI). We find oxygen in such a stripped-down state only if it is in a very high-temperature gas, where collisions between the oxygen atoms and fast-moving electrons or high-energy photons knock most of the electrons away from the oxygen atom. Later, telescopes sensitive to extreme ultraviolet radiation on the *Apollo-Soyuz* test project, launched for other purposes in 1976, and on *Voyager* showed that the light from distant hot white dwarfs was not absorbed by neutral hydrogen atoms, a finding indicating that the gas between us and these stars is very hot. At least half the space between the stars is filled with this million-degree gas.[10]

The Orion nebula, a cloud of interstellar gas. A very hot form of this gas fills most of the space between the stars. (Lick Observatory photograph.)

## Superbubbles

What heats the interstellar gas to a temperature of a million degrees? Another discovery of a remarkable object from space astronomy may provide part of the answer. Webster Cash, Philip Charles, and Stuart Bowyer were analyzing the data from the first of the *High Energy Astronomical Observatories* (*HEAO*s) when they came upon a new X-ray source. When they looked at adjacent parts of the sky, they found out that this structure stretched across most of the constellation of Cygnus and was far larger than any such structure seen before. If you could see this structure with your eyes, it would be ten times the size of the full moon. This object was christened a *superbubble*.

The most reasonable explanation of this superbubble is that it is the result of a chain reaction of stellar explosions. When massive stars run out of nuclear fuel, they explode as supernovae. When a star goes supernova, it increases its brightness a million times and sends its outer layers blasting off into space at a speed of roughly ten thousand kilometers per second. We have not seen a bright supernova in our own galaxy since 1604, six years before the telescope was invented. A nearby supernova would be bright enough to see in the daytime. The shock wave sent through space by a supernova explosion heats the interstellar gas, creating a hot, expanding bubble. It also triggers star formation, and some of these new stars also go supernova, heating and expanding the bubble still further. This chain reaction creates the superbubble. This scenario is also supported by some other evidence, including some direct evidence from meteorites that a supernova explosion triggered the formation of the Solar System.[11]

## Infrared Galaxies

The IRAS infrared satellite surveyed the entire sky in the infrared part of the spectrum. Some relatively bright IRAS sources turned out to be identified with galaxies, huge collections of hundreds of billions of stars. These galaxies were very feeble emitters of light but were emitting virtually all of their energy in the infrared part of the spectrum, unlike all other previously known galaxies. The news media dubbed them giant *furnaces* in the sky. Why a galaxy should be so peculiar is unknown. The most reasonable explanation is that these galaxies are surrounded by a huge cloud of dust that absorbs the starlight and reradiates it as heat. But what made the dust? Why do only some galaxies show this "furnace" phenomenon? These questions remain unanswered.

## Gamma-Ray Bursts

One unusual discovery from space astronomy came not from astronomical satellites but from spy satellites. In the late 1960s, a number of satellites called *Velas* were launched in order to detect clandestine nuclear explosions. These satellites could pinpoint sudden bursts of gamma rays. They may have discovered a nuclear explosion off the coast of South Africa; the nature of this particular event remains controversial. What is not controversial is that they accidentally discovered nuclear explosions on distant stars, visible as sudden blasts of gamma rays that came from far off in space.

The cause of these gamma-ray bursts remained mysterious for about ten years. I've said before that we often need more than one telescope to do astronomy, and it was true here, too. In this case, it was the Harvard Observatory's archives that played an important supporting role. Harvard had been running a sky patrol program for nearly a century. If you want to find out whether there was an unusually variable star in some particular location in the sky, the archives are the place to look. Bradley Schaefer, a graduate student at MIT, patiently searched through the Harvard archives, looking for something unusual at the site of several gamma-ray bursts. He discovered that, on several occasions, these gamma-ray bursters also emitted a sudden burst of light. Someone looking in the right place at the right time would have seen a star suddenly flare up to naked-eye visibility for about thirty seconds or so and then fade. (Just such a flareup was spotted recently by a network of Canadian meteor observers.)[12]

Schaefer was able to show that the gamma-ray burst sources repeatedly flared up, every thirty years or so, and also produced optical emission. His finding supports the idea that the gamma-ray bursts are produced by a rapid dumping of material onto the surface of a neutron star by a binary companion. Because it is superdense, a neutron star exerts a strong gravitational pull. As mentioned earlier, if a pin were dropped onto a neutron star, it would release five Hiroshima bombs' worth of energy when it hit the surface and stopped. The powerful gravity of neutron stars can easily produce all of the energy seen in gamma-ray bursts.

These new objects, and others noted earlier like black holes, high-energy solar activity, and the hot gas clouds in clusters of galaxies, cover almost all fields of astronomy. Interest in them goes beyond the astronomical community; at least three of the objects listed here (superbubbles, protoplanetary dust clouds, and furnace galaxies) made news in national TV broadcasts or news weeklies when their discovery was announced. There are many more discoveries that I haven't mentioned. Many involve not new objects or phenomena, but new ways of looking at previously known objects. All this has happened as a result of relatively few missions in the past, as space astronomy got started well after some of the other space science disciplines. What's up next?

## GREAT OBSERVATORIES OF THE FUTURE: SPACE TELESCOPE

Before *Challenger* blew up, the first of a series of great observatories in space was to have been launched in the fall of 1986. Because many investigations require measurements at many wavelengths, NASA has designated a number of spacecraft as "great observatories" which, when combined, will provide a comprehensive look at the distant universe and its fascinatingly diverse contents. The space telescope is a ninety-inch telescope that will be launched into earth orbit. It is called an *observatory* because we anticipate that it will last for half a century at least, with the prospects for in-orbit refurbishing and upgrading. This general-purpose instrument will play a major role in almost all fields of astronomy. The Hubble Space Telescope (HST—of course, there has to be an acronym) is one of a very small number of projects in all of the space program that will be set back only two or three years by the shuttle disaster. Its high priority in NASA's thinking, as well as the tremendous cost of keeping it on the ground, has indicated that it will be the fifth payload launched when the shuttle program gets rolling again, according to the shuttle manifest. The manifest gives a date of 17 November 1988; I wouldn't be surprised to see this date slip into 1989.

What will the space telescope do? In view of the many unexpected results from space astronomy (and from ground-based astronomy) in the last twenty years, it is likely that the most exciting discoveries from HST will be those that we don't expect at all. Yet, an instrument like HST is not designed in a vacuum. The scientists who spent years of their lives designing, selling, and building this project had a number of definite ideas in mind. Some of these ideas have taken shape in the form of "key projects," high-priority programs that will require lots of telescope time. It is likely, but not certain, that these key projects will be given some of the time they need. Three key projects identified so far all have the study of the origin of the universe as their prime target, so I'll have to digress a bit for a quick tour of modern cosmology, the study of the evolution of the entire universe.

Perhaps one of the deepest questions that astronomers—or any human beings, for that matter—can ask is "Where did it all come from? How did the universe begin, and how did it develop over time?" Cosmology is the name given to the answers to these questions. To an anthropologist, *cosmology* refers to creation stories, stories that are part of tribal folklore but are unsupported by any scientific evidence. Until the twentieth century, the answers to these questions given by Western culture, even by scientists, basically fell into the same category of stories. But since 1965, we have developed a remarkably complete outline of cosmic evolution, the Big Bang cosmology, which is scientific cosmology because there's evidence to support it.

Let me try to compress twenty billion years of cosmic evolution into one paragraph. Somewhere between ten and twenty billion years ago, everything that we can see in the observable universe was squashed very close together, a hunk of very hot hydrogen gas. It's unfortunate that we can't pinpoint the time scale any better than that, but a range of ten to twenty billion years for the age of the universe represents the current state of our knowledge. This stuff expanded very rapidly and cooled. Nuclear reactions, occurring in the first twenty minutes of cosmic evolution, transformed about one quarter of the primordial material from hydrogen into helium. Any hot gas contains radiation, and this primordial gas was no exception. The radiation banged around, bumping into gas particles, until the universe was a million years old and cooled to the point where its gases became neutral and transparent. Soon thereafter, in some unknown way and at some unknown time, this gas condensed to form galaxies. The first stars in the first galaxies were pure hydrogen and helium; when they evolved and exploded, the heavier elements, like the carbon and oxygen in their interiors, were splattered out into space, to become part of subsequent generations of stars, planets, and living creatures. On a rather ordinary yellow star, located on the outskirts of a rather unremarkable galaxy, some of these living creatures acquired the gifts of intelligence and curiosity and began to wonder what made it all happen.[13]

The evidence in favor of this scenario is compelling yet incomplete. First of all, we see that the universe is expanding, pushed outward by the heat of the Big Bang. This expansion was discovered in the early part of the twentieth century by Edwin Hubble, a Rhodes scholar who passed up careers in heavyweight boxing and in the law for his first love, astronomy. The space telescope is named in his honor. The Big Bang is the easiest way to explain the expanding universe. There's more evidence. In the first million years of cosmic evolution, we expect one quarter of the universe to be transformed from hydrogen into helium, by nuclear reactions. Because a hot primeval universe contains radiation, and because the radiation persists until this day, we expect the universe to be filled with radiation. These particular cosmic fossils—the helium and, most important, the radiation—have been found in the real universe. It's hard to account for them in any other way.

The evidence supporting the Big Bang cosmology is incomplete. The most important gap is understanding how the structure in the universe formed—how the galaxies, the stars, and the planets that we live on came to be. Another problem that has gone well past the nuisance stage is that we don't know exactly how old the universe is. The ages of the oldest star clusters, fifteen billion years, are rather close to the age of the entire universe. If these clusters turned out to be older than the universe, the whole Big Bang picture would have to be scrapped, and a new cosmology would have to be invented.

What can the space telescope do that ground-based telescopes can't do? The Hubble Space Telescope can, like other telescopes in space, cover a broad range of wavelengths and is not limited to the narrow optical window visible from the ground. It is also high above the turbulent atmosphere, which blurs out the images of stars and swamps them in the background light from the night sky. The HST can photograph and study objects that are a hundred times fainter than the faintest objects seen on similar ground-based photographs.

What can the HST do for cosmology? As a start, it can tell us how old the universe is. A particular type of star, called a *Cepheid variable,* turns out to be a unique cosmic yardstick in that, if we measure how bright it is and how fast it varies, we can tell just how far off it is. But with ground-based telescopes, these variable stars can be spotted only in relatively nearby galaxies, too nearby to help us probe the cosmic expansion cleanly. The HST will be able to surmount the problem, locating and measuring these variables in very distant galaxies. If we measure how fast these distant galaxies are moving away from us, we can tell how long it's been since those galaxies and we were in the same place at the time of the Big Bang. Additional observations with the HST can pin down the ages of the oldest star clusters in the universe and can find out whether these clusters are indeed younger than the universe as a whole, as they have to be if the Big Bang scenario is valid.

Doing cosmology with the HST is alluring, but some astronomers working in other fields will be using it to make their own exciting discoveries. For example, many astronomers who won the privilege of using the telescope by agreeing to spend years of their lives designing and building instruments are seeking planets or planetlike objects around nearby stars by a variety of methods. It's such a flexible instrument that virtually any field of astronomy can benefit from its existence.

## BEYOND THE HUBBLE SPACE TELESCOPE

Beyond the HST, many astronomy missions have been planned, approved, and started, though flight dates for many are uncertain. The *Gamma Ray Observatory* (GRO) will do the first deep, detailed study of gamma-ray sources and is the gamma-ray counterpart of the Hubble Space Telescope in NASA's group of four great observatories. The *Cosmic Background Explorer* will look for the signs of primeval galaxies in radiation from the million-year-old universe. The *Extreme Ultraviolet Explorer* will open up the last unexplored window in the electromagnetic spectrum. The *X-Ray Timing Explorer* will explore in detail the inner regions of accretion disks surrounding black holes, neutron stars, and white dwarfs. ROSAT, an unusual collaborative project in which the West German space agency is the lead agency and NASA is a junior partner, will follow up the X-ray discoveries made by *EINSTEIN* and *EXOSAT*. The European HIPPARCOS will measure stellar distances with unprecedented accuracy.

Another set of projects is only in the planning stages; the money for the construction of these telescopes is yet to come. The Advanced X-ray Astrophysics Facility (AXAF) is a very large X-ray telescope that is the third great observatory. The HST is the first of them, GRO the second, and AXAF the third. The successes of the IRAS mission, which discovered the protoplanetary clouds and infrared galaxies mentioned earlier, call for a follow-up mission. The Satellite Infrared Telescope Facility (SIRTF), originally designed to fly on the shuttle but redesigned as a free-flyer when it turned out that the shuttle environment is a powerful source of infrared radiation, is the fourth of the series.

It's not clear when any of these missions will fly. The HST and the Gamma Ray Observatory are on the space shuttle manifest and look reasonably secure. The manifest does not address the fate of smaller payloads like many other astronomy missions. Some may well be launched on rockets, requiring costly redesign. The community feels frustrated because of the delays and cancellations. The space shuttle was all set to pay off for astronomy in 1986. I've heard the greatest concerns from the people who build instruments, people who have devoted years of their lives to one particular project. A friend recently groused to me about designing "nonexistent instruments for nonexistent payloads." Peter Boyce, executive officer of the American Astronomical Society, expressed his worry that some of the key scientists and technicians on these projects will be lured away to other fields, unwilling to hang on, not knowing when, or whether, their projects will be launched.[14] A similar concern has been mentioned by two blue-ribbon advisory panels.[15]

Looking toward the twenty-first century, the future of space astronomy may well become bright again if we can survive the lean years. In contrast to planetary scientists, astronomers can benefit greatly from the capabilities of the space station, which can, in principle at least, significantly reduce the cost of almost any astronomical mission. Most of the cost of an astronomical mission is the cost not of building the telescope, but of building the satellite to house it, point it, cool it, power it, and send the data back to earth. Part of the space station concept is a collection of platforms that will provide these necessary support services. If the potential of the space station is realized, astronomers of the future should be able to design and build a new telescope for one-tenth of the present cost of a full-fledged mission. This telescope would be lifted to the astronomy platform, bolted to it, and run for a short time to be checked out. If necessary, it could be unplugged, brought back to earth, modified or refurbished, and then put in space again.

## THE EUROPEAN COMMUNITY: NEW LEADERS IN SPACE SCIENCE?

Even before the *Challenger* explosion, the European Space Agency (ESA) began asserting itself in the space sciences. An ambitious report, "European Space Science—Horizon 2000," outlined four cornerstone projects that could, as the authors put it, give "Europe the means of being an equal partner in a worldwide prospectus in space science, while honoring its cultural heritage and scientific tradition."[16] ESA, a consortium of a number of European countries, or a European counterpart of NASA, has its own launch capability in the form of Ariane.

Relations between American space scientists and their European counterparts have changed dramatically in the last ten years. When I first started doing space astronomy in the mid-1970s, the American attitude was quite paternalistic. We would welcome the Europeans on board but anticipated that they would play a secondary role. I was accustomed to seeing European scientists visiting American universities and space centers, grateful for the opportunity to obtain precious moments of observing time. The situation has reversed itself quite rapidly, with Europeans becoming equal partners and, in some disciplines, leaders.

For example, in 1984, Rob Petre (of NASA's Goddard Space Flight Center) and I wanted to continue to obtain some X-ray observations of hot white dwarf stars to follow up on some work we had done with NASA's *EINSTEIN* X-ray observatory. The only place where we could go was the European *EXOSAT* observatory. Now the Americans had to wait a couple of years until ESA let outsiders like Rob and me apply for observing time. American X-ray astronomers were welcome, but as junior partners. (Rob and I were lucky and did get some observing time, and *EXOSAT* started on our project before it spiraled into the ocean in the spring of 1986.)

*    *    *

The ability to send telescopes above the atmosphere permits astronomers to use these robot eyes to probe the entire electromagnetic spectrum. This much broader view of the universe has revealed a number of peculiar phenomena, such as ultrahot stars, black holes, protoplanetary disks, and very hot interstellar gas. Our own sun, seen from space, is far from the placid yellow disk seen from the ground. Gas and magnetic fields around the sun seeth with activity, occasionally erupting in a solar flare, producing particles that could endanger astronauts.

The vast number of stars in the universe suggests that somewhere there is some other intelligent creature who may be looking out into space and discovering this enormous variety of cosmic phenomena. The search for other life in the universe is a quest that comes deep from the human spirit. Although it is a relatively small part of NASA's effort, the exobiology program appeals to many individuals as an unusually important way of fulfilling part of human destiny.

# CHAPTER 12

# Intelligent Life in the Universe

Are we alone? This intriguing question has perplexed philosophers and scientists for millennia. For most of human history, the question was addressed in a philosophical or theological context: Did the Creator want to create a host of intelligent species, a few of them, or only one—us? This question is not a scientific one, as there is no way of using experiments or observations to reveal the Creator's intentions. About four hundred years ago, astronomers, physicists, and other thinkers accepted that the earth was one of many planets in the Solar System, and that the sun was one of many stars in the Milky Way Galaxy. But we could only speculate and wonder whether whatever it was that produced us could have produced other intelligent creatures somewhere else in this vast universe that we live in. It has been only in the last two or three decades that scientists have obtained the data and insights needed to ask, as scientists, this elusive question: Are we alone?

Our understanding of the genesis and development of life on the earth indicates that the natural processes that occurred here could certainly have occurred elsewhere in the universe. The scientific evidence cannot exclude the possibility of a divine role in the evolution of complex life forms or, for that matter, in the creation of life itself. However, natural processes certainly can account for all of what we see in the biosphere, though many important gaps in the evolutionary chain are not fully understood. If it could happen here, it could happen elsewhere.

Accepting this conclusion, many astronomers have decided that it is worth making some modest efforts, at least, to search for life elsewhere in the universe. Physicists Giuseppe Cocconi and Philip Morrison brought the search for extraterrestrial life into the world of science in 1959 by pointing out that radio is the uniquely rational way for extraterrestrial civilizations to communicate. The smart way to worry about extraterrestrial life was not to become embroiled in the wild, woolly, and frustrating controversy about unidentified flying objects (more about these later). Radio searches for extraterrestrials have continued, at some level of effort, since the 1960s.

Although there is a general consensus in the astronomical community that a search for extraterrestrial life is worthwhile, there is less agreement on the prospects of success, on the likely abundance of life in the universe. Some astronomers, most notably Carl Sagan, visualize a universe that is almost teeming with life, waiting for us to get our radio telescopes working so that we can join the galactic club.[1] Others point to the many gaps in the evolutionary chain leading from an inanimate universe to the complexities of human intelligence and contend that an equally strong case can be made for the viewpoint that we may well be alone in the universe.[2] Most astronomers, including me, believe that life is likely to be relatively abundant, and that is the position that I shall concentrate on here. Even the skeptics, however, believe that the rewards of a search for life are so fundamental that the quest is worth it, especially considering that it can be done very economically.

## WE ARE NOT ALONE

If you can find a dark spot somewhere, far from the glare of the city or suburban lights, go out in some season when it's comfortably warm after dark, and wait a few minutes for your eyes to get used to the lack of light. Unless it's spring, you will see a band of light stretching across the sky. (In the spring, someone in mid-northern latitudes cannot see the Milky Way because it happens to be located near the horizon, in all directions. It gets lost in the ground haze.) This band is the Milky Way—the visible sign of the huge galaxy that we live in. In particular directions in space, thousands of stars lie along your line of sight, so many stars that your eye cannot separate their images into single stars. These stars blend together into a strip of luminous smoke, the Milky Way. Galileo pointed his telescope at this band in 1610 and discovered that it was made up of countless numbers of stars.

No one has counted all the stars in the Milky Way Galaxy. Astronomers can enumerate what is visible in particular directions, and we can analyze the motions of the stars that we do see. It's now clear that we live in a huge, flat disk containing well over a hundred billion stars. We are in the plane of the disk. When our line of sight is directed in the plane of the disk, we see thousands of stars. When we look up out of the disk, we see many fewer stars and a dark sky. This explains the phenomenon of the Milky Way.

When we first realized what stars were, numerous articles and books mused on the philosophical and theological consequences of the possibility that there were other worlds with intelligent inhabitants.[3] There was little scientific thought in much of the early writings. However, Cocconi and Morrison's finding that we could, with present technology, send intelligible radio messages from one star to another surfaced around the same time that biologists began real research on the origin of life. The search for extraterrestrial life has become scientifically respectable. It's not a big item in NASA's budget, but it plays a major role in stimulating the human intellect and imagination. It has to have a name and an acronym, of course: SETI, the "search for extraterrestrial intelligence." A small number of courageous souls have embarked on a difficult, but important, human venture.

There are hundreds of billions of stars in the Milky Way Galaxy, and there are millions of galaxies within reach of our telescopes. This superabundance of life sites is both good and bad for SETI. With so many possible places in the universe where life could flourish, it is almost inconceivable that we are absolutely alone in the universe. But these hundreds of billions of stars are a huge cosmic haystack in which a much smaller number of extraterrestrial civilizations are possibly hidden. How much of the cosmic haystack must we look through in order to find an extraterrestrial? Asking this question opens up a host of questions regarding the origin and evolution of life on the earth, questions that are fascinating in their own right.

What we need to do is to think about the likelihood that an intelligent, communicative civilization exists out there, and to estimate what fraction of the hundred billion stars in the Milky Way Galaxy might be the sun that provides its energy to a civilization that might contact us. Consider first what life needs. All living systems that we know of use a liquid medium in their cells in order to push chemicals around from one cell organ to another. The liquid found in all terrestrial organisms is water. This liquid medium also plays a central role in moderating the chemical reactions that occur when cells grow. Every time a bunch of amino acids link together to form a protein, water molecules are given off. So life as we know it requires liquid water.

It's certainly possible that some other chemical could be the liquid of life, but there's no way of deciding for sure whether such an idea is idle speculation or a real possibility. Ammonia has been widely mentioned as one possible substitute. For the moment at least, let's set such speculations aside and restrict ourselves to water-based life. Allowing for ammonia-based life could only increase the number of possible life sites in the galaxy.

To estimate the number of life sites in the galaxy, consider what different types of stars there are out there, and what stars are sufficiently sunlike so that they could nurture a planet like ours sufficiently well. Think of these as good suns. In order to have liquid water on its surface, a planet needs to orbit at the right distance from its parent star. This requirement does not, at first, rule out any stars, for a planet could be a long way from a hot, powerful star or huddled very close to a cool, dim star and still have a surface temperature that is within a tolerable range. But a little more thought does start to limit the scope of the search for life, to pick out the good suns.


## GOOD SUNS

Some of the stars in the Milky Way Galaxy are very hot. Most of these hot stars, though only somewhat more massive than the sun, use up their nuclear fuel at a prodigious rate, shining thousands of times more powerfully than the sun. Such stars will use up their nuclear fuel much faster than the sun will and will last in their present state for only a few million years or so. It's taken 4.5 billion years for life on the earth to originate and become complex. On planets around these hot stars, there's probably just not enough time for intelligent life to develop. Before life gets started, and certainly before life becomes more complex, these stars will run out of nuclear fuel, either boiling or freezing any living creatures that exist on planets orbiting them. It's possible that the planets around these stars will be vaporized completely. If we require that a star remain in its present state for say, a billion years in order to become a good life site, we're restricted to stars with masses less than 40% larger than the sun, that is, stars with masses less than 1.4 times the mass of the sun.

The cool stars in the Milky Way Galaxy pose a different set of problems. These stars, smaller than the sun, survive on hydrogen fusion in much the same way that the sun does. These stars will remain in the present stage of their life cycle for plenty of time. But a planet would have to huddle very close to such a star in order to be warm enough so that liquid water could exist on its surface. If a planet is too close to its parent star, tidal gravitational forces will cause the planet to rotate very slowly on its axis, perhaps keeping one face permanently pointed toward its sun in the same way that one side of the moon always faces the earth. In such a case, conventional wisdom has it that the sunward side of the planet will always roast and the back side of the planet will always freeze. Exactly how cool does a star have to be for this catastrophe to occur? Estimates differ, but it does seem that a star with half a solar mass or less is certainly ruled out. About 60% of the stars in the galaxy are such cool stars, and this conventional interpretation would rule them out as good suns. (These types of stars are called *M dwarfs,* named for their spectral class [M] and their size.)

There are a couple of other types of stars that remain to be disqualified. Somewhere between 5% and 25% of the stars in the galaxy are white dwarf stars, dying cinders. These stars cool reasonably rapidly, from very high temperatures to $10,000°K$ in a billion years or so. Any planets close enough to a white dwarf to be habitable would probably have been vaporized in earlier stages of stellar evolution, so we'd have to rule them out. A large fraction of the stars in the sky are double stars, star systems in which two or more stars orbit each other.

According to conventional wisdom, star formation processes can produce either a multiple star system or planets but not both, and so only single stars are good suns. Depending on whether we want to rule out all multiple star systems or only those in which the stars are very close together, this restriction cuts out about 50%–85% of all stars in the Milky Way Galaxy.[4]

The conventional wisdom could be wrong, and (in my opinion, at least) M dwarfs and some double stars *could* be good suns. We know two planets that have very long days and that have substantial atmospheres: Venus and Uranus. The days on Venus are about 120 earth days long, and the days on most of Uranus are 84 earth years long because of the slow rotation of Venus and the bizarre tilt of the axis of Uranus. On both of these planets, the temperature of the nighttime side is similar to the temperature of the daytime side. Now, perhaps it takes an uninhabitably thick atmosphere to produce a reasonable temperature on a slowly rotating planet, and we'd still have to rule out slowly rotating planets (thereby ruling out the M-type stars). Another way of including the M-type stars is to allow for the possibility of life that could survive at very low temperatures, where ammonia (or an ammonia–water clathrate?) could take the place of water.

Also, I've never been very comfortable with the idea that double stars are necessarily bad suns. We don't know enough about the formation of planets to rule out the possibility that another stellar object nearby could prevent planetary accretion. After all, mighty Jupiter didn't sweep up all the stuff in the solar system and prevent the earth from forming. Clearly, we can rule out a double star system in which the distance between the two components is about equal to the distance that a watery planet would have to be from either of them. But consider the sun's nearest neighbor, Alpha Centauri, where a remarkably sunlike star coexists with a somewhat cooler star orbiting in the vicinity of Neptune. Think about this situation as seen from a planet in an orbit like the earth's, close to the sunlike star. This cooler companion would be about a hundred times brighter than the full moon, lighting up the night skies when a planet was between them. But its effects on the weather would be trivial because it would be several hundred times fainter than the sunlike star. And the third star in the system is so far away that it doesn't matter at all. I should point out in all honesty that most other scientists who write about this topic do rule out double stars as good suns, but it seems to me that at least the well-separated double stars should still be left in the running.[5]

Where does all this leave us? Let's start with a group of 100 typical stars and use the conventional wisdom (now that I've mentioned my unconventional ideas about double stars and very cool stars, I'll let them drop; in the end, these ideas would have a relatively minor effect on the final answer anyway). About 60 of those stars are too cool, and one is too hot, to have habitable planets orbiting them. About 15 of the stars are white dwarfs, where habitable planets have been alternately fried and frozen in the last billion years. We're left with 14 sufficiently sunlike stars, and if we toss out the 60%–80% that are double stars, we're left with 3 to 5 good suns out of the original 100. This sounds like a small number. But there are not just a few hundred stars in the Milky Way Galaxy; there are a few hundred billion stars, and of these few hundred billion, there are about ten billion (that's 10,000,000,000) that are good suns.

Now, how many habitable planets are there? We don't know how planets are spaced in other planetary systems, and we also don't know what it takes to make a planet habitable. We know the earth is habitable, but just how earthlike does a planet have to be in order to support life? We have to throw scientific caution to the winds to get anywhere, so let's press on. Venus is clearly not habitable. Mars is also too cold, but it was warmer in the past. There was liquid water on the surface of Mars one billion years ago. Perhaps the problem with Mars is not just that it's too far from the sun, but also that it's too small and therefore has an atmosphere that's too thin. So let's count Mars as being almost habitable; it seems that if Earth had been located halfway between the present orbits of Earth and Mars, it would be colder than it is now but still livable. We don't know about the architecture of other solar systems, but based on these arguments, it would seem that if there is a random collection of small rocky planets in orbits more-or-less like the orbits we have, at least one of them would be in an orbit that would allow liquid water to exist on its surface. A skeptical reader may feel that my logic in this paragraph is rather glib (and it is). A number of writers have described similar arguments in more detail and have come up with similar answers.[6] So our ten billion good suns in the Milky Way are surrounded by something like ten billion habitable planets.

The estimation of this probability, the probability that a planet will find itself at the right distance from its sun, is one place where a diversity of views exists in the scientific community. We do not know how perilously balanced the earth's climate is between the extremes of a tropical oven like Venus and an Ice Age. Because our understanding of the earth's climate is limited, it is not possible to resort to computer models to settle the question definitively.

## HOW LIFE BEGAN

Is there life on these habitable planets? Will life evolve on any planet with a watery ocean on it? These questions are basically equivalent to asking one of the deepest questions in biology: How did life get started in the first place? We don't know the whole answer, but we do have some reasonable ideas. Our basic conclusion is that natural processes can, in principle, lead to the spontaneous development of living cells, given the right environment and enough time. Although there are a number of gaps in our reasoning, gaps where primitive-earth experiments cannot be run for long enough to investigate directly what went on, there is no reason to assume that the development of life required very special circumstances that would be infrequent in the universe.

Let's start by asking what the primeval earth was like. Sunlight (in which the ultraviolet was not blocked by an ozone layer), lightning, and volcanic eruptions poured energy into the atmosphere and oceans. The terrestrial atmosphere was quite different from the present one. We don't know what was in the primitive atmosphere, but we do know that the oxygen that makes up about one-fifth of our current atmosphere could not exist if green planets weren't around to sustain it. If all the plants on the earth were killed, the oxygen in our atmosphere would rapidly combine with the soil and soil particles in water to make stable oxides (in the way that iron plus oxygen forms rust), leaving little to breathe. Whatever the other uncertainties are, scientists are quite sure that the primeval atmosphere had no free oxygen in it. In fact, we have direct evidence, from iron deposits in rocks, showing that the atmosphere contained 1% oxygen about two billion years ago.[7]

Nearly thirty years ago, graduate student Stanley Miller, working under the leadership of geochemist Harold Urey, asked what kinds of chemicals would be produced on the primitive earth. He put a mixture of ammonia, methane, and water in a suitable vessel and subjected it to electrical shocks and other energy sources. In a few days, more complex organic molecules were formed. Remarkably, many of these complex molecules are the same molecules that are formed in living systems. Two amino acids, the building blocks of proteins, were formed in this primordial soup. This experiment has been repeated, with variations, hundreds of times, and the results remain the same. Complex chemicals like those found in living systems arise from natural reactions under almost any circumstances on a habitable planet.

But it's a long way from a primeval ocean with a few organic molecules in it (called a "hot, dilute primeval soup" by J. B. S. Haldane) to a living cell, a bag of chemicals that can do those remarkable things characteristic of life, such as reproducing, mutating, eating, excreting, and growing. The gap between chemical evolution (where natural chemical reactions result in the production of increasingly complex molecules) and biological evolution (where organisms control the chemistry taking place inside them and reproduce) is a big one. But we have some ideas about how this particular gap can be bridged.

It might have happened in a tidepool at the ocean's edge. Currently, the favored location is the sticky surface of claylike minerals, where the amino acids and other building blocks can bond to the clay and to each other.[8] The Russian biochemist A. I. Oparin proposed that coacervate droplets, little globs of water inside spherical oily films, either were the initial site or played a significant role in forming the wall of the first living organism, the membrane that separated the living interior from the nonliving exterior. Wherever it happened, the initial steps in the scenario must have been something like these: Amino acids and nucleic acids found themselves in an environment where natural processes produced a high concentration of them. Pairs, then chains, of these simple building blocks began to form. At some point, a special larger molecule, a chain of amino acids, formed. This special molecule was one that catalyzed, or speeded up, the production of other molecules. A growing, concentrated witch's brew of chemicals began to form at this site.

And somehow—we don't know how and probably won't know

how for a long time, if ever—this bag of biochemicals crossed the dividing line that separates the living from the nonliving. It developed the ability to metabolize and reproduce, becoming the first living organism. It then reproduced and spread very rapidly. An organism that divided only once a day could spread over the primeval oceans in a matter of months because there would have been no predators or competitors for food to inhibit its colonization of new territory.

The precise mechanisms that produced life remain to be described in detail. Some writers argue that, because it's hard to put all of the right chemicals in the same place at one time, natural processes can't explain the origin of life. Such arguments miss the point. A key step is the development of chemicals, or groups of chemicals, that can catalyze their own reproduction.[9] It's a little as if you were cracking a safe with a complicated lock with a twenty-number combination. If you had to try all possible combinations at random before you hit on the one that opens the door, it would take forever. But if you could listen to the tumblers click into place when you had a few of the numbers right in sequence, you could open the safe much faster. When a small group of amino acids and nucleic acids gets together in the right place and begins catalyzing its own reproduction, it is like having the tumblers click into place because this particular combination of chemicals does not have to get together by chance ever again. It's now there, sitting on the clay substrate, waiting for the remaining amino acids to come along.

Our picture of the origin of life is quite generalized. Scientists have produced monomers (the building blocks of life) and short chains of several monomers in the laboratory. Scientists have not made living organisms in the laboratory but have thought of a number of ways in which they could have occurred in nature. Nature had hundreds of millions of years to create life, and lab experiments don't last that long. Based on what we know about the origin of life on earth, most scientists believe that it is reasonable to estimate that life evolved on most, if not all, of the ten billion habitable planets in the Milky Way Galaxy.

However, we can't establish communication with a one-celled amoeba in the primeval slime on a planet surrounding Alpha Centauri or any other nearby star. On the earth, the processes of Darwinian evolution produced a variety of different organisms, starting with the time when the first living cell became sufficiently numerous so that its food supply was limited, and when a different cell that could digest a different chemical was produced by random mutations. It is quite possible that, on many of the ten billion habitable planets, there has been neither the time nor the circumstances to allow more complex life forms to evolve. To establish communication, we need more than complex life out there. We need a creature who can build radio telescopes, who wants to build them, and who will put some effort into communicating with other cosmic creatures.

Now it becomes rather difficult to estimate the probabilities that the development of complex, intelligent organisms will naturally occur as a result of evolution. For instance, a number of evolutionary biologists have argued that the probability of Darwinian evolution's producing an intelligent species is quite small.[10] The basic argument is that the evolutionary track that connects us to the first living cell involves millions of branches and is not a straight line of "progress" upward on a tree. Of the millions of possible evolutionary paths, only one led to us. Were similar evolution to occur on another planet, the chances that precisely the same branches would be followed is very small, of the order of one in a million.

However, this argument misses the point, to a great extent. This calculation shows only how improbable species with the body form of human beings are, not how improbable intelligence itself is. We all agree that creatures that look like humans won't evolve on other planets. There are some traits, however, that any organism will have. Sight is valuable for survival, and organs that provide sight—eyes—have evolved many different times during the earth's history. Vision will apparently develop anywhere because the eye offers a tremendous reproductive advantage to any organism that has one or, better yet, two. Is intelligence a similar, inevitable consequence of the processes of Darwinian evolution? Basically, we don't know.

A second uncertain issue is the likelihood that an intelligent extraterrestrial species will develop the technology needed to communicate over interstellar distances. The human experience suggests that a technological civilization may emerge relatively infrequently among intelligent species elsewhere. Human beings have established nearly a hundred cultures on this planet. The high-tech culture of the United States and Europe can build—and has built—radio telescopes. But the !Kung bushmen of the Kalahari are perfectly happy to shun such technological marvels. Even in technologically sophisticated countries, groups of people who disdain the use of technology flourish. The Amish are content with horses and buggies. However, it isn't just the West that has embraced technology: the Chinese and Japanese certainly would have developed the ability to build radio telescopes by themselves.

There's one last factor to consider. If we are to communicate with an extraterrestrial being, he or she must be out there *right now*. It may well be that, at one time, there was a flourishing technological civilization on a planet around Alpha Centauri that self-destructed millions of years ago. The rusting remains of their radio telescopes would be mute now, unable to send or receive messages. The first radio telescope was built by Karl Jansky in 1933, and twelve years later we built, tested, and used the first nuclear weapons, devices that could easily annihilate all intelligent life on the earth. The technological sophistication that is needed to communicate over interstellar distances is the same level of development that has, on earth, built weapons of mass destruction. How long will a technical civilization last, on the average?

It's easy to be pessimistic about the lifetime of our technical civilization. You can take your choice of nuclear annihilation, environmental catastrophe, or population explosion as a possible cause of our destruction. But speaking personally, I see reasons to hope that our civilization will indeed last for a long time. More than forty years have passed since the bombing of Hiroshima and Nagasaki. There have been two major East–West conflicts (the Korean and Vietnamese wars). Countless other conflicts could have escalated to embroil the superpowers. Other incidents like the Cuban missile crisis in the 1960s or the shooting down of Korean Airlines Flight 007 in the 1980s could also have served as an excuse for some trigger-happy leader. But despite all of these international tensions, leaders with varying amounts of competence in international politics have still avoided nuclear war. I find that there is reason for optimism, reason to believe that technological civilizations can last for a long time.

Anyone can consider the previous discussion and come up with just about any set of defensible numbers regarding the fraction of the ten billion stars in the Milky Way Galaxy on which at least one type of organism has surmounted the hurdles needed to become a civilization capable of interstellar communication. These hurdles require that a species on one of these stars develop the necessary complexity, intelligence, and technology to communicate. Furthermore, the civilization has to last long enough so that the probability that we will find them there right now is reasonably large. Someone who is wildly optimistic could argue, quite defensibly, that there is no reason to suppose that any of these hurdles is impossibly difficult, that at least one species on every one of the these ten billion habitable planets has reached the stage of development that we have. A pessimist could make an equally strong case that some of the hurdles could be quite difficult to surmount.

In order to estimate the difficulty of the search for extraterrestrials, some numbers are useful and even necessary, however uncertain they are. A number of authors have gone through this process.[11] Although the estimates vary, reasonable numbers look something like the following: The evolution of complex life forms seems, based on terrestrial evolution, to be quite natural and is likely to occur on most of the ten billion habitable planets. Intelligence is another issue; it would seem reasonable that at least one intelligent species would evolve in 10% of the cases, leaving us with one billion planets with intelligent life forms. Some form of technology has developed in something like 10% of human cultures, and if these statistics apply to the entire universe, we would have 100 million planets where radio telescopes were built at some time. We can only guess at the likely lifetime of a technical civilization; our survival after forty years in the nuclear era suggests that it's not unreasonable to suspect that 1% of all civilizations can survive for a long time, say, the lifetime of their star; thus, we would have one million technical civilizations in the Milky Way Galaxy. This number is an estimate—nothing more than that.

## WHERE ARE THEY?

One result from the above line of reasoning is considerably more certain than the estimate of the number of technical civilizations in the

Milky Way Galaxy. If life is abundant in the universe, then those civilizations that have developed have to have lived a reasonable fraction of the lifetime of a star, which is several billion years. If technical civilizations last for only a short time, then there will be only rusting radio telescopes out there, not living creatures whom we can communicate with. The result is that if extraterrestrial civilizations are abundant in the Milky Way Galaxy, most of them must have lived a long time. This makes us the youngest and dumbest of them all.

Skeptics who believe that we may be alone have seized on this argument and turned it on its head. The question goes beyond argument about the probabilities, which rapidly turns into a futile exercise. It can be summarized quite briefly: *Where are they?* The Italian physicist Enrico Fermi, who developed a well-deserved reputation for asking questions that get to the heart of the problem, was supposedly the first to ask this question. The argument has since been restated and reworked by many others.[12] If extraterrestrial life is abundant and worth looking for, it has to be able to send us radio signals. If it can send us radio signals, it can also develop interstellar travel. If this hypothetical civilization had developed interstellar travel, it would colonize the Milky Way, either by traveling as physical organisms or by constructing self-reproducing machines. If this had happened, they would be here now. They are not here now, and therefore, extraterrestrial life does not exist.

This argument has withstood some of the more obvious attacks that can be made on it, but I nevertheless find it unconvincing. The usual motive for colonization is the provision of more living space. A species intent on obtaining more space to live in would find a space colony much cheaper than a trip to Alpha Centauri. Interstellar travel is sufficiently difficult so that mere curiosity would probably not provide enough motivation for sending physical objects instead of photons across space. I recognize that, in summarizing these arguments, I have skipped over some major subtleties; interested readers can find more detailed discussions elsewhere.[13] "Where are they?" is not a question that can be easily dismissed, but I don't think that it proves that we are alone.

## SHOULD WE LOOK FOR EXTRATERRESTRIAL LIFE?

With all these uncertainties surrounding any estimate of the abundance of life in the universe, is there any point in looking for it? Scientists differ in their estimates of the chances of success, but almost all agree that the search should take place, especially considering how inexpensive it turns out to be these days. The consequences of finding extraterrestrial life will be profound. Because the prize is so valuable, even those who think our chances of success are small believe that we should go on looking anyway.

Perhaps another way of asking whether we should search is to ask how much money should be allocated to the search. Several years ago, I participated in a "Delphi" poll in which about a hundred astronomers, biologists, and physicists were asked questions about SETI and were then given a chance to revise their answers on the basis of the opinions of the group. I was surprised at the degree of unanimity about the effort that should go into SETI: a few million dollars a year. The poll was taken before NASA's SETI program started in the early 1980s, and it happens that a few million dollars per year is just about what we are spending on SETI. Compared to the cost of other things our country does, that's not very much. A few million dollars a year, these days, will buy a single high-priced basketball or baseball player. In 1986, my hometown baseball team, the Philadelphia Phillies, paid Steve Carlton a million dollars to pitch for other baseball teams.[14] A million dollars is less than 0.10% of NASA's budget, not to mention the Pentagon's. A few million dollars a year is a small price to pay for a tremendous human venture, one that uplifts the human spirit toward the stars.

## SEARCHING THROUGH THE COSMIC HAYSTACK

If there are a million civilizations in the universe, only one star in 100,000 has a civilization around it. These 100,000 stars constitute a tremendously large "cosmic haystack" in which there may be a tiny needle. The numbers, to be sure, are uncertain, but they illustrate the difficulty of the task. How do astronomers even begin to approach this big job?

To start with, consider what scientists are looking for. In 1959, physicists Giuseppe Cocconi and Philip Morrison came up with a good framework for answering this question.[15] An extraterrestrial civilization and technology could be in just about any imaginable form. However, we and this civilization live in the same universe and have the same basic laws of nature. These circumstances provide some cosmic limitations on the likely nature of any interstellar signal.

What would you use to carry the information in interstellar communication? Stars are very far apart: something flying at the speed of a typical jet airplane would take five million years to get to Alpha Centauri. If you want to send something faster, you need to use energy, and this limitation suggests that you want to communicate information using as low mass a carrier as possible. You want something that travels very fast and is easy to produce and capture. These very basic considerations suggest that a photon, the basic unit of electromagnetic radiation, is the best way to communicate over interstellar distances. It takes only (!) a little over four years for a photon to go between here and the nearest star, and we know how to generate photons. You are looking at photons of visible light right now when you read these words.

What wavelength photons should be used? I've mentioned radio telescopes several times, but why? If the cosmic creature out there has eyes, as is likely, she or he may well see different colors, possibly using X-ray vision, though that's unlikely. We have to look beyond differences in physiology, atmospheric composition, and the like to find fundamentals. We want to pick a type of photon that can easily travel over interstellar distances and that is the easiest to separate from other types of photons that pervade the universe.

Surprisingly, these very general considerations suggest a uniquely rational way to communicate over interstellar distances: radio waves. It takes a lot of energy to manufacture a high-energy photon, like an X-ray, and so these would seem unlikely. Such photons don't travel through any conceivable planetary atmosphere, and X-ray detectors have to be placed in space, so they would be expensive. Lower energy X-ray photons can't travel through interstellar space so easily, and so we can rule them out as well. When we get to the ultraviolet, visible, and infrared part of the spectrum, our signal is fighting to be seen against the radiation from our star. These fundamental considerations, which depend only on the physical properties of atoms and molecules, rule out large portions of the electromagnetic spectrum as likely media for interstellar communication.

Radio waves are subject to none of these disadvantages. The small energy per photon makes radio much more economical as a communications link. Radio waves travel unimpeded through the interstellar medium, and the background babble of the universe is smaller in the radio part of the electromagnetic spectrum than it is anywhere else. Radio has been accepted as being by far the best way for two civilizations to communicate over interstellar distances.

A great many books have been written about the possibility of physical contact between extraterrestrials and us. Science fiction writers need such contact to produce a credible scenario; authors can't write interesting plots about people who can only send radio messages to each other. For the last forty years, a small band of writers has done their best to persuade their readers that funny lights in the sky are actually spaceships piloted by intelligent beings. This is not the place to write in detail about the UFO controversy. I know of no UFO sighting where the evidence is so strong that a natural atmospheric phenomenon, a balloon, or some other prosaic explanation can be ruled out as the cause for what was seen. The "best cases" have always gone away when they were scrutinized in detail.[16]

Since 1960, radio astronomers have acted on the suggestion that radio is the rational way to communicate over interstellar distances and have spent some telescope time looking for signals. Pioneer Frank Drake pointed the newly constructed 26-meter-diameter antenna of the National Radio Astronomy Observatory at two nearby stars in 1960. One of the stars, Tau Ceti, is just a little cooler than the sun, twelve light-years away. The second, Epsilon Eridani, is a bit cooler still and is eleven light-years away. Drake had to pick a frequency to look at; he followed Cocconi and Morrison in selecting a magic frequency of 1420.405 megahertz, which is emitted and absorbed by the most common element in the universe: hydrogen.

Drake's experience reflects many of the problems that have arisen in subsequent SETI programs. Drake had to build special equipment to analyze the signal that was broadcast by the radio telescopes. One expects a signal to be concentrated in a very narrow-frequency band, in much the same way that terrestrial radio stations broadcast at only one frequency. If you twirl the button on your car radio, the signal will change when you go through the frequency emitted by a nearby radio station. However, the equipment routinely used by radio astronomers is insensitive to such narrow-band signals because natural objects don't give off that kind of radio emission.

This first-ever search for extraterrestrial intelligence found no signal from Tau Ceti. But when Epsilon Eridani rose above the horizon, there was something there! Drake did not seriously believe that something would be found on the second star that we looked at and waited to reexamine Epsilon Eridani a second time. The signal was never seen again. The best guess is that some classified military plane was emitting radio signals, and that Drake's equipment picked them up.

In the years since this beginning effort, well over forty separate programs have involved a total of several thousand hours of observing time. Only several hundred stars have been looked at directly, well short of 100,000, which is the usual estimate of the size of the cosmic haystack. Most of the time, the signals are recorded on magnetic tape and are analyzed later by computer. The human work involves the frustrating task of trying to chase down what might have caused the signal picked out by the computer.

We have not found intelligent life in any of these searches. But what's been heard is not the uniform, uninteresting hiss of the natural universe. Jill Tarter, the astronomer who has probably spent more time on these searches than anyone else, summarized: "Most of us in this business have a drawer somewhere containing a list of 'birdy' coordinates and frequencies that we reobserve at every opportunity, hoping to reproduce a prior detection."[17] The trouble is that conducting a search for extraterrestrials amidst the bustle of human activity is a bit like trying to hold a whispered conversation in Yankee Stadium in the middle of a World Series game. Humanity is much closer than extraterrestrials, and human activities produce narrow-band, regular signals that can easily mimic an extraterrestrial communication.

Tarter described one particular experience that provides an instructive example of the difficulties of SETI. Astronomers Tarter, Tom Clark, Jeff Cuzzi, and David Black recorded hundreds of magnetic tapes in a five-day observing run with the large, 300-foot antenna of the National Radio Astronomy Observatory in Green Bank, West Virginia. One particular star showed very strong signals every time it was looked at. This discovery would probably have remained as one of the hottest possibilities for a real extraterrestrial civilization if one of the members of the observing team, Tom Clark, had not been a radio ham. Clark noticed that the frequencies being picked up were the same pattern as the frequencies produced by CB radios, raising suspicions that the signals came from nearby CB operators rather than from non-human creatures. Another clue to solving the puzzle came from the limitations on the 300-foot telescope itself. This telescope cannot be moved in both directions, and so any particular part of the sky can be observed only at the same time of day, when it is due south of Green Bank. The interesting star was always observed at just about 8 A.M., when the telescope operators changed shift, and virtually all of their cars and pickup trucks were equipped with CB radios. No one can be absolutely positive, but it seems likely that the CB radios were the sources of the signals that the team received.

In this case, the team was lucky because the clues to solving the mystery of the origin of these ''birdy'' signals were available. Most of the time, the investigator lacks such clues and has to use scientific judgment in deciding just where a particular signal came from. For example, Drake was unable to investigate his hunch that classified airplane flights were the source of the signals that looked as if they were coming from Epsilon Eridani because he could not penetrate the curtain of military secrecy. SETI investigators have accepted the standard of reproducibility that has evolved through centuries of scientific tradition. For scientists to believe that a purported message is real, it must be detected twice, at different times of day, with different telescopes.

## THE IMMEDIATE FUTURE

In the early 1970s, NASA sponsored a summer study program to design a system that could scan 100,000 stars in a decade. The design, called Project Cyclops, was a megaproject on a grand scale.[18] Hundreds of 300-foot dishes, manufactured cheaply by mass-production techniques, would be lined up cheek by jowl to cover several square miles of territory. The cost of such a system could only be guessed at and would be likely to be immense. It is now clear that it would be pointless to build something like that on the earth because of terrestrial radio interference. If Cyclops were ever to be built, it would be on the far side of the moon.

The Cyclops study seemed to kill completely the idea of a realistic SETI program because it seemed to indicate that any project with a realistic chance of success would be unbelievably expensive. Throughout the 1970s, the principal efforts in SETI were isolated projects by radio astronomers who managed to obtain a few days' observing time on radio telescopes that were constructed primarily to do radio astronomy. But the Cyclops study was completed just before the revolution in computer technology that produced cheap personal computers. These computers made it much easier to look for extraterrestrial signals and brought the cost of SETI down to its current reasonable level.

One can look for a weak radio signal in two ways. The Cyclops approach was a brute-force one: build a radio antenna that is so big that virtually anyone can hear a narrow-band *beep-beep-beep* by listening to a star on a loudspeaker. The other approach, which has revitalized SETI research, is to use a computer to recognize the characteristics of a weak signal from an extraterrestrial intelligent being amidst the much louder static. Because the necessary computers are now quite inexpensive, all one needs is a small radio antenna to do the job.

The result is that scientists don't need to spend the gross national product on a Cyclops-style megaproject to do SETI. An example of a small, ongoing effort is the dedicated search taking place at the Ohio State University Radio Observatory. Ohio State had built a radio telescope to survey the sky, paid for by the National Science Foundation. When the sky survey stopped in 1973, radio astronomers John Kraus and Robert Dixon continued to operate the telescope on their own time in a SETI project. Kraus and Dixon ingeniously designed some receiving equipment that operates unattended for the most part; the result is a very low-budget operation.

Paul Horowitz of Harvard has also developed some equipment

that can do inexpensive searches for extraterrestrial life. The key to his project is to use a computer to scan a number of frequencies simultaneously. You do a single-channel frequency scan whenever you switch on a car radio and turn the tuning knob, waiting to hear a signal from a radio station that you like. A scientist looking for an extraterrestrial message is looking for any signal at all over the entire range of the dial, and it would be much faster to analyze all the frequencies at the same time. Horowitz began with a project called Suitcase SETI, a portable computer that he took to the 1,000-foot radio telescope at Arecibo, Puerto Rico. Next came project Sentinel, where 128,000 channels were scanned simultaneously. In the mid-1980s, Horowitz is analyzing eight million frequency channels simultaneously. Interestingly enough, filmmaker Steven Spielberg, who produced two movies involving contact with fictional extraterrestrials (*E.T.* and *Close Encounters of the Third Kind*), contributed $100,000 to the project. Unusual as it might seem to have a filmmaker underwriting science, it's been done before. The German movie director Fritz Lang was one of the early supporters of the German rocket pioneers in the 1930s.[19]

With all this activity, what's left to be done? The cosmic haystack is so big that a few astronomers working in their spare time can pick over only a small piece of it. Jill Tarter and Ben Zuckerman, two radio astronomers active in SETI, have estimated that searches so far have involved $10^{-17}$ of necessary search space in terms of frequencies, directions in space, sensitivity, and so on.[20] That means that there is 100,000,000,000,000,000 times more work to be done. Because much of SETI was done B.C. (before computers), there is hope that the pace of searching the cosmic haystack can accelerate rapidly. This early work was not wasted, nevertheless. Every project since Drake's pioneering efforts has helped demonstrate what we need to do in order to make the search more efficient, to let the computers rather than the astronomers throw out the demonstrable false alarms.

In the next ten years, NASA plans to search $10^7$ times as much of the cosmic haystack as has been searched so far. This sounds like a great achievement, but it still means that when we're done we will have searched only $10^{-10}$ of it. The hope is that we're looking in the right place, either because we're clever or because we're lucky. Many people have tried to play the guessing game of figuring out what "magic frequencies" an extraterrestrial civilization would use to send an attention-getting beacon to us. It's a little like trying to figure out where you would meet someone in Washington, DC, if you had not prearranged a meeting. Several "universally popular" locations come to mind: the Washington Monument, the White House, the Capitol. Can we be clever enough to find the Washington Monuments in the frequency spectrum? Some people like the "water hole," a frequency region between the 1,421-megahertz transition of hydrogen and the 1,667-megahertz transition of OH. (H + OH is, of course, water, and where have desert dwellers met since antiquity? At the water hole, of course.)

It may not be necessary to guess where to look. If we can continue to increase the efficiency of our search by a factor of $10^7$ every decade, we will be able to search the whole cosmic haystack in thirty years. Most of the searching will be done at the end: from 1985 to 1995, we will search $10^{-10}$ of the haystack; from 1995 to 2005, we will search $10^{-3}$ of it; and then another improvement of $10^7$ will allow us to search it $10^4$ times over from 2005 to 2015. These numbers are not meant to be taken too seriously. Computers have been getting better and better, the power of computers of comparable cost has been improved only (and I use the word *only* in a loose sense) improving by 100-fold per decade from 1960 through 1985. If we are to search the whole cosmic haystack by 2015, SETI researchers must do much better than computer companies in improving the efficiency of their operation.

*     *     *

Are we alone? Astronomers, physicists, and biologists have developed two ways of approaching this question. The first is to consider just what set of circumstances led to the development of a set of two-legged creatures thriving on a green world orbiting an otherwise unremarkable yellow star in the outskirts of the Milky Way Galaxy. From what scientists can tell, purely natural processes can account for each of the essential steps leading from a gas cloud that contracted to make the sun, to living creatures looking up at the sky and wondering if we human beings are alone. If it happened here as a result of natural processes, it can happen elsewhere.

The second approach to the question of the existence of extraterrestrial life is a more pragmatic one: searching for evidence of its existence. General cosmic limitations suggest that the most economical way for intelligent creatures to communicate over interstellar distances is to send radio signals to each other. A number of intrepid radio astronomers have developed clever schemes in which a modest investment of a few million dollars per year can support a realistic search program. Something like 100,000 stars need to be examined at a wide range of radio frequencies before we will detect a signal, even based on rather optimistic estimates of the abundance of life in the universe. The search may be difficult, but the goal—detection of an extraterrestrial signal—is a deeply rewarding one.

# Summary of Part III

Space has opened up the entire universe for human exploration. Although astronauts have progressed no further than the moon, automated space probes have cruised by all the planets except Neptune and Pluto and have orbited many of the nearer ones. Many whole new worlds have been transformed from mere dots into fascinating landscapes. The largest volcano known to human beings is on the planet Mars. Right now, there are probably half a dozen sulfurous volcanoes erupting on Jupiter's satellite Io. Planets as varied as the hot oven of Venus and the frozen snowballs that are satellites of Saturn and Neptune are now part of the human experience. These insights stimulate the human intellect and imagination, to be sure; they also provide a cosmic laboratory that will help us to understand some of the global processes that make the earth work.

Space provides a number of new windows on the more distant universe, too. The earth's atmosphere is completely opaque to many wavelengths of electromagnetic radiation, and scientists' ability to hurl telescopes above the atmospheric curtain has opened up a new universe to us. The sun has been transformed from a placid yellow disk into a whirlwind of activity, where gases and magnetic fields swirl around and interact with each other, occasionally producing the phenomenon known as a *solar flare*. Astronomers using telescopes in space have discovered black holes, ultrahot stars and interstellar gas, protoplanetary dust clouds, and many other brand-new cosmic phenomena. A very small part of NASA's efforts go into investigating one of the deepest human questions: Are we alone?

The past achievements present a significant contrast with the actual situation of American space science in the late 1980s. The planetary exploration program has seen no new satellites launched since 1978. Many astronomical missions, seeking to explore the more distant universe, were lined up and ready to fly in 1986. The purportedly routine access to space provided by the space shuttle program was to pay off at last for astronomy. However, this golden year was not to be, and astronomers face another decade of waiting, surviving until the space station comes along.

The future requires us to press onward and to overcome the *Challenger* tragedy. Ten years from now, the space station will be in orbit, and humans will be playing a much greater variety of roles in the space program. Twenty years from now, when the twenty-first century has dawned, the present difficulties will be a tragic memory. The last part of this book looks toward the future.

# PART IV

# *The Future*

The mid-1980s mark a critical juncture in the American space program. The *Challenger* explosion highlighted many of the ways in which the space program has changed since the *Apollo* days and presents its own set of immediate problems with a crowded launch schedule for the next ten years. We must look beyond the *Challenger,* for many of the decisions that are made now will have their impact well into the twenty-first century.

The construction of the first space stations has meant, and will mean, a considerable expansion of the many roles that human beings can play in the space program. The astronauts in space will acquire more roles as they become repairers, builders, scientists, engineers, teachers, journalists, and someday, ordinary tourists. A growing diversity of people will gain access to space; scientists, engineers, and members of Congress are among the list of those who have already gone up. The American space station itself will present a number of opportunities for private industry to become involved, possibly at a considerably lower cost than is required at present.

But what will the astronauts of the twenty-first century do in space? If the cost of access to space is dramatically reduced by future technological developments, many of the blue-sky dreams can become reality. If the cost remains roughly as high as it is now, the opportunities will still remain but will be more limited. There are many areas of space activity: communications, weather forecasting, military, earth observations, materials processing, planetary science, astronomy, and the search for extraterrestrial intelligence. All of these can benefit greatly from increased access to earth orbit and beyond.

However, turning even some of these dreams into reality will require us to press onward, despite *Challenger*. Previous human explorations have faced hard times; Columbus spent ten years collecting rejections before he finally persuaded Queen Isabella to support his trip across the Atlantic. The future beckons. Do we have the vision to follow the high road?

# CHAPTER 13

# The Human Roles in Space

Traditionally, the subject of this chapter would be expressed as the usual question: Should the space program be manned or unmanned? The traditional way of asking this question is both sexist and misleading. Human beings play an essential role in any space mission, whether or not the mission involves astronauts in space. There is no such thing as an "unmanned" space mission, because humans control and interact with the machines we put in space in a growing variety of ways. The focus of this chapter is really on the variety of roles that people can play in the space program, particularly people who actually travel into space. So far, humans have been spacecraft operators and controllers, test objects, pilots, scientist-explorers, experimenters, repairers, symbols, and ambassadors. Soon to appear will be space travel writers (a role that will include the teachers and journalists) and builders. In the longer run, we can see people as tourists, and possibly even as colonists. Studies of the human response to the zero-gravity environment show no fundamental limitations on what humans can do in space in the short run, though there may be some effects that will limit long-duration existence in space (such as would be required for a human expedition to Mars).

The debate over the human role in space often evokes comparisons with the human exploration of the earth by Columbus, Magellan, and others. Fervent advocates of manned space flight often use the importance of these expeditions in human history as justification for sending people everywhere that people could conceivably go. Although there are significant similarities between the Age of Discovery and the Space Age, there are significant differences as well. Space exploration costs considerably more than the Columbus expeditions did, and we can gain a significant amount of knowledge about the territories to be explored without sending people there. The similarities as well as the differences are important issues that must enter any rational consideration of the future roles that human beings should play in our exploration of the near and distant universe.

## SPACECRAFT CONTROLLERS

The "unmanned" space program is one of the most unfortunately named human activities that I know of. To most people, doing something in an "unmanned" way means sending some sort of self-programmed, self-propelled, unfeeling robot out there into the cosmos. The implication is that once this robot craft has been launched, humans don't interact with it anymore and don't share in the sense of drama and adventure that has been part of all human exploration. But the unmanned space program doesn't work that way.

Any spacecraft that is up there is being controlled by human beings on the ground a great deal of the time. Much of the time, the spacecraft controller is preoccupied with "housekeeping" chores like monitoring the spacecraft's location, power consumption, attitude (which way it's pointing), and so on. But for a probe that is exploring, there is always the important adventure of deciding where it should go and what it should look at.

In many cases, especially with somewhat older spacecraft, the sense of adventure can be submerged because of the large amount of machinery or bureaucracy between the person who actually decides what to do with the spacecraft and the spacecraft operation. My first involvement with the space program, in the mid-1970s, was as a guest investigator with the *Copernicus* satellite. I traveled to Princeton University and spent a couple of days in offices and computer centers, consulting with the scientists who built the satellite to decide how I should use the twenty-four hours of on-target time that I had been awarded. These places looked just like other offices and computer centers and seemed far from the adventurous, romantic space program. I had to prepare a set of instructions telling the instruments exactly what to do and then check these out on the Princeton University computer, which had a program that simulated the spacecraft operation. I needed to make sure that my instructions were, in fact, a correct translation of my wishes. Although I could have traveled to the Goddard Space Flight Center to be there when my stars were looked at, I wasn't encouraged to take advantage of that opportunity because my observing program was quite straightforward and there was no need for me to be there. When the observations were complete, I returned to Princeton to analyze the data. I had little or no contact with the spacecraft itself, and the idea that I was using something that was up in space was something I thought about only in idle moments.

But the 1980s are different, and people on the ground can have a much greater involvement with the machinery that they are using. Developments in computer technology can allow for a greater sense of "hands-on" control of the spacecraft—and a greater feeling of actually being up there in space—than was possible before. When I use the International Ultraviolet Explorer satellite, I do travel to the spacecraft control center at Goddard and sit right next to the telescope operator who actually controls the motion and operation of the satellite. The telescope is operated from afar, or "teleoperated" in NASA jargon. A recent NASA Study, *The Human Role in Space* (it has an acronym, of course: THURIS) identified six different categories of "man–machine" interaction:

- *Manual*, where a human, in a spacesuit or not, uses simple hand tools like wrenches and screwdrivers.
- *Supported*, where supporting "machinery," such as rocket-powered backpacks, helps the person get around.
- *Augmented*, where human capabilities are magnified (e.g., by a microscope or a telescope).
- *Teleoperated*, where remote-controlled "sensors and actuators" like the robot arm on the shuttle are used.
- *Supervised*, where a computer controls the process, and a human, in space or on the ground, watches over it.
- *Independent*, where the piece of machinery is left to itself for an extended period of time.[1]

In the early days of the space age, the choice was between equipment that sat at either end of the spectrum. In the 1960s, we could not even send instructions to an ''unmanned'' space probe; it was independent of us once it was launched. For example, the early particle detectors, like the one that discovered the Van Allen belts, was built, was launched, and then just returned data. If we wanted to do something more complicated, we had to send an astronaut up there to do things manually. By the 1970s, we could send simple instructions to spacecraft and supervise what they did, but there was still a reasonably large gap between the simple tasks that machines could perform and the complex tasks that astronauts could do.

In contrast, now we have systems that can do rather complex tasks that fall in the middle of the spectrum. These permit people to perform tasks that would formerly have required someone in a spacesuit, without leaving more comfortable environments like the interior of the space shuttle. In the case of the International Ultraviolet Explorer, the operator doesn't even have to go up into orbit. Space shuttle astronauts can grasp satellites and build space stations while using the Canadian-built remote manipulator arm on the space shuttle, avoiding the hassle of a ''spacewalk'' or an EVA (extravehicular activity). It takes time to put an astronaut into a spacesuit, to allow his or her body to adjust to the different atmospheric pressure in the suit, and to get the astronaut out the door and back in again. On the other end of the spectrum, the development of such things as the manned maneuvering unit allows astronauts to range farther away from the space shuttle in order to retrieve and repair satellites.

The future may bring ''humanoid'' teleoperated systems complete with armlike and leglike appendages that the operator can use as a remote extension of human capabilities: when the operator moves an arm, hand, and finger to grasp something, the machine does the same thing. In late 1986, the design of the space station was significantly modified so that teleoperated systems could do more of the construction work. Teleoperated systems can be much simpler than genuine robots because they don't need the complex feedback systems that fully automated systems need in order to grab something without crushing it.

This broad spectrum of interactions between humans and machines demonstrates that a broad middle ground between ''manned'' and "unmanned" space activity is beginning to appear. It's no longer a simple question of whether to send an idiot machine or an expensive, spacesuited astronaut into space to do something. Rather, the question is (or should be): What role should human beings play in this particular space activity? In the past, humans have played a variety of roles in space, and the importance of these roles goes well beyond the simple hardware questions of what a person has to do in order to accomplish some particular task.

## PAST ROLES OF ASTRONAUTS IN SPACE

Even before computers and robots came along, human beings in space played a wide variety of roles ranging from test objects to repairers to ambassadors. Had *Challenger* not blown up, Christa McAuliffe would have added another role, that of teacher. In the future, we can anticipate that humans will function as builders, journalists, tourists, and even colonists. An exploration of these roles can demonstrate the versatility of the astronaut, a versatility emphasized by veteran space-policy analyst Marcia Smith.[2]

Initially, the astronaut was just a test object that was hurled aloft in a can that might just as well have had monkeys in it. (Some of the early tests of *Mercury* space capsules did indeed involve monkeys.) The astronaut's role was to test the human response to low gravity: Would he be completely incapacitated and be incapable of operating a spacecraft at all or could he function as a normal human being? Initially, there wasn't even a window to look out of. The astronaut couldn't climb out of the spacecraft unassisted. The capsule splashed down and wallowed in the water, with the astronaut stuck helplessly inside. A gang of U.S. Navy sailors had to unbolt the hatch and pull the astronaut out.

As is the case with many of the other roles of people in space, this one is still with us, though astronauts don't have to put up with the indignities of the early days. A major activity on any space shuttle flight is the continuing investigation of the response of human beings and other living organisms to the effects of zero gravity. One reason for doing these investigations is to determine just what the human factors in spaceflight are, what limitations there are on who can go up into space. But another major factor is to use the microgravity environment as an additional variable in some fundamental investigations of human physiology.

Later on in the *Mercury* program, the astronauts began to take a more active role in controlling their spacecraft. John Glenn, the first American to orbit the earth in February 1962, manually controlled the small rockets that kept the spacecraft from tumbling. But the astronauts still played a very minor role in spacecraft control until the *Gemini* program, a precursor to *Apollo*. In Gemini, astronauts in two-person capsules practiced rendezvous in orbit, and the astronauts became the pilots that they had always wanted to be.

With the lunar landing, the astronaut became an explorer, a scientist who headed off into the depths of space to prospect, to do experiments, and to tell the rest of us what was there. Harrison Schmitt was the only one of the *Apollo* astronauts who was professionally trained as a geologist, but the rest of the astronaut crew picked up an impressive amount of geological knowledge during their astronaut training.

A closely related role is that of experimenter, a person who goes into space to manipulate apparatus and make things. Many materials-processing experiments had been done on the *Skylab* and *Salyut* space stations in the early 1970s. The role of experimenter became firmly established when Charles Walker of McDonnell Douglas became a frequent traveler on the space shuttle. Walker had developed the electrophoresis apparatus for purifying drugs in space, and a question arose about who was going to operate this machinery in orbit. NASA could either send an astronaut to Walker's lab to get training in electrophoresis or send Walker to the astronaut training program. It was decided to train Walker as an astronaut. Many other ''payload specialists'' have been flown in space because of their expertise in using particular types of equipment that are carried aloft on space shuttle missions.

Does one need a scientist-astronaut in order to do science in space? Most often, the answer is no. Remotely operated instruments have grown sufficiently capable so that observations of the earth, the planets, and the distant universe can be done without people. It's best to avoid involving astronauts if possible, because the equipment has to be designed and built much more carefully if its failure could cause loss of life. Astronomers are particularly interested in getting away from people because the environment of a manned spacecraft is severely contaminated by urine dumps. Human movement in a lightweight space station could jar telescopes that are pinpointed on celestial targets trillions of miles away. Even materials science experiments, which usually require human monitoring, can be affected by other human activities. One of the crystal-growing experiments was done on the same space shuttle mission that rescued two communications satellites, and the continual acceleration and deceleration of the shuttle could have been one of the causes of the experiment's lack of success.

But occasionally, a human scientist can make a lot of difference. Earth scientists had listened to astronaut accounts of subtle differences in ocean color for years. The astronauts repeatedly said that photographs taken from orbit, spectacular as they seemed, were only a pale, washed-out rendition of what could be seen by the human eye. Oceanographer Paul Scully-Power seized the opportunity to go on a routine shuttle mission in 1984 when a payload change resulted in an available seat, on three months' notice. He discovered spiral eddies that covered the entire Mediterranean Sea, as well as cloud structures that acted as reliable tracers for ocean currents.[3] Astronomers on board *Skylab* were able to use superior human perception to make excellent drawings of Comet Kohoutek. Enthusiasts of manned spaceflight can go too far. I have heard people using the successes of these missions as a basis for arguing that all future oceanography or astronomy missions must have scientist-astronauts in orbit. Only occasionally do circumstances call for a scientist-astronaut.

The most dramatic addition to the growing set of human roles in space came at the start of the *Skylab* mission in May 1973. *Skylab* was seriously damaged during launch, and the first crew that went to it spent a great deal of effort repairing it. One of the solar panels was torn off and the station's cooling system didn't work properly. The crew pieced together an umbrella of sorts out of spare parts to shade the vehicle and deployed the second solar panel, which didn't work correctly at first. *Skylab,* the first space station, was rescued and was home for three crews that lived in it for a total of six months.

The shuttle astronauts were called on many times to fix and re-trieve errant satellites. The Solar Maximum Mission satellite, a space science satellite designed to study the sun in detail around the peak of the solar cycle when solar activity and flares are most numerous, needed repair when the guidance system failed. In April 1984, the satellite was brought into the shuttle bay after heroic measures slowed its spin down so that the remote manipulator arm could grab it. Shuttle astronauts fixed it and sent it on its way. Only a few months later, a more complicated repair mission succeeded in rescuing the *Westar* and *Palapa* communications satellites, which had failed to reach earth orbit. *Westar* and *Palapa* were not equipped with grappling pins, and astronauts Joe Allen and Dale Gardner had to fly out to them, grab on to them with a specially designed tool that fastened firmly to the satellite's rocket nozzle, and then be pulled by the manipulator arm into the cargo bay.

Allen's gripping description of this rescue mission illustrates two of the intermediate modes of astronaut–machine interaction discussed earlier in this chapter.[4] Allen and Gardner had to use machines in a supporting role when they used the backpacks to fly out to the satellite. But they couldn't guide these massive satellites into the cargo bay by themselves. Rather, they linked up with the remote manipulator arm, which was being teleoperated by another astronaut inside the space shuttle. They then fit the satellites into special fittings and returned them to earth, where they were refitted for launch aboard a later shuttle flight.

Humans who are test objects, pilots, scientists, engineers, or re-pairers are doing things in space, but one of the most important roles of humans in space is to serve as symbols. The symbolic role of people was made clear from the beginning, as Yuri Gagarin, the first person to orbit the earth, was paraded through Red Square to celebrate the Soviet Union's victory over the United States in the space race. Kennedy's response to *Sputnik* had been to send people, not machines, to the moon. When the astronauts landed on the moon, precious moments that could have been spent in exploration were instead devoted to unfurling flags and talking baseball with President Nixon.

Astronauts can use their symbolic roles to sally forth into the cosmos, carrying their national flags where no human has gone before.

The Soviet Union and then the United States exploited this particular human role in space by welcoming astronauts from other countries on various space missions in the 1960s and 1980s. The Soviet Union's ''guest cosmonaut'' program involved sending astronauts from other Soviet bloc countries (and eventually from other countries as well) in what would otherwise have been an empty seat on the *Soyuz* flights that replenished supplies in the orbiting *Salyut* space station. James Oberg has made it reasonably clear that these people were sent primarily as passengers, who may have played a secondary role in entertaining and cheering up the crew, which was enduring a long six-month stay on board *Salyut*.[5]

Astronauts have acted as aggressive, even antagonistic, symbols of national pride and power. But in one case, at least, they served as ambassadors of detente, illustrating how different nations, with different value systems, could work together toward a common goal. In 1976, an *Apollo* lunar module docked with a Soviet *Soyuz* spacecraft in a mission that was simultaneously a symbol of detente, a test of the American and Soviet astronauts' ability to rescue each other, a first step toward possible Soviet-American cooperation in space, and a scientific mission as well. Critics derided this mission as a multi-million-dollar handshake. Senator Spark Matsunaga saw it as a first step on the road toward global cooperation, a step that was not followed up because the ambassadorial role of astronauts can be a double-edged sword. The Carter administration wanted to punish the Soviet Union for its suppression of the Polish Solidarity movement, and a convenient way to do this, supposedly, was to cancel the talks that were planning a space shuttle mission to the Soviet *Salyut* space station. Evidently, the Carter administration viewed Soviet-American cooperation in space as something that was purely symbolic, seeing the multiple dimensions of the space program as unimportant or nonexistent.

One vision of the future sees a joint U.S.–USSR mission to Mars as the focus of joint exploration efforts by both countries. Although part of the justification for such a mission is a purely scientific interest in the exploration of Mars, the science by itself could surely be done less extensively by a series of robot landers and sample-return missions. Many writers advocating the Mars mission see a joint mission to

Mars as a gleaming symbol of what can be done with American-Soviet cooperation. Senator Spark Matsunaga described space in general and a joint Mars effort in particular as a "primary, but not exclusive, opening wedge" in a new approach to relations between the two super-powers.[6]

## FUTURE HUMAN ROLES IN SPACE

Christa McAuliffe was slated to extend the number of roles that humans play in space when she was launched toward orbit as the first teacher in space. Tragically, her career as spacefarer and teacher was cut short in the fireball that destroyed the space shuttle *Challenger.* Her mission was attacked by some as a public relations gimmick, but I believe that the role she was to play was an important and largely overlooked one. Her job on the space shuttle was to involve American schoolchildren in the space program. Her job after the shuttle mission would have been vastly more important, for she was to bring the space experience back to the American people. After the explosion, the many teachers who had competed for her seat into orbit went out into the school systems and did their best to stimulate interest in space.

My work on this book has taken me into the University of Delaware library many times, and I have always been struck by the plethora of travel books about the various expeditions that opened up the globe to western European eyes. Most expeditions across the globe spawned several books by the participants, who were able to share their experiences with others. This hasn't happened with the space program; most astronauts have been mute or inarticulate about their experiences, at least in print or in the mass media. Joseph Allen's magnificent account of the retrieval of the *Westar* and *Palapa* satellites is a conspicuous exception.

Someday, another teacher will go into space and bring back the experience to the average person. Journalists, poets, and musicians may well follow. If these people are carefully selected, they can play the role that has, as yet, no name. Because astronaut Joe Allen has come closest to filling it with his book, I'll call it the *space travel writer,* though someone who really fills the role well will be able to communicate through the broadcast media and in person, as well as in print.

In the immediate future, astronauts will also become builders. NASA plans to construct a space station in orbit, starting in the late 1980s, with an operational station scheduled for 1994. This structure will be carried up in pieces and will be put together like a giant erector set. Initial configurations required many spacesuited astronauts carrying screwdrivers and wrenches on spacewalks, connecting the various modules together with long trusses. With the growing recognition of the capabilities of teleoperated systems like the remote manipulator arm, the station was recently redesigned so that more of the construction could be done by such systems, avoiding the need for spacewalks.

The difficulty with spacewalks is that long preparations are required to acclimatize the astronauts to the different environment inside the spacesuit. The air pressure in the spacesuit (4.3 pounds per square inch, or psi, in the American suits, and 5.8 for the Soviet suits) is considerably less than the cabin pressure, earth-normal 15 psi. The "air" in the suits is pure oxygen, rather than the 20% oxygen–80% nitrogen mix in the cabin. Higher pressure inside the spacesuit would make it hard for a spacesuited astronaut to move because he or she would have to push all that air around to bend an elbow. As is the case for underwater bridge builders, a change in atmospheric pressure can have nasty consequences, called the *caisson disease,* or the *bends.* (Bridge builders work in pressurized compartments called *caissons,* and one of the symptoms of the disease is excruciating pain in the joints, causing the victims to double up or bend over in agony.) What apparently happens is that the nitrogen gas, normally dissolved in body fluids like blood, fizzes out in gaseous form in the same way that gas fizzes out of a soda can when it's opened. The remedy is to allow some time for the body to acclimatize to the changing pressure.

Preparations for an EVA are complex and time-consuming; astronaut Mike Lampton told me that sending someone on an EVA is much more like launching a spacecraft (with a live human being in it) than it is like putting one's clothes on and going outside on a cold wintry day. To prevent the bends, the cabin pressure is dropped to 10 psi at least overnight before an EVA. Two hours before an EVA, the astronauts strap on oxygen masks and breathe pure oxygen at a lower pressure of

3.3 psi. The astronauts then don a liquid cooling and ventilation garment that is worn under the suit. The suit itself comes in three pieces, which can be roughly described as pants, shirt, and helmet. The astronaut puts the suit on, connects the parts together, and puts on gloves, the communications module, and the backpack containing the oxygen supply, then goes out of the airlock and hooks herself or himself to the rocket-powered backpack, or MMU, which is stored in the cargo bay. Two and a half hours after starting, the astronaut is now ready to go to work as a builder. She or he has at most five hours to work before having to go through the process in reverse order (though unless the astronaut develops symptoms of the bends, he or she doesn't have to do the two hours of oxygen breathing again). The whole process is time-consuming and exhausting, and if it can be avoided by increased use of teleoperated systems, so much the better.[7]

## HUMAN ROLES IN SPACE IN THE TWENTY-FIRST CENTURY

Looking ahead to the more distant future, I'll speculate a bit about the roles that human beings are likely to play in the future. The existing roles of test object, pilot, scientist-explorer, experimenter, and repairer will continue. The role of astronauts as pilots may diminish some because the space station will allow for more extended missions. In addition, *Challenger* demonstrated that it is senseless to send a bunch of people up in a space shuttle to launch communications satellites when unmanned rockets could do the job. Teleoperation can also be applied to remotely piloted vehicles, and I think it's likely that an unmanned vehicle, derived from the present space shuttle, will be developed to do the straightforward trucking work. The number of experimenters in orbit will very likely increase because the space station will allow for longer missions, particularly in the materials-processing area.

Another rather widely publicized possible human role in space is that of tourist. Society Expeditions of Seattle, Washington, has collected 265 reservations for an orbital sightseeing tour; the company expects that it will cost $50,000 for a ticket on a twelve-hour orbital flight.[8] However, this company doesn't yet have a launch vehicle to take its passengers into orbit because the space shuttle is too expensive ($1 million per passenger) as well as unavailable for ten years. On a small scale, we've actually had tourism in space, if we count the "guest cosmonauts" in the Soviet space program and the flights of Senator Garn and Congressman Nelson on the space shuttle (though none of these people had to pay for their flights into orbit). If trans-oceanic flight is done in a hypersonic transport that nearly goes into orbit like an ICBM, some fraction of the millions of passengers carried by airplanes each day may be carried into orbit rather than simply around the globe. The history of the air transport industry is grounds for long-term optimism. (Think back: flying in an airplane was simply not done a hundred years ago and was a real adventure fifty years ago, and it's routine now.) No doubt, in the years to come, some entrepreneurs will think of clever (or bizarre?) space-based tourist attractions. Space hotels? Honeymoons in space? Ballet theaters in microgravity? Zero-gravity racquetball? Low-gravity swimming pools? Who knows?

## SPACE COLONIES

The most speculative role for human beings in space is colonization or long-term residence. Princeton University physicist Gerard O'Neill initially promoted the vision of space colonies in an article in *Physics Today*.[9] Several conferences, summer studies, and books followed in the 1970s. As a result, O'Neill's particular concept of a space colony has been fleshed out rather considerably with reasonably complete, if schematic, designs.[10]

The proposed space colony occupies one of the stable Lagrangian points in the moon's orbit. Joseph-Louis Lagrange calculated two centuries ago that there were five special places in a two-body system where the gravitational forces from each body were balanced, in which an object that was placed there would not be pushed away. Three of these points, named *L-1*, *L-2*, and *L-3*, lie along the line between the earth and the moon. Something orbiting in one of these places will be in unstable equilibrium, in the same way that a pencil balanced on its point is in unstable equilibrium. A colony at L-2, for example, could remain where it was as long as it was *exactly* at L-2, but any motion away from L-2 would tend to become amplified. However, L-4 and L-5, the two Lagrangian points located in the moon's orbit around the earth, but sixty degrees ahead of and behind the moon, form an equilateral triangle with the earth, the moon, and the Lagrangian point. L-4 and L-5 are stable Lagrangian points, and material that is in the vicinity of these points tends to stay there. Thus, the original proposal was to put the colony near L-5 or, more precisely, in orbit around L-5 (as the sun would eventually pull a colony away from L-5). Although there is still some debate about whether L-5 is the best point, it has been sufficiently well identified with the space colony so that a group of people supporting the space colonization concept has adopted L-5 as the name of their society.[11]

The scheme is to build a giant colony shaped like a wheel with an outer radius of 0.8 kilometers. The inner edge of the wheel, where an inner surface would be built to hold the air in, is 65 meters inside the outer edge. The consensus of the NASA summer study was that artificial gravity was needed because the long-term prospects for living in zero gravity seem dim. As a result, a rotating system is part of the design. Science fiction writers and visionary scientists have proposed other rotating space colonies, consisting of cylinders, dumbbells, spheres, or toruses in some combination. Planners visualize that 10,000 people could live in such a colony, giving 49 square meters of residential area per person, a figure somewhat in excess of the 38 square meters per capita that is available on Manhattan Island. The colony would have to be surrounded by a relatively massive radiation shield, which would simply be a layer of lunar rock.

The new idea in O'Neill's colony is to use the moon as a source of much of the bulky material that will be needed to build the colony. Hauling stuff up out of the earth's gravitational field requires a great deal of rocket fuel; a much smaller velocity change is needed to move material from the lunar surface to L-5. Since the first of the summer studies (which is the basis for much of the popular literature on these space colonies), it has become clear that, in velocity terms, many earth-approaching asteroids are much closer to L-5 than the surface of the moon is, and that they might be even better sources of raw material. If the material were to be obtained from the surface of the moon, an electromagnetic "mass driver" would be used to boost it to L-5. Small vehicles containing superconducting coils (which conduct electricity with no resistance) are accelerated by carefully timed magnetic fields, generated by "drive coils" that surround the buckets.

There are two basic reasons that the space colony idea still remains a vision on paper: cost and utility. Such a colony would, despite the economies of getting material from the moon instead of hauling it out of the earth's deep gravity well, be expensive. The summer studies led by O'Neill led to a cost of $190.8 billion 1975 dollars, or about $400 billion 1985 dollars. However, these cost figures are based on a shuttle launch cost to low earth orbit of $440 per kilogram, or $200 per pound, which is less than one-tenth the actual cost we are experiencing.[12] Taking the original summer study estimates at face value gives a cost of $40 million per colonist for a 10,000 person colony.

The colonists would have to do something quite worthwhile in order to justify such an investment. In the late 1970s, the proposed activity of the space colonists was to have been the construction and operation of the solar-power satellite system. The vision here is that solar photovoltaic cells in space would generate electrical power, which would be transmitted to earth by a microwave beam. Solar power was a hot energy option in the 1970s, partly because of high oil prices and partly because it was heavily promoted by the Carter administration. The collapse of the OPEC cartel has reduced the economic attractiveness of solar energy. In addition, the Reagan administration has dismantled the tax incentives and the direct government spending programs that supported the study of solar energy in general.

The solar-power satellite system still has a number of problems, even when solar power generated in space is compared with solar power generated on the ground. The total amount of solar energy in space, 1.3 kilowatts per square meter, is only somewhat higher than the total amount of solar energy reaching the ground in a good site like Phoenix (1 kilowatt per square meter). A study by the Congress's research arm, the Office of Technology Assessment, concluded that 26 square kilometers of solar cells at a site in Arizona could produce the same amount of power, on a continuous basis, as 10 square kilometers in space. Thus, we would gain only a factor of 3 in total power output by putting solar cells in space. In order to compete with ground-based solar power, the entire solar power satellite system has to be no more than three times as expensive as a comparable ground-based solar system. In addition, there are a number of potential environmental hazards that the solar-power satellite system has to overcome.[13]

For these reasons, there has been relatively little discussion of the solar-power satellite system for several years. However, a space colony needs something like the solar power satellites in order to justify its existence. Scientific research alone, even if humans were involved much more extensively than has been the case so far in the space program, would not justify a huge colony of ten thousand persons. Science is the ostensible reason for maintaining a continuous human presence in the Antarctic, and the stations there generally house a small number of people (about 18 at the South Pole, and about 125 at McMurdo Sound) during the Antarctic winter. The National Commission on Space, which no one has accused of lacking vision, visualized only (!) comparatively small settlements, with a few dozen people at most, as coming within its fifty-year time horizon.

## THE HISTORICAL ANALOGY

Some space enthusiasts would disagree with my pessimistic assessment of the prospects for space colonization. Frequently, they invoke the Age of Discovery to support their belief that exploration must inevitably lead to colonization. After all, the historic voyage of Columbus was shortly followed by the Spanish conquest of ancient Mexico and the establishment of Spanish colonies in the New World. However, a number of significant differences between the human exploration of the earth and the exploration of the Solar System must be considered. Although I don't always agree with James Van Allen's perspective on the space program, I think he was perfectly correct, if a bit shrill, in stating:

> Fervent advocates of the view that it is mankind's manifest destiny to populate space inflict a plethora of false analogies on anyone who contests this belief. At the mere mention of the name of Christopher Columbus they expect the opposition to wither and slink away. If reference to

Columbus is made in an offhand, thoughtless way, it is incompetent; but
if made with full knowledge of the facts, it is deceitful and fraudulent.[14]

The simple fact is that, by the standards of space exploration, Columbus's expedition was cheap. Samuel Eliot Morison cited the cost of this expedition as two million *maravedis,* which Morison equated to $14,000 in 1942 dollars.[15] Prices have increased by about tenfold since then, so we're talking about something like $100,000 to $200,000, something which an individual could finance a substantial piece of. In fact, Columbus and his friends did pay for about a tenth of the cost of his trip. For comparison, the cost of the *Apollo* program was about $40 billion 1985 dollars, 200,000 times the cost of Columbus's expedition.

Freeman Dyson has done some similar calculations to try to estimate the costs of the *Mayflower* expedition that brought the Pilgrims to Massachusetts and the migration of the Mormons to Utah led by Brigham Young. He estimated the costs (in 1975 dollars) to be $6 million and $15 million, respectively, which correspond to $11 million and $28 million in 1985 dollars.[16] Both of these were colonizing expeditions, unlike Columbus's exploratory expedition. But still, by comparison with the *Apollo* program and O'Neill's L-5 colony, these expeditions were cheap.

In fairness to Dyson, I should point out that he regards *Apollo* and L-5 as examples of an extravagant style of space colonization. The large cost of *Apollo* and L-5 illustrates the extreme difficulties of getting out of the earth's deep gravity well. Dyson went on to point out that someone living on L-5 who wanted to go onward to the asteroid belt could do so at very low cost, citing O'Neill's figure of $1 million per person. However, the high cost of getting away from the earth is not just due to the need for a government project like *Apollo* to triple-check everything to be sure that human lives are lost as rarely as possible. It is simply true that living in space requires a lot of hardware and that lugging that hardware from the gravity well of the earth costs a lot of money.

Columbus had much that we lack in the early days of the space age. He had an ocean on which his ships could float, rather than empty space. His ships sailed through an atmosphere which supplied rainwater to drink and wind to push his sailing ships around. His crew could breathe this air and could find food on the shores of the distant land that was his destination. Going into space, we have none of these creature comforts, so that it is harder and more expensive to get there.

Yet, we have something that Columbus didn't have: knowledge. We can send our robot probes to Mars, Venus, or wherever and find out what it's like there. Columbus had only driftwood from Central America, which fetched up on the shores of the Azores, as a clue to hint at land westward. Perhaps he even had access to reports of the Leif Ericsson expedition in the Vatican (as has been recently alleged); perhaps not. But no matter, he knew little or nothing of what lay to the west. In contrast, we need not rely on explorers to sail up and down faraway coasts and send back reports, often exaggerated, on what's there (as was the practice in the hundred years after Columbus). We can find out what's on Mars and find that it is far more forbidding than the Antarctic.

As a result, there is a distinct difference between our situation about thirty years after the opening of the space age and the situation of Spain thirty years after Columbus, in 1522, awaiting the return of Magellan's expedition around the globe. We are a wealthier society, to be sure, but each expedition can consume much greater amounts of resources. We know much more of what we will encounter in the depths of the Solar System. The planets are fascinating worlds, but worlds that are hard for human beings to live on. The challenge of establishing outposts on Mars will exceed the challenge of the Antarctic by a considerable factor.

I should not leave Columbus before recognizing the similarities between the Age of Discovery and the present situation. Space is a multidimensional activity,[17] and there are some respects in which the astronauts' role as space pioneers parallels that of Columbus. The symbolic role of the early human explorers of the earth, carrying their native or adopted national flags to distant lands, was an important part of their ability to persuade their governments to finance their expeditions. Similarly, the prospects of the hammer and sickle fluttering in the thin winds of Mars when the first Soviet manned expedition lands on the Martian surface may yet prove to be the stimulus that will lead to the adoption of a human expedition to Mars as a national goal of the United States.

Where does this leave us in considering the future human role in space? The analogy with Columbus is imperfect at best. The costs of space exploration are much greater, but not so great that governments cannot bear them. Automated and teleoperated equipment can tell us a lot more about what's out there, so that we do not need to be in such a rush to send people. Yet the need for a symbolic ambassador of humanity is still one that is rooted deep within the human soul. So, to conclude this chapter, let's toss cost questions aside and ask what fundamental limitations there are on the human capability to venture into space.

## LIMITATIONS ON THE HUMAN ROLE IN SPACE

Space is a hostile environment. There is no air. In the sun, one fries, and in the shade, one freezes. Water evaporates very rapidly. Beyond the protective shield of the earth's inner magnetosphere, radiation doses from solar flares could kill an unprotected human being, and the fragile shield of a spacesuit is little protection. How can one possibly visualize people as test objects, pilots, scientists, explorers, engineers, repairers, builders, tourists, and possibly colonists or settlers? What limitations are there on the roles that people can play in space?

As is the case with much of the space program, the procedures adopted in the early days cast a long shadow in spite of all that we have learned since. The original seven astronauts selected for Project Mercury went through a very severe screening process. Members of the military were used to taking physicals, but this was a bit more than the usual. Doctors stuck electrodes into their thumb muscles and then zapped their hands to test reflexes. They were indeed treated like "lab rats," to use Tom Wolfe's term.[18] The seven that emerged were regarded as prize physical specimens, and the myth that astronauts have to be exceptionally fit has persisted to this day.

However, it has turned out that the rigors of zero gravity are not all that bad in the short run. There are a number of minor inconveniences. Half of the astronauts get spacesick, succumbing to what NASA delicately calls the "space adaptation syndrome," or SAS. Gravity no longer drains fluid to their feet, so that their faces and eyes

get puffy, distorting facial expressions and the usual nonverbal communicative clues that we rely on in trying to interpret what someone
says. People in space get taller (about half an inch) and leaner, losing
about 3.4% of their body mass in flights lasting up to two weeks. But
none of this is debilitating, and NASA's physical requirements have
been considerably relaxed. One still needs to be in reasonably good
shape to go up, but anyone who can pass a normal physical exam can
probably pass the one given to payload specialists. The shuttle in
particular makes the job of being an astronaut easier, for takeoff involves only a maximum, momentary acceleration of three times the
earth's gravity just before the external fuel tank is discarded. Rockets
were far worse.[19]

There are still a number of potential difficulties in long-term
missions that need to be faced if we are thinking seriously about
extended stays in space. The time required for a trip to Mars varies a
bit depending on how it's done, but nine months in space on both
outbound and return journeys is typical for most mission scenarios. A
variety of problems, ranging from the purely physical to the purely
psychological or sociological, remain to be solved before we can confidently say that human beings can stand the rigors of the kind of long-
duration spaceflight that is required if we are to send people to Mars
and beyond, to settle the inner Solar System.

American astronauts have had little experience with the rigors of
extremely long-duration spaceflight; this is one area in which the Soviet space program is clearly well ahead. The longest tour of duty on a
space station was the three-month stint for the last crew that stayed in
*Skylab.* However, many Soviet astronauts have spent nearly eight
months in the *Salyut* space station. One cosmonaut, Valeriy Ryumin,
was up for six months and then went back to the same station on the
next mission for another six-month tour. Even when the American
space station is in orbit, current plans still call for three-month tours of
duty for the American astronauts.

There is one purely physical problem of long-duration spaceflight
that could prove fatal to missions to Mars and beyond. The critical
mineral calcium is lost from bone mass, and there is no indication that
this loss levels off over time. Two to four percent of the calcium was
lost from the bone mass in the *Apollo* and later *Skylab* flights. Calcula-

tions based on the *Skylab* data indicate that a one-year flight might result in the loss of 25% of the calcium in the body pool.[20] Once the astronaut is returned to an environment with gravity, the lost calcium is regained, rather slowly. But because no one has followed a one-year spaceflight, it is quite possible that long-duration spaceflight could result in irreversible skeletal damage.

Long-duration space missions would benefit greatly from the development of a system that would allow us to recycle more of the stuff that is used by humans on a space station. Buckminster Fuller used the phrase "spaceship earth" in the 1960s to sensitize the environmental movement to the closed nature of our ecosystem. On earth, nothing is thrown away. Unfortunately, this phrase evokes a picture of a spaceship that has nothing to do with reality, for contemporary spaceships throw a lot of things away. We are far from having an "earthlike spaceship." In one day, a person in space requires roughly 0.6 kilograms of food (dry weight), 0.9 kilograms of oxygen, 1.8 kilograms of drinking water, and 2.3 kilograms of sanitary water, for a total of 5.6 kilograms. In most spacecraft, additional water is needed simply to keep the spacecraft going; a recent report cited a figure of 16.8 kilograms per person per day.[21] Outputs per person per day typically consist of 1 kilogram of carbon dioxide, 0.2 kilograms of feces, and a lot of water, which comes out as perspiration, exhalation, and something like 1 kilogram of urine.

In many spacecraft, all of this stuff was simply launched into orbit, used as needed, and brought back to earth or dumped over the side. On the earth, these materials are cycled through the biosphere in various cycles that transport water, carbon, oxygen, and foodstuffs around and around again. The challenge to a spacecraft designer is to build equipment that will allow us to "close the loop" as much as possible, to recycle these materials in somewhat the same way that the biosphere does.

We are far from closing any of the life support loops in current spacecraft. Once again, the United States experience with long-duration spaceflight is confined to the *Skylab* mission; the space shuttle missions are so short that making any great effort to close these loops is not cost-effective. *Skylab* was a completely open-cycle system; 3 tons of water, 2½ tons of oxygen, nearly 0.75 tons of food, and 0.5

tons of nitrogen were sent up on the first mission and consumed as needed. The shuttle does a limited amount of "recycling," in that water for the spacecraft thermal-control system and for drinking is supplied by the fuel cells that power the spacecraft. (In a fuel cell, hydrogen and oxygen are combined to make pure water.) This is the only time that water is used twice on the shuttle; all the water used by humans and collected from the air in the spacecraft is then dumped over the side, creating a little cloud of contaminants around the spacecraft. Virtually all of the American investigation of closed-loop life-support systems is at a very early stage.

The Soviet Union has built some partially closed systems on the *Salyut* space station; for example, there are reports that they use the condensate produced from the cabin air for drinking water.[22] Because of a lack of routine communication between Soviet and American space scientists, we know little of the details of their operations, and on the basis of what we do know, their cycles are only very partially closed as well. *Soyuz* spacecraft continually fly into orbit to resupply the *Salyut* space station with food, water, and other consumables. Buckminster Fuller's analogy, poetic as it is, could not have been more wrong. If we are to go to Mars, we have to get much more practical experience with closed-loop life-support systems. Perhaps the easiest way to get experience with such technology is to cooperate more with the Soviet Union, which has had some real working experience with partial recycling.

Another problem of long-duration spaceflight that is partly physical and partly environmental has to do with exercise. The loss of lean body mass persists throughout a spaceflight. This corresponds to a deterioration in the condition of the muscles, which don't have to work to hold the body up in space. There is less concern about the possible permanence of such loss because it's easy enough to gain the weight back once the astronaut returns to earth. But on something like a Mars trip, the astronauts couldn't spend several months on Mars as torpid, debilitated creatures, waiting for their bodies to get back in shape. Exercise on board would have to keep up the muscle tone.

Exercise in space has proved to be relatively unpopular with the astronauts. Part of the problem is that there is no room to do anything more than ride a stationary bicycle or walk on a treadmill, but another difficulty is that, once they work up a good sweat, it is hard to get really clean. Soviet astronaut Ryumin wrote in his memoirs, ''Besides being simple hard work, it was also boring and monotonous. . . . In space, drops of perspiration do not drip off the body, but take on a spherical shape without coalescing. These water 'peas' are collected with a towel.''[23] The problem is made even worse by the inadequate shower facilities in space, which don't allow the astronauts to wash the sweat off easily. Water doesn't run down to the floor, so some kind of vacuum system has to be devised to collect the water and prevent it from running all over the cabin. *Salyut* and *Skylab* had similar showers, consisting of long plastic cylinders with vacuum holes in the bottom to suck the water out. The cylinder has to be dried carefully or it will mildew when it is put away.[24]

A related hygienic problem is the difficulty of devising a toilet that works well enough in zero gravity. Catheters and plastic bags could work for the brief all-male missions of the *Apollo* program. The shuttle has a toilet that works for both sexes, after a fashion. From the outside, it looks like a standard facility with foot restraints and a seat belt to keep the astronauts from floating away. Suction pulls the waste down toward the floor. When the toilet is flushed, a set of tines shreds and slings the fecal matter to the side of the toilet bowl. Hence the name of this device: the slinger. (Or, to put it more accurately, the shit hits the fan.) The ''flushing'' is finished when what remains in the toilet is vacuum-dried to its walls. Obviously, the gradual buildup of junk on the walls can go on for only so long, and this design wouldn't work for a Mars mission. Apparently, this design has given serious problems in actual use and has required a good bit of cleaning.

Anyone who has been in a confined space where the toilet doesn't work knows how irritating the smell is. A space environment is particularly confining and has its own set of problems. The windows of the cabin on the space shuttle can't be opened to air the space out if the toilet perpetually stinks. Furthermore, people can't burp in space because gas does not naturally flow to the top of the digestive tract, and so all the gas comes out the other end, adding to the smells. It would be surprising if toilet design turns out to be an insurmountable obstacle to extended spaceflight, but it needs more work if the space environment is to be made pleasant enough so that people can stand it for long periods of time.

## SMALL GROUPS IN
## CONFINED PLACES

These hygienic problems are partly physical, partly engineering, and to a great extent psychological. In the past, the American space program has paid little attention to the psychological side of the space program, partly because the original "right stuff" astronauts thought of the psychologists as nosy nerds who couldn't fly a paper airplane six inches, but who did have the power to bump astronauts off the flights that they really wanted to go on. In addition, people can stand just about anything for a week or two, and except for the three *Skylab* missions, all of the American spaceflights have been short. NASA is looking to the future, though, and has established a human factors team at the NASA-Ames Research Center to consider the psychological problems that will face crews living in the space station for three months and that may face the people who go on a more extended mission to Mars.[25]

Submariners and people who winter over in the Antarctic face an environment somewhat similar to the environment of space. About eighteen people live at the South Pole Station in the Antarctic; this station closes in February and opens in November of each year. For nine months, these people have only each other to live with. The Bartol Research Foundation of the Franklin Institute, which has offices in the same building that I work in, regularly sends a cosmic ray observer to the South Pole. I talked to Don Kent of Bartol about the insights that he has derived from helping to select those cosmic ray observers.

Some of what we've learned from the Antarctic is not very surprising. "Almost any stresses are magnified," Don pointed out to me. "Things you wouldn't worry about in a normal environment have to be considered in amplified terms. When you interview candidates, you have to take into account any irritating habits they might have."[26] NASA planners have discussed the development of an "annoyance questionnaire" that could be used both ways: to prevent selecting people with traits that will irritate others, and to prevent selecting people who are so finicky that any human frailty will drive them up the wall.[27]

An interesting aspect of the Antarctic experience that may or may not be applicable to the space program is the blurring of roles that takes place over the nine-month Antarctic winter. About half the eighteen people at the South Pole are scientists, selected by independent organizations; the remainder are support personnel provided by companies with government contracts. I would have expected that a division between the scientists and the support personnel would persist for a period of time, in the same way that the payload specialists are not seen (and do not see themselves) as "real" astronauts like the pilots and the mission specialists, who know much more about the ins and outs of the space shuttle. However, Don Kent tells me that cliques tend to form and dissolve over the nine-month Antarctic winter, with no split between the scientists and the support people. Perhaps this happens because the menial chores like snow shoveling are rotated so that the scientists do them just as anyone else does. It's possible that, in space, where people are trained for specific tasks, crews may remain split even on long-duration missions. However, the need to train people as backups may cause roles to be blurred as the mission progresses.

A small group engaged on a mission like a stay on a space station or a flight to Mars needs to have people engaged in a number of different roles. Sociologists recognize a clear distinction between a "task leader," who parcels out various chores to help the group get the job done, and a "socioemotional leader," who promotes harmonious relations within the group. Only rarely does the same person fill both roles. In the Antarctic, a station manager is appointed as the task leader. The identity of the socioemotional leader or nurturer varies. Sometimes, it's the cook: food takes on more than a nutritional role on such expeditions. Sometimes, the physician is the socioemotional leader. Additional roles in small-group interactions require additional people. One widely mentioned role is that of the clown or jokester, who can keep up morale, remind people not to take things too seriously, and occasionally use his or her position to challenge the conventional wisdom.

So far, space crews have generally consisted of individuals chosen separately who have happened to be thrown together on a particular mission. Last-minute flight assignments in both major space programs illustrate that interactions between crew members receive little consideration. Ryumin, who made two six-month flights back to back, was assigned to the second flight at the last minute because the cosmonaut who was supposed to make the flight was injured. The American space shuttle program has been plagued by last-minute changes in flight manifests and corresponding changes in crews. Perhaps, the relative homogeneity of the astronaut corps has produced no ill consequences from the lack of concern with group dynamics. Perhaps, task leaders, socioemotional leaders, jokesters, and the rest naturally emerge from any particular group of suitably qualified people. My friends at the Bartol Research Foundation make their own decision about whom to send to the South Pole with no knowledge of the makeup of the rest of the group, no idea of what role their chosen candidate will fill in the group's dynamics.

For a year-long mission like the trip to Mars, another issue in group dynamics emerges: sex. Public discussion of this issue in the context of the space program has been muted because the official Soviet and American cultures are puritanical and because the circumstances in existing space missions have discouraged, if not effectively prohibited, sexual expression. However, as space psychologist Yvonne Clearwater wrote, "In space, regardless of crew gender composition, if we lock people up for 90-day periods, we must plan for the possibility of intimate behavior."[28] When I talked to Don Kent about the Antarctic experience, the issue of sex rarely came up; he seemed to feel that it wasn't his business to be overly concerned about how things worked out between consenting adults. I agree, and I feel that the possibility of sex in space has simply to be accepted and then ignored as much as possible. There's little mission planners can do about it anyway. Assigning single-sex crews to long voyages will not eliminate the possibility of sex between crew members, and college students (and others) have demonstrated that no one can design living quarters in such a way as to make sex absolutely impossible.

In summary, then, the only insurmountable barrier to long-duration space travel seems to be the gradual loss of calcium from astronaut's bones. There are a number of other problems relating to exercise, hygiene, and group dynamics that haven't been solved yet, but that seem to be the sorts of problems that can be worked around in time. *If* we can get around the bone problem, the other barriers to long-duration space expeditions seem to be less fundamental.

*    *    *

Human beings have played a variety of roles in space. In the past, we humans have been spacecraft controllers, teleoperators, test objects, pilots, scientists, explorers, experimenters, and repairers. Builders and travel writers will venture into space in the near future to accomplish their respective tasks. It is possible, though less certain, that the cost of space travel will be reduced to the point where space tourism will be economically attractive.

More speculative are the prospects for space settlement and colonization. Large-scale colonies like those suggested by Gerard O'Neill seem to be quite far away. If there is to be settlement of the Solar System in the next fifty years, it will probably take place in the Antarctic mode of a few dozen people living in a space station or outpost. These people will not be permanent settlers but will be temporary space dwellers, out there for a shift of a year or two. Most visions of the near future place such settlements in near-earth space, either on a space station in orbit or on a lunar base on the moon, or both.

# CHAPTER 14

# Permanent Stations in Space and on the Moon

The comparison between the costs of Columbus's expedition and the *Apollo* program highlights the most significant barrier to wider use of space: its high cost. Although the costs of the American civilian space program are less than 1% of the entire federal budget, these costs are a very significant barrier to private participation in the space program. The costs must come down if space is to be developed in a free-enterprise mode.

It costs a lot to haul people and equipment up into orbit and to bring them back down again. Is there some way that we can avoid sending people and machines up and down the highway to space, back and forth, practically commuting into earth orbit at a cost of hundreds of millions of dollars per trip? The apparently simple solution is to leave people, machines, or both in orbit for longer periods of time, even indefinitely. The earliest space pioneers visualized that some kind of permanent structure or space station would be constructed in orbit—or would be constructed on the ground and left in orbit—at a fairly early stage in the development of a space program.

The construction of a space station and associated orbiting platforms is to be the next major step in the American space program. President Reagan established this as the next relatively short-term goal for NASA in his 1984 State of the Union message. Development of the space station, like everything else, has been slowed by *Challenger*. However, because there is no fixed design or size for the space station, its configuration may be adjusted to take account of the changing circumstances and can still allow it to be deployed on schedule beginning in 1992. The current goal is to have a station that can be permanently inhabited starting in 1994, and that will continue to grow in size thereafter.

NASA's vision is that the space station will open the door to a much broader participation in space activities. Some of the people using the space station will be those who are currently using shuttle flights or free-flying satellites to do the variety of space investigations described earlier in this book. If that's all that the space station is used for, it will fall far short of its potential. The real motivation for constructing the space station is to encourage a much broader range of space activities by reducing the cost, the lead time, and the expertise required to participate in the business of space.

## WHY BUILD A SPACE STATION?

If access to space is to become more routine, the difficulty of getting into space and working in space needs to be reduced. Currently, someone wishing to do something in space needs to build a set of hardware, often completely from scratch, and then compete with others to obtain a scarce launching opportunity on the space shuttle. Let's take materials processing in space as an example. The bulky electrophoresis apparatus, constructed at great expense by McDonnell Douglas, has to be launched into space at a cost of several thousand dollars a pound. (Although it is true that NASA now pays for the launch cost because the apparatus is being used for research, the arrangement between McDonnell Douglas and NASA calls for McDonnell to pay the launch costs when the apparatus is used for production.) The apparatus is used for a week and is then brought back to earth. When the results of that run are fully digested, it is launched once more, again incurring a cost of several thousand dollars a pound, and once again, it is used for a week.

McDonnell Douglas recognizes that there is no way to run a production facility under such circumstances. Some kind of structure is needed that would allow them to leave the processing apparatus in space and send astronauts and materials to it from the ground. The structure needs to have a volume of space sufficient to contain the furnace and the operator. In the jargon of space stations, this means that "pressurized volume" is needed. McDonnell is gambling that some type of space station or platform will be available when the time comes to produce commercial, marketable quantities of drugs in space.

Let's take astronomy as another example. When hundreds of millions of dollars are spent to launch a one-meter telescope into earth orbit, the cost is not for the telescope itself. Most of the cost is for the spacecraft, consisting of a pointing system, a power system, a data-handling system, a radio antenna that will relay the data to earth, a thermal control system, and so on. When the mirror coating degrades so the telescope becomes useless (as was the case on the *Copernicus* satellite), the spacecraft is no longer used. The pointing system and all of the other stuff still works. What would make space astronomy much easier is an orbiting platform to which a telescope could be bolted, plugged in, and let work for several years.

These two illustrations indicate the kinds of things a space station can do. They come under the general category of *space infrastructure*. On earth, the word *infrastructure* is applied to things like roads and bridges, which all of us use for common purposes. The space shuttle is an example of that kind of infrastructure, a transportation infrastructure. Because we can't simply pitch a tent in orbit and live in it, the provision of the appropriate rentable living and working space also falls into the same category of infrastructure.

These two examples also illustrate the large number of different tasks that people will ask for from space infrastructure. Materials processing requires substantial amounts of pressurized volume and, for the early stages at least, continuous human presence to monitor the furnaces. Space astronomy, in contrast, simply requires an orbiting platform, which should be located at some considerable distance from an inhabited structure because the environment around something with people in it is not at all the environment wanted for astronomy. Water is being dumped out of a pressurized volume with people in it all of the time, and the motion of astronauts inside the station makes it jitter around and disrupts the precise pointing that is one of the reasons for doing astronomy from space in the first place. Other space sciences and applications have different needs as well. The communications business requires ways of getting to and from geosynchronous orbit to refuel and refurbish communications satellites. Earth watchers want to be in polar orbits so that their cameras and radars can view the entire earth.

This variety of activities that could fake advantage of a permanent space infrastructure illustrates that there is no such thing as *the* space station. Examples of permanent space infrastructures could range from simple, astronaut-tended platforms to extensive, permanently inhabited multielement volumes. From this perspective, the Hubble Space Telescope could be thought of as a kind of space station because it is intended to be permanently in orbit, to be used indefinitely, and to be refurbished by shuttle missions as needed. However, given the *Apollo* heritage, and given the special role of people in space, the term *space station* is generally conceived of as meaning something that includes pressurized volume with people living in it. Reagan's State of the Union message specifically mentioned a manned space station.

A number of manned space stations have been operated by both of the major space countries. NASA's space station, as now conceived, has capabilities that range far beyond those of past and current stations. It will be worth our considering the types of stations currently in use as a historical background to further space station projects.

## PAST AND PRESENT SPACE STATIONS: *SALYUT*, *SKYLAB*, AND *MIR*

The first space station launched into orbit was the Soviet Union's *Salyut 1*, placed in orbit in 1971. *Salyut* is a cylindrical space station, 13 meters long and 4.2 meters in diameter at its maximum, with 100 cubic meters of usable volume. Astronauts are carried to and from *Salyut* in a *Soyuz* spacecraft. The *Soyuz 10* astronauts failed to get the hatch open on *Salyut 1*, but *Soyuz 11* carried three astronauts who spent three weeks in orbit in 1971. Unfortunately, these three astronauts were killed on their way back to earth, and the Soviet space station program was set back several years. The Soviet space station program resumed with *Salyut 3*, an apparently military space station launched in 1973. Since then, a series of four additional *Salyut* space stations have been launched at intervals, the last of the series being *Salyut 7*, launched in 1982.[1]

On the later *Salyut* missions, the crew stayed up in orbit for a very long time. Resupply flights were made by a spacecraft called *Progress*, and frequently, subsequent *Soyuz* missions flew to the *Salyut* space station, bringing some fresh supplies and a couple of additional astronauts, who would stay in orbit for a week or so and then go home. These visiting *Soyuz* astronauts contributed to the overall mission in a number of ways. They could carry supplies into orbit and bring back some of the results of orbital sciences, such as film packs and samples produced in the materials-processing facilities. They could leave a new *Soyuz* in space, attached to the station, and come back to earth in an old one. (The *Soyuz* spacecraft apparently has a limited lifetime of ninety days.) The *Soyuz* flights provided the Soviet Union with an opportunity to fly astronauts from other countries as a diplomatic gesture of international friendship. The visitors also had a role in providing a slightly more varied environment for the two *Salyut* astronauts, who had only each other for company for six months and more. *Salyut* 6 and 7 had two docking ports, in contrast to the single port for *Salyut 1–5*, which had only a single docking port.

Astronauts on board *Salyut* carried out a number of space science and applications experiments. A number of telescopes measured radiation from stars and galaxies in the ultraviolet and X-ray spectral ranges. The cosmonauts made a variety of earth resources observations and high-resolution photographs, often coordinated with aircraft flights. Large crystals and well-mixed alloys were made in a furnace. Plants and fruit flies grew in a microgravity environment. Despite this admirable variety of space activities, the achievement of *Salyut* that has had the most scientific impact outside the Soviet Union has been the experience of very long-duration spaceflight. If humans are ever to travel to Mars, the grueling endurance tests of *Salyut* will provide some critical experience with the response of human beings to prolonged weightlessness.

Three weeks after the *Challenger* accident, on 20 February 1986, the Soviet Union launched a new type of space station called *Mir*. It's not much larger than *Salyut*, being 17 meters long rather than 13 meters. However, there are a number of indications that *Mir* may be the nucleus of a significantly larger space-station complex. The dining area can accommodate six people, and two small rooms like dormitory singles can give the astronauts rooms of their own. Its size alone hides the fundamental difference between *Mir* and *Salyut*.

The launching of *Mir* was apparently the well-publicized event that was to mark the twenty-fifth anniversary of Yuri Gagarin's orbital flight and to highlight the difference between a Soviet space program that was slowly forging ahead and an American space program that was on hold. Soviet television broadcast the ascent of *Mir*'s first crew as it was happening on 13 March, in a rare display of Soviet openness. The American press virtually ignored *Mir*. Perhaps it was mired deep in the gloom of *Challenger*. More likely, the distinction between *Mir* and *Salyut* is subtle enough to slip by newspaper editors, who decide what is printed and what isn't.

Many people who work in the news media seem to be obsessed with scorekeeping, reporting on such things as satellite size and weight. I sometimes feel that, if they can't reduce an achievement to something that is as simple as a baseball score, they view it as being not comprehensible, and it's therefore not reported. *Mir* does have six docking ports, as compared to *Salyut*'s two.

The difference between *Salyut* and *Mir* is much greater than the score in docking ports of 6–2 might indicate. Mir may represent a fundamental change in Soviet space station technology. Most of the time that *Salyut* was up there, it was the primary if not the only piece of usable machinery in orbit. All the scientific equipment that was used for space station science was inside, making it very crowded. The two docking adapters on *Salyut* 6 and 7 allowed for only a minimal amount of expandability. These ports were largely occupied by the *Progress* and *Soyuz* craft, which resupplied the station with consumables and astronauts but did not increase the number of things that could be done with or inside it. However, on *Mir,* the six docking ports will apparently permit the attachment of a variety of modules for living and working. Astronauts Kizim and Solovyov built a fifty-foot tower on the first *Mir* mission, further suggesting that *Mir* could be the nucleus of an expandable, modular space station.[2]

While the *Salyut* program was moving ahead in the 1970s, the

United States orbited one single space station: *Skylab*. The competition between the United States and the Soviet Union in the 1970s and 1980s has often been likened to the race between the tortoise and the hare. This analogy is far from perfect because there are many dimensions of space activity, and the two nations are often pursuing different goals. The notion of a single "space race" to a single goal is one of the many legacies of the *Apollo* program that are simply not applicable to the present era.

If you restrict your attention to the goal of building a space station, the tortoise–hare analogy is really quite apt. *Skylab* was well ahead of the Soviet *Salyut* program in measurable things like the amount of pressurized volume. It was launched in 1973, was visited three times, and was then forgotten. If the American space station program meets its timetable, twenty years will have passed between the last visit to *Skylab* in 1974 and the first crew to visit the American station in 1994. In the meantime, the Soviet Union was sending up *Salyut* after *Salyut,* visiting them repeatedly with two-man astronaut crews, and gradually accumulating a great deal of working experience.

The *Skylab* space station was a derivative of the *Apollo* technology and included an *Apollo* module as the way of transporting the crew to and from earth orbit. The station itself was a modified upper stage of the giant Saturn V rocket, which launched the *Apollo* astronauts to the moon. The leakproof hydrogen fuel tank was transformed into living and working quarters. It was huge by comparison with *Salyut*. Including the *Apollo* command and service module, the multiple docking adapter, the airlock module, and the orbital workship, the entire assemblage was 36 meters long and 6.6 meters in diameter and contained a working volume of 339 cubic meters. The corresponding figures for *Salyut* 6 are 29 meters long (with two *Soyuz* spacecraft attached; the station itself is only 15 meters long), 4.15 meters in maximum diameter (with two compartments being only 2 meters and 2.9 meters in diameter), with about 100 cubic meters of working volume (even the two *Soyuz* spacecraft add only about 8 cubic meters of space). For comparison, a good-sized living room (16 by 20 feet with an 8-foot ceiling) has a volume of 75 cubic meters, a little smaller than the entire *Salyut* complex.

The 171 days in which three crews occupied *Skylab* were tremendously productive ones. The *Apollo* telescope mount program returned a wealth of data on solar physics and solar activity. The first crew had to repair *Skylab* because one bank of solar cells was ripped off in launch and the other did not deploy correctly. This repair dramatically demonstrated this particular human role in space. The *Skylab* astronauts suffered no serious physiological or psychological effects as a result of their 84-day longest mission. We developed some of the support systems, such as toilets and showers, that would be needed to build a more permanent space station.

But *Skylab* didn't last. Any space station orbiting at the altitude of 435 kilometers (270 miles) will eventually have to be continually boosted to maintain its orbit, because atmospheric drag will cause it to do a slow death spiral toward earth. It's possible that the space shuttle could have rendezvoused with *Skylab* and attached a little rocket to it that could boost it into a higher orbit or, alternatively, intentionally dump it into an uninhabited area like the Pacific Ocean. The shuttle program suffered delays. In addition, the sun produced an unusually large number of flares as the solar cycle moved toward its peak in 1979–1980. Solar activity heats the earth's upper atmosphere, causing an increase in density at high altitudes like that of *Skylab*. As a result, no shuttle was ready in the summer of 1979, when *Skylab* finally descended to orbit. Most of *Skylab* fell into the Indian Ocean, but a number of pieces hit the thinly populated western region of Australia.

The *Skylab* program gave America a jackrabbit start that led rapidly to limbo. By the time it flew in 1973 and 1974, it was clear that NASA had the resources to carry on only one major project, and that was the space shuttle. The *Skylab* legacy is very much with us as we look to the development of a space station in 1994 because it is the only American experience with long-duration stays in orbit. Our ability to benefit from the Soviet experience is limited because there is currently no framework for formal American-Soviet cooperation in space, and we lack the sort of detailed knowledge of their experience that would be necessary if we were to use it as a basis for constructing a space station.

## GENERAL CHARACTERISTICS OF THE SPACE STATION

The design of the space station is not yet fixed. The *Challenger* disaster highlighted some of the concerns that were being voiced within NASA regarding the number of space shuttle flights and spacewalks (EVAs) that would have been required to build the space station as it was originally conceived. The term *spacewalk*, widely used in the media, makes an EVA sound deceptively simple, something like going out on a cold day, where it takes, say, ten minutes at most to suit up. However, the complicated prebreathing procedures make an EVA an endurance test, not a "walk."

Another reason for taking some time to design the space station is that there is an enormous variety of uses that it can have. Many early conceptions of the space station apparently drew very heavily on the early literature on space travel, when computers were basically glorified adding machines and teleoperation or automatic control of instruments had never been heard of. In these visions, the entire space infrastructure would be taken up with habitable, pressurized modules. The only things outside these modules were a few accessories sticking out of the impenetrable walls. One reason that the scientific community was initially unenthusiastic about the space station was that the initial conceptions of the space station did not take the special needs of some scientific users into account. Astronomy, for example, is difficult to do if the instrument is located ten meters or so from a manned module that is dumping water and other contaminants overboard all the time, creating its little cloud of an atmosphere around the inhabited module. Microgravity experiments could be seriously upset if the space station had to maneuver to go and fetch a satellite in an errant orbit.

During the early 1980s, many of these conflicting needs and uses of a space station were gradually worked out and a number of features that any space station will have, have begun to be established. The first characteristic is that the station will be modular and expandable. Here, the space station envisioned by NASA is a great contrast to *Salyut* and *Skylab*, where the can that was initially launched into orbit was the same can that astronauts lived and worked in. Because only a small station can be put into orbit with a single launch, the size of the station was restricted, and the American *Skylab* astronauts lived in a king-sized can and the Russians on *Salyut* in a pint-sized can. But cans they were, and the living and working space was quite limited. The *Mir* space station may turn out to be expandable in the same way that current designs of the American space station will be.

The basic concept for the construction of the American space station is that several shuttle flights will be required to haul the pieces into orbit and put them together into an initial space station. A recent proposal suggests the possibility that expendable rockets can be used to launch some components during the assembly process, and the shuttle flights would be used primarily to put the astronauts in orbit so that they could put the entire erector-set-like structure together. The number of shuttle flights required to establish an operating space station is not set yet and will depend on whatever is decided about an initial crew size. I rather suspect that these decisions will be governed primarily by budgets and timetables. Reagan's 1984 State of the Union message set "within a decade" as the goal for space station operations, and an $8-billion budget ceiling for the cost of the initial space station was initially set by the political process. Clearly, some kind of large space station, hundreds of meters in its outer dimensions, can be constructed for that kind of money. A report in late 1986 suggested that something like seven or eight shuttle flights would be required to establish an initial manned capability.[3] I suspect that the initial crew size will be around four people. The cost may well go up.

A number of kinds of modules are required for this initial operating capability. We need an orbiter docking module so that we can get people and consumables to and from the station. The remote manipulator arm and some manned maneuvering units would be sent up early to help with assembly; robots or teleoperated assemblers may also be sent up in early stages so that much of the assembly can be done without EVA. A utility module including solar arrays will provide electrical power, thermal control, data-handling capability, and attitude control for the station. The habitability module is the place that the crew will live in, containing bunks, a galley (nautical terms are always found in space, and for landlubbers this means the kitchen), a wardroom (dining room), an exercise area, the bathroom (delicately called the "personal hygiene compartment"), and other places to store stuff. This module may well be separate from the laboratory module, where microgravity or life science experiments would be carried out. The station may eventually include many laboratory and habitability modules, even if it starts with one of each. Logistics modules would be carried up to the station from the shuttle and would contain the consumables needed by the crew. They would be very similar in function to the containerized units that are loaded onto commercial airline flights with dinners; these are loaded onto the airplane, emptied and served to passengers, filled up again with dirty dishes, and then taken off of the plane as a unit. The logistics modules would arrive, would be emptied of the useful stuff, and would be filled with trash and sent back down again.[4]

This modular structure allows for a very natural mode of international collaboration. NASA has actively sought partners in the space station effort, and these various partners can contribute particular modules to the space station. These modules could call on the station for electrical power, thermal control, data handling, and docking but could be relatively independently functioning modules otherwise. As of the end of 1986, Japan and the European Space Agency will probably contribute one laboratory module apiece, and Canada will contribute a remote manipulator arm that can move around the station's superstructure.[5]

The overall structure of the station has been one of its most changeable features, and it may change again before it is built. (Because so many changes have taken place, I decided not to include an artist's conception of the space station in this book. Such conceptions are often quickly outdated.) The initial "power tower" configuration had a giant, 200-meter-long truss as its central core, with a pair of solar panels attached to either side of it like wings. In early 1986, this design was scrapped for a dual-keel design, in which two parallel beams flank the station's central region and give the structure its name. Further redesign was prompted by *Challenger,* allowing for eventual expansion to the dual-keel concept but viewing the initial station as a single, 360-foot-long truss with solar panels on either end and a bunch of modules attached along it.

One idea that has been around for a while is using the shuttle's giant external tank (affectionately known as ET) in orbit. In a normal shuttle launch, the ET is carried almost all the way into orbit, separating from the shuttle eight minutes after launch and plunging into the sea. It contains two fuel tanks with a total of 2,000 cubic meters of pressurized volume, six times the volume of *Skylab* and twenty times as big as *Salyut.* When it separates, it contains, on the average, about 4,500 tons of fuel and oxygen.

The puzzle is to figure out how to get rid of the fuel and what to do with the tank. When the tank is in orbit, the fuel will not naturally sink to the bottom and flow out of an open tap; it has to be scavenged. The most obvious thing to do is to use it for what it is: a storage tank for fuel. It could simply be a mounting structure that telescopes would be bolted onto. The most radical proposals involve modifying the ET in the same way that the upper stage of a Saturn rocket was modified in order to become *Skylab*. However, one must keep in mind that astronauts in spacesuits cannot do as much work as people on the ground can, and that sending astronauts out in spacesuits takes time, astronaut energy, and ultimately money.[6]

At some time in the development of the space station, a number of additional types of modules and platforms will be required, largely because of the station's anticipated role as a service facility. The space station will initially be in a low orbit, visualized as about 300 nautical miles above the earth's surface, inclined by 28.5 degrees to the earth's equator. This is the natural orbit to put stuff in from Cape Canaveral, which is at a latitude of 28.5 degrees. However, various people involved in earth watching want a platform in an orbit that goes over the earth's poles. Launching instruments into such an orbit from Cape Canaveral is impossible because the booster rockets would fall onto populated areas in the Caribbean and in South America, and so the station itself cannot be in a polar orbit. People in communications want to be able to service hardware in geosynchronous orbit. As a result, the station will need to have a docking module so that various sorts of orbital transfer vehicles can move things to and from it.[7] A teleoperated orbital transfer vehicle could go and fetch things in low orbits. More challenging is the task of servicing and retrieving satellites in geosynchronous orbit, which is considerably farther away in miles and in the velocity change needed to get there. Substantial operation of such service vehicles would require the space station to include a number of modules that would store rocket fuels and oxidizers.

The platforms associated with the space station would be physically separate from it. The usual distinction between a *station* and a *platform* is that a station is continuously or almost continuously inhabited by astronauts, containing pressurized modules that people can work in. A platform is a structure that is basically unmanned, though it can be visited by astronauts as necessary. Astronauts could replace film packs, retrieve data tapes, repair or replace equipment that is not working, or even remove a machine and replace it with an upgraded version. The scientific instruments on the Hubble Space Telescope, for example, are designed to be easily removable so that a "second generation" of scientific instruments can replace the first one. A platform can provide the common elements, like pointing, power, data handling, and data transmission, that at present are designed and built for each free-flying satellite. Many such platforms and other elements of space infrastructure exist now, as part of past or planned shuttle missions, and may become part of the space station program, although they would be physically separate from the station.

## EXISTING SPACE PLATFORMS

One space platform has already flown on two space shuttle missions. The German company Messerschmitt-Bolkow-Blohm has developed a platform called the Space Pallet Satellite, or SPAS. It is a truss of graphite–epoxy tubes (which are strong and lightweight) and fits snugly into the bay of the space shuttle. On the seventh shuttle flight, it was released from the shuttle, was operated for ten hours, and was then retrieved. It flew again on the tenth shuttle flight, remaining in the cargo bay and carrying equipment for several commercial users into orbit. It has only battery power and uses compressed cans of gas to push it around, so it is suitable only for short-duration missions. But it's cheap; the basic platform costs less than $4 million.[8] (However, if you want to use SPAS, you have to pay for a launch, which cost $12 million before *Challenger,* and for the equipment that you put on it.) Another primitive platform that has already flown is NASA's Long Duration Exposure Facility (LDEF), a coffin-shaped structure that housed fifty-seven experiments designed to test how materials and machines respond to the environment in low orbit. Boeing has designed a somewhat more expensive $10-million platform called MESA (Modular Experimental Platform for Science and Applications), intended for launch on ESA's Ariane rocket.

Several platforms that are considerably more sophisticated than

SPAS or LDEF are in the advanced development stage. The European Space Agency has developed a relatively sophisticated platform called the European Retrievable Carrier (EURECA), originally designed specifically for microgravity research. Pre-*Challenger* plans called for a 1988 mission in which the EURECA would be released by the shuttle, would transfer under its own power to a higher orbit 525 kilometers above the earth's surface (contrasted with the 300-kilometer altitude of the shuttle), would remain in orbit for six to nine months, and would then return to shuttle altitude and be retrieved. The latest plans call for a 1991 mission. EURECA contains a solar array to provide power, an attitude control system so that the satellite can be pointed roughly, and a communications system to the ground. It can handle a 1,000-kilogram payload, which can occupy 8.5 cubic meters of space. The first EURECA mission was planned for a number of microgravity experiments. Because it is a free flyer, any required shuttle maneuvers or astronaut perturbations will not push the experiments around. NASA and ESA are investigating the possibilities of using EURECA for missions in other fields, like space astronomy.

A number of more sophisticated platforms, which would cost more, would carry more payload, and would remain in orbit permanently, are on the drawing boards. Fairchild, an American aerospace company, has designed a platform called LEASECRAFT. Its mission profile is rather similar to EURECA's, but it can stay in orbit much longer and can accommodate a 10,000-kilogram payload. The Mc-Donnell Douglas electrophoresis experiment is often mentioned as a possible customer for LEASECRAFT. Space Industries, Inc., a relatively new commercial space company, has proposed a permanent industrial space facility, a platform that would house microgravity factories. A possible customer (which could include a space science experiment within NASA or ESA) would buy an entire platform, or space on a platform, from one of these companies and attach an experiment to it. Because the platform would have to be designed only once and could be "mass" produced, this mode of operation would, in principle, be cheaper than doing things the way we do them now. At present, a group wanting to do something in space starts by designing an instrument. The necessary support structure, power systems, communications, and so on are then designed and built especially for that instrument, at great cost. Often, the cost of the instrument is only one-tenth of the cost of the entire structure.

Three groups of users have particular needs for platforms associated with the space station. Astronomers on the ground put observatories in remote areas to escape from city lights and urban air pollution. Astronomers in space have the same desire for remoteness and isolation and want an astronomy platform that would be located some distance from the space station, but in a similar, low-inclination orbit. Earth watchers want to send cameras and radars into orbits that go over the earth's poles, requiring a separate platform in polar orbit. Sun-synchronous polar orbits, which are arranged so that they cross the equator at the same time each day, are particularly attractive to this group. The communications satellite industry has been talking about yet a third type of platform in geosynchronous orbit. Here, the platform would be refuelable and could remain on its station indefinitely. Different antennas in the same place in orbit would point at different spots on the ground and would use different frequencies.

If we want to use a platform in geosynchronous orbit, we need to be able to get there easily. Another piece of space infrastructure would be some kind of reusable orbital transfer vehicle that could take stuff from the space station orbit into geosynchronous orbit, come back to low orbit, be refueled at the space station, and go up again. An interesting possibility that could save fuel is that the vehicle could enter the earth's atmosphere and use aerobraking to slow it down when it came from high orbit to low orbit. Although aerobraking has been used to bring manned capsules back to earth ever since the days of Yuri Gagarin, it's a new idea to use the atmosphere to slow something down and then put it back into space.

You may have noticed that a number of new names have crept into this discussion of space platforms. For most of this book, it has been government agencies that have built space hardware: NASA, the Soviet Union, the Japanese, ESA, and so on. Platforms have been built or designed by a number of private companies using their own resources. In the space station era, a number of private companies will be involved in the space business, partly as suppliers of space hardware and eventually as possible users.

The key ingredient that makes the space station scheme work is that one nation or company will make the capital investment in infrastructure and then rent it at low cost to end users. The ability to use infrastructure that has been built by someone else can reduce costs rather dramatically. The only past opportunity for gaining such low-cost access to space occurred when a small, lightweight satellite could be launched at the same time as a more massive satellite was launched. A group of amateur radio operators took a great deal of advantage of this opportunity and built a reasonably large enterprise from their initial efforts.

## INEXPENSIVE SPACE ACTIVITY: $65 AND UP

No, that's not a typographical error. The whole point of building a space station is to make it possible for people to try out things in space without having to spend $100 million as an entry fee. There is an example from history in which a group of people, amateur radio operators, became involved in the space program by launching a satellite for something like $65 in 1961. They hitchhiked on someone else's rocket flight in the same way that, we hope, entrepreneurs of the future will be able to try out ideas in a small piece of the space station.

The story of this $65 satellite appears to have a definite beginning in February 1959, when an amateur named Don Stoner wrote the following in a column in *CQ* magazine: "Currently being tested is a solar powered 6–2 m repeater which could be ballooned over the southwest. Can anyone come up with a spare rocket for orbiting purposes?" What is a repeater, and why would anyone want to put one in orbit? The name *repeater* comes from the early days of the telephone system; then, it was a device that picked up a weak phone signal, amplified it, and sent it on to its destination. In the context of amateur radio, a repeater is a communications satellite that receives a message at one frequency and retransmits it, amplified, at another frequency. Amateurs sought to put one in orbit to make long-distance communication possible. The long-wavelength radio waves used in ham radio bounce off the unpredictable ionosphere, the electrically conducting upper layer of the earth's atmosphere, to allow ham operators to send messages across the country. A high-altitude repeater, or one in orbit, would be much more reliable than the flaky ionosphere.

Fred Hicks, a worker employed by a large missile contractor, read this column and used his contacts to find a slot on a forthcoming launch of a U.S. Air Force satellite. The two satellites would be launched at the same time and then separate. Hicks, Stoner, and others conceived, built, and launched *OSCAR 1*. The satellite was built in garages and basements, out of spare parts in people's junk boxes, at an estimated cost of $65. No one added up the cost of the bits and pieces; this figure was quoted by Roger Soderman in a talk at a scientific meeting.[9] *OSCAR 1* was then tested in vacuum chambers to ensure that its capacitors wouldn't explode in space, and it was launched on 12 December 1961. (*OSCAR* stands for "orbiting satellite carrying amateur radio.") It was a simple satellite, just a beacon that transmitted "Hi" in Morse code. It lasted for twenty-two days before it plunged into the earth's atmosphere, but it galvanized interest in satellites among the worldwide amateur radio community.

*OSCAR 1* was the first of a series. Stoner's dream of an orbiting repeater was realized quite early with *OSCAR 3,* which allowed amateurs to set some distance records and to communicate between Massachusetts and Germany, for example. The technology was always rather modest; at one point, a carpenter's ruler was used to make a folding radio antenna. The current amateur radio satellite, *OSCAR 10,* is a long-lived satellite that permits worldwide amateur radio communication.

The *OSCAR* story illustrates entrepreneurship at its best. Clearly, the initial $65 satellite didn't do a great deal by itself. It was only the pilot project that opened the door, and that demonstrated that there were a whole bunch of amateur radio operators out there who were interested in space and willing to invest time and money in their own satellite. The reason that *OSCAR 1* cost only $65 was that the radio amateurs took advantage of a launch opportunity provided by someone else, and that they invested their time and energy at low cost (in this case, at no cost at all). Space entrepreneurship isn't going to involve $100-million satellites. Rather, if it happens, there must be an opportunity to try out ideas at low cost. Many of these ideas won't work. But some will, and these $10,000 investments could grow into the multi-million-dollar businesses of tomorrow.

The space shuttle has provided a few of these low-cost opportunities for access to space, and a wide range of people have shown interest in them. One of the most popular has been the Get Away Special, known as a *GAS can* partly from the acronym and partly because of the shape of the payload. A GAS can is a cylindrical can about 40 inches in diameter and either 14 or 28 inches high. GAS cans are stuffed in the side of the space shuttle bay. If you fly a GAS can, you can ask the astronauts to turn your instrument on and to operate two toggle switches during flight. You deliver your instrument to NASA two or three months before launch and get your package back two weeks after the shuttle lands. Paperwork and cost are minimal; prices range from $3,000 to $10,000. Your package does have to pass a safety check, but that's about all.

Customers of the GAS program vary enormously. GTE has done some experiments on light bulbs in a GAS can. College and high-school students have designed experiments that have flown as GAS cans. A seed company flew twenty-five pounds of vegetable, herb, and flower seeds in a GAS can. Astronomers studying the ultraviolet background put some telescopes in two adjacent GAS cans and the controlling avionics in a third, adjacent canister. The complex of three cans flew successfully in early 1986. Before *Challenger,* about fifty GAS experiments were planned per year. I suspect that something like two or three dozen GAS experiments per year will continue, as they take very little extra space and weight in the shuttle. The GAS program is basically a formal way of instituting the same process that allowed the ham radio people to fly *OSCAR 1* as an extra satellite in a rocket launch.[10]

Scientists, especially space scientists in training, need opportunities for low-cost access to space. For the last twenty years, NASA has run an extensive sounding rocket program that gives graduate students an opportunity to fly an experiment at low cost. Well-established scientists have also used the rocket program, as an idea can be tried out relatively quickly, without the risk of much money. NASA has developed some more sophisticated ways of getting into space at low cost, on board the shuttle. The simplest of these is called a "SPARTAN," a box that is deployed as a subsatellite of the space shuttle and that is retrieved after a several-day mission. Unfortunately, one of the SPARTANs was aboard the *Challenger* when it blew up, and there may not be many SPARTAN opportunities in the near future.

The most sophisticated way of getting science payloads into space without the hassle and high cost of building and flying an independent satellite has been the spacelab program. In a spacelab mission, the cargo bay of the space shuttle is filled with a combination of pallets exposed to the space environment and pressurized cans that astronauts can move around in. The pallets are used for astronomy and space science experiments; materials-processing and earth-watching experiments are typically found in the pressurized module. Four spacelab flights were flown before *Challenger;* only three more are scheduled before 1990.[11]

The shuttle program had just begun to open the door to low-cost access to space, and the *Challenger* accident apparently slammed that door shut again. Once the effects of *Challenger* work their way through the system, the door will open again, and people with ideas will be able to get into space to try them out inexpensively. I'm convinced that, in the space station era, this kind of opportunity will be much more abundant than it is now, and that those who have the patience to stay with the space game through the lean years ahead will be able to take advantage of cheap access to space.

## BEYOND THE SPACE STATION: THE LUNAR BASE

A number of people in the space business see a base on the moon as the next evolutionary step beyond the space station. For years, people within NASA had been very cautious in talking about the lunar base (or, indeed, in talking about anything beyond the space station), partly because the space station had not been firmly emplanted in NASA's future agenda. Now that the space station plan is apparently firmly in place, discussion of a lunar base has become much more respectable, and such a base was mentioned in the National Commission on Space report.

What could we do on a lunar base that we can't do on a space station? Exploration of the moon itself is certainly one possibility, although it's true that the moon is one of the less interesting terrestrial planets when looked at strictly from the perspective of a planetary scientist. It's a great place to do space astronomy from because there is a firm platform to fasten telescopes to, and lots of dollars and engineering effort need not be spent on systems that can point a free-flying satellite. Radio telescopes on the far side of the moon would be well shielded from terrestrial interference. However, thinking about a lunar base is still at the early stages, and little has been written that goes beyond these generalities.[12]

A more likely role for a lunar base is that of a mining outpost. A wide variety of future space activities may well require bulky materials. If we go to Mars, for example, the spacecraft bound for the red planet will require a heavily shielded "storm cellar" where the astronauts can retreat and protect themselves from the radiation of a solar flare. The shielding can be of any bulky material, and it takes a lot less rocket fuel to lift a ton of rock from the moon than it takes to lug that same ton of stuff from the earth. Whether there will be enough demand for bulk lunar material to justify the front-end costs of establishing a "base" remains to be seen.

As is true of many space activities, no one reason for constructing a lunar base could, by itself, justify the cost of doing so. The cost need not be very great, as the peculiar way of reckoning distances in space means that, in terms of the cost of getting there, the moon is not much farther away from the earth's surface than geosynchronous orbit is. A lunar base may not cost much more than the *Apollo* program: $100 billion spent over twenty years. However, if the base happens, it won't be until the twenty-first century.

## THE HISTORICAL ANALOGY

In April 1817, the legislature of New York State appropriated the money required to build the Erie Canal. In the 1980s, this waterway is best known as a mosquito-breeding ground, but in the early part of the nineteenth century, it was the bridge that connected the rapidly growing American West with metropolises like New York City. The investment of $7 million was repaid by toll collections alone within nine years after the canal's opening in 1825. Freight rates between Buffalo and New York were lowered from $100 to $15 per ton. New York City, which now had the good fortune to be located at the end of a route through the Appalachian mountain chain, soon surpassed Baltimore, Boston, and Philadelphia to become the primary commercial center on the East Coast.[13] (The other cities tried to build canals, too, but found that they would have to build more complex canal systems to penetrate the mountains.) The coming of interstate highways in the 1960s and 1970s was a more diffuse example of the same pattern repeating itself later on, leading to the development of the American suburbs.

Is the space station the Erie Canal or the interstate highway system of the twenty-first century? Its most ardent proponents will implicitly or explicitly cite the building of national road systems as a historical precedent and will assume that the case has been made. As usual, there are differences between expansion in space and on the earth. In the Erie Canal case, cities like New York knew that there was money to be made by trading with the American West. New Orleans was getting all of the trade in bulky items, and there was a market for smaller manufactured goods that were carried by wagon train across the mountains. The New York legislature knew a lot more about how the Erie Canal would be used than we know about the potential of space commerce. The space station is much more of a gamble than the Erie Canal was.

*    *    *

President Reagan committed NASA to the space station program in his State of the Union message in 1984. Although there is still some uncertainty about the type of space station we will have, it will happen, and it will be much more expandable than previous space stations like *Skylab* or *Salyut*. Although the centerpiece will be a single structure, the space station concept includes a flotilla of associated platforms and orbital transfer vehicles that will open the road to a variety of activities in space, hopefully at a significantly lower cost than at present. The ability to repair space machinery and the existence of platforms that can provide essential, commonly needed services like power, data handling, and pointing will lead to a fundamental change in the way we do business in space. Except for communications, where all the equipment is in geosynchronous orbit and will be inaccessible to the space station until a sophisticated orbit transfer vehicle is developed, and planetary exploration, where the automated probes are far beyond the station's repair capabilities, there is no area of space activity that will not be affected profoundly by the development of the space station.

The great changes that the space station will bring make it difficult to look beyond the era of the space station to the twenty-first century. But look ahead we must.

# CHAPTER 15

# Space in the Twenty-First Century

## WARNING: SPECULATION LIES AHEAD!

So far in this book, I have played the role of a reporter, telling you what we have been doing in space, why we have been doing it, and what's planned for the near future. As far as I know, I've stuck with the facts, basing my predictions of the future on what we're doing now and on immediate plans. In this last chapter, I will throw caution to the winds and make some potentially outrageous predictions. It's fun to think about the role of space in our lives fifty years from now, and I think that my experience with the space program gives me at least a reasonable base of knowledge to work from. I have deliberately avoided putting my own opinions into previous chapters; in newspaper terms, you can think of this chapter as the "editorial page" of this book.

The next ten years look lean. Even if *Challenger* hadn't exploded, there is no way in the world that NASA could have launched all of the payloads that were on various space shuttle manifests. The loss of an orbiter along with a two-year hiatus in shuttle flights has made the situation much worse. Many of my friends are very pessimistic, seeing various projects that represent the investment of years of work put on hold or canceled because they can't fit into a reasonable launch schedule. I feel the effects of the hiatus, too; my bottom drawer is filled with proposals to use various space instruments that I will fish out five years from now to see if they still make sense. However, I feel that all of us must look beyond the contemporary gloom.

By the year 2000, we'll have a space station, a private launch industry, and another orbiter. The space station could revolutionize the way in which we do business in space and could open the door to broader participation by private industry. Space science could become easier and cheaper, allowing for ambitious expeditions as well as low-cost tinkering and experimenting. It's likely that space scientists will make more unexpected discoveries like the complex ring systems of the giant planets, black holes in space, and the ethane oceans of Titan. The space station may be the key to the microgravity business, allowing more time in orbit and significantly reduced costs for these operations. We will be looking for what to do next, casting our eyes toward the moon and Mars.

The more distant future is harder to forecast. Despite all the optimistic discussion of a trip to Mars, there are still a couple of fundamental barriers that have to be overcome. If humans can go to Mars, we will go, certainly sometime in the twenty-first century. Such a trip can serve as a long-term focus for the present space program.

## THE HAZARDS OF PROPHECY

Ever since Arthur C. Clarke predicted communications satellites in 1945, people associated with the space program have tried to emulate prophets. We have to; it takes so long to implement most space projects that we space scientists have to have a long-range perspective. NASA is one of the few government agencies that I know of that pays people to think about what will happen beyond the next election. It's fun to dream, but this means that the rest of this book will take on a different cast from what you have read so far. I will distinguish between dreams, reasonably solid plans, and the vast territory in between.

Just so that you don't take my predictions (or anyone else's) too seriously, it's sobering to recall the success of previous attempts to predict the future. In his path-breaking book *Profiles of the Future*, Arthur Clarke recounted the sad history of a number of astronomers who blandly predicted that certain things wouldn't happen. William H. Pickering, a few years after the first airplanes, scoffed at the prospect of "gigantic flying machines speeding across the Atlantic and carrying innumerable passengers in a way analogous to our modern steam-ships.'' I can't think of a better description of a 747 packed to the gills with sweating, snoring humanity, winging its way from New York to L.A. Richard Woolley, Astronomer Royal of Britain in the mid-1950s, said, "Space travel is utter bilge.''[1] I have specifically selected instances in which astronomers ended up with egg on their face to remind you not to take me too seriously, as that's my trade, too!

Based on these past failures, Clarke formulated his oft-quoted aphorism that every major technological development goes through three stages, represented by the prevailing attitudes of the scientific establishment:[2]

1. It can't be done.
2. It can be done, but it isn't worth it.
3. Why weren't we doing it all along?

Ever since Clarke predicted satellite communications in 1945, every-body who writes about the space program has been deathly afraid of looking stupid, like the astronomers I just quoted, and the fashion now is to be wildly optimistic in predictions about the future. At the end of the *Apollo* program, many writers had us on Mars within fifteen years. In the mid-1970s, people were still estimating a cost of a hundred dollars per pound or even less to launch things aboard the space shut-tle, one-twentieth of the actual cost. Some writers saw space indus-tries, colonies, and power satellites coming in the next decade (mean-ing the 1980s).

I'll try to steer a middle ground. Technological progress has been so explosive that there's not much that people have thought about that can't be done. I doubt that something that costs the gross national product to build will be done, at least in the foreseeable future.

## THE NEXT TEN YEARS: DARKNESS BEFORE THE DAWN

During the next ten years, NASA will fix the space shuttle, get it flying again, and rebuild a four-orbiter fleet. In the fall of 1986, NASA announced a tentative date of 18 February 1988 for the first space shuttle flight. According to the schedule, shown in Table 10, five

Table 10.    Space Shuttle Payload Manifest[a]

| | | |
|---|---|---|
| 1988 | 18 Feb. | *TDRS* (NASA communications satellite) |
| | May | DOD geosynchronous mission |
| | July | DOD large imaging reconnaissance satellite? |
| | Sept. | *TDRS-D* (second TDRS) |
| | 17 Nov. | *Hubble Space Telescope* |
| 1989 | Jan. | *Astro-1* ultraviolet telescope |
| | March | DOD geosynchronous payload |
| | April | *Magellan* (Venus radar mapper) |
| | June | DOD *Navstar* satellites plus materials science pallet |
| | June | SDI Spacelab Mission |
| | July | DOD |
| | Sept. | DOD (2 flights) |
| | Nov. | *Galileo* (Jupiter probe) or *Ulysses* (solar polar) |
| | Dec. | Spacelab life sciences |
| 1990 | Jan. | *Gamma Ray Observatory* |
| | Feb. | DOD |
| | April | International materials science spacelab |
| | May | *Navstar;* McDonnell Douglas electrophoresis (EOS); space-station heat-pipe test |
| | May | DOD |
| | July | DOD |
| | July | British Skynet-4 military payload |
| | August | DOD |
| | Oct. | *Galileo* or *Ulysses* |
| | Oct. | U.S.-Italian tethered satellite (atmospheric sounding); *Insat* (India); *Navstar* |
| | — | Retrieve Long Duration Exposure Facility (which will have spent five years in space, not just one; Syncom-4) |
| 1991 | Jan. | Spacelab pallet (atmospheric data) |
| | Feb. | *Navstar;* materials pallet |
| | March | DOD |
| | April | EURECA (European Retrievable Carrier) |
| | 2Q[b] | Japanese spacelab |
| | 3Q[b] | German D-2 spacelab |
| | | Refurbishment of Hubble Space Telescope |
| | 4Q[b] | Retrieval of EURECA |

[a]This table refers to major payloads only. It assumes a flight rate that many people think is higher than can actually be achieved. *Mars Observer* is set for launch in 1990 or 1992.
[b]2Q = second quarter; 3Q = third quarter; 4Q = fourth quarter of the year.

flights in 1988 will launch two TDRS satellites, two U.S. Defense Department payloads, and the Hubble Space Telescope. The flight rate is scheduled to slowly increase to ten in 1989, eleven in 1990, and sixteen per year after 1991, when a replacement orbiter will enter the fleet. The NASA manifest lists payloads through the end of 1994, with nearly 41% of these being national security missions, 47% being NASA science and applications payloads, and only 12% being scheduled foreign and commercial payloads.[3] Payloads through mid-1991 are assigned to specific shuttle flights; after that, most of the flights are assigned to space station assembly (beginning in 1993), some commercial flights, and U.S. Defense Department payloads. I honestly doubt that the schedule can be met, and Admiral Richard Truly, head of NASA's Office of Space Flight, emphasized that "there will be *no* attempt to meet an arbitrary flight rate." In other words, I doubt that the dates should be taken very literally. It seems to me that it is a safe bet that the NASA manifest includes all the major payloads that will fly in the next decade, and that something like at least 80% of them will actually be launched.

The shuttle manifest (which refers only to major payloads) does indicate that a number of missions will be delayed from their original launch dates by two to three years and sometimes more. A number of critical science missions are on the list, including many mentioned earlier in this book. The major difficulty will be with commercial and foreign customers, who will have to fly on expendable rockets launched by private companies unless they are so configured that they can fly only on the shuttle. NASA has other tasks on its plate for the next ten years. Construction of the space station is slated to begin in 1993, and an initial space station should be built by the mid to late 1990s. Expendable rockets will launch many of the routine payloads into orbit, and a private launch industry will become established.

As a result of all these mission delays, many of my friends in the space science business became quite gloomy during 1986. At a recent conference, a senior space scientist grumbled to me, "My advice to anyone is to stay out of the space business." I hear the words *shambles* and *disarray* used very often to describe the current state of the space program. Yet, at the same time, we have the bold report of the National Commission on Space, asking America to lead the exploration and development of the space frontier, advancing science, technology, and enterprise, and building institutions and systems that make accessible vast new resources and support human settlements beyond Earth orbit, from the highlands of the Moon to the plains of Mars.[4]

Which vision of the future is correct—the gloomy one or the optimistic, visionary one? It's a matter of time scale. It will be very tough for the next ten years. Some people find it hard to look beyond that, especially older people whose productive professional careers may not last until the promise of the space station becomes a reality. Many senior space scientists entered the business around 1960, when they were, say, in their early thirties; thus, they are in their late fifties now, and they will be pushing seventy when the situation improves. I find it easy to understand their gloomy perspective, but I don't share it (perhaps because I'm in my late thirties). Many of us in the space business will be active when *Challenger* is a distant, tragic memory. I find it possible, even necessary, to look beyond the next ten years and to see a much happier long-term future. I believe that NASA still has the resources and the will to continue to support the space science and applications communities, which will develop some things for the space shuttle and the space station to do after 1995.

Two particular communities within the space program will be especially hard hit. Planetary scientists have designed a low-cost program for planetary exploration based on frequent launches (one per year) of simpler spacecraft. If the schedule in the manifest is kept, even approximately, only three planetary missions—*Magellan* to Venus, *Galileo* to Jupiter, and the *Mars Observer*—will have been launched in the seventeen years between 1977 and 1994. This is hardly the frequent launch schedule that had been hoped for. The repair capabilities of the space station will be little help for most planetary missions except for a sample return mission to Mars.

Astronomers exploring the distant universe share the problem of dashed expectations. Two weeks before Challenger, NASA administrator Sam Keller told the American Astronomical Society that 1986 was to be "our" year, with the *Astro-1, Ulysses, Galileo,* Space Telescope, and *Astro-2* major astronomy-related missions all scheduled for launch. The Space Shuttle was to finally pay off for astronomy. Of course, in 1986 *none* of these missions flew, and the *Astro-2* reflight of the *Astro-1* set of telescopes is not even on the shuttle manifest. Those that will fly will face delays of several years.

Much of the gloom in the space science community stems from the experience of the late 1970s, when cost overruns in the space shuttle program ate up all available funds within NASA. Will a similar fate befall the space station? I think and hope not, because there is a crucial difference between the station and the shuttle. The space station is modular, and if it turns out that you can only build 80% of what's currently visualized as "the" space station and remain within a limited budget, you have the option of building 80% of the current version of space infrastructure which is called "the space station." To be more specific, you could, for example, build four modules instead of five. The Space Shuttle could not be scaled down at the last minute in the same way; 80% of a space shuttle won't fly, but 80% of a "space station" is just a smaller space station.

Columbus, too, faced a bleak decade in the '80s—the 1480s. The history books you read in school tell little of the story of the 10-year struggle behind the Columbus expedition. To the extent that they think about it at all, most students probably have the impression that Columbus simply went to Queen Isabella of Spain, asked for money, got it, and discovered America. It was definitely not that easy. He first tried to sell his transatlantic expedition in 1484 to King Joao II of Portugal, and failed. Two more failures with Queen Isabella of Spain followed, along with final confirmation that Portugal was simply uninterested in sailing across the Atlantic because Bartolomeo Dias discovered that ships could reach India by sailing around Africa. Finally, Columbus found a friend, Santangel, who interceded with Queen Isabella. Santangel and Columbus had to use their own money to finance one quarter of the cost of his first expedition—an old version of what's now called *cost sharing*. Table 11 shows the ragged funding history of this magnificent human adventure, and it should offer hope to people whose projects face similar difficulties.

Curiously, all of the people who turned Columbus down, listed in the table, rejected his proposal for good reasons. Columbus justified his expedition as a way of sailing to Japan and opening up trade routes. By underestimating the size of the earth by 20%, overestimating the

Table 11.   Funding History

Project: Transatlantic Expedition
Principal Investigator: C. Columbus

| Date | Agency | Result |
|------|--------|--------|
| 1484 | Joao II (Portugal) | Rejected. |
| 1486 | Isabella (Spain) | Sent to committee for four years. |
| 1488 | Portugal | Final rejection. Dias found the African route to India and Portugal was no longer interested in Atlantic Ocean routes. |
| 1490 | Isabella | Committee rejects after four years of deliberations. |
| 1491 | Isabella | Rejected. |
| 1492 | Isabella/Santangel | Accepted, with Santangel and Columbus putting up 25% (cost sharing). |

size of Asia, and overestimating the distance from China to the offshore island arc of Japan, Columbus came up with some numbers that put Japan 2,400 miles west of Spain, about where the east coast of North America is now. These numbers now seem ridiculous, and even by the standards of five hundred years ago, Columbus was, at best, stretching the data to make his point. Various advisers to the monarchs on the Iberian peninsula spotted these fudges and understandably recommended rejection. Were it not for America, Columbus would have run out of consumables (food and water) long before he reached Japan. The lesson to be drawn is that Columbus was persistent and had the staying power to last through some lean years.[5]

## CHANGES AT NASA

During the next ten years, the nature of human activity in space will change, and the public perception of NASA and the agency itself will have to change along with it. To a great extent, NASA is an agency put together to go to the moon, but we're not doing that anymore. Many policy analysts commenting on the current scene see three basic issues as dominating the debate about where the space program is going or should go: space industrialization, international cooperation and competition, and space militarization. I think there is another issue that NASA has also been confronting: the need to adapt to a balanced space program.

Space industrialization has been much discussed, but little has happened in the recent past. One policy decision made in the summer of 1986 may, at long last, make the hoped-for participation of the private sector in the space business (beyond communications satellites) become real. President Reagan directed NASA to stop using the shuttle for launching commercial satellites, payloads that could be launched on expendable rockets. It is not surprising that private industry found it so hard to sell launching services to commercial customers. Two government-subsidized behemoths, NASA and ESA, have engaged in the same sort of fare wars that have brought down Freddie Laker, People Express, and other airlines. NASA seemed to want to become the Pan Am of the skies, a trucking company instead of a government agency on the cutting edge of science, engineering, and exploration.

NASA's withdrawal from the commercial market may stimulate much greater private-sector participation in the provision of launch services, and in the space program in general. A Business-Higher Education Forum study of the space program was relatively unimpressed by the private-sector participation in the space program, largely because few companies outside the aerospace industry have chosen to work with NASA. The Forum study called for the establishment of ''a culture supportive of entrepreneurial efforts.''[6] Perhaps the development of a private launch vehicle industry can lead the way. If NASA can withdraw from launching commercial satellites, it will have the opportunity to return to what it is good at: pushing outward on the cutting edge of space science, exploration, and technology. Private industry can run the trucking company.

A second issue before NASA and before the country in the next decade is our international posture in space. During the Apollo days, ''pride and power''[7] was the driving force behind the space program. Because there was only one space race, Americans could visualize NASA as an agency that could put the United States in a preeminent position in a magnificent human endeavor, to use the haunting word first used by President Kennedy in 1962. Many advisory groups, including NASA's own Advisory Council, the National Academy of Sciences' Space Science Board, and the National Commission on Space, have argued that a budget increase for NASA is necessary if the United States is to maintain a preeminent position.

In a time when there is great concern about deficit spending, it nevertheless seems that a budget increase for NASA is indeed possible. There is a broad base of support for the space program. Opinion surveys taken before *Challenger* showed that about 10% of the adult population are interested in space exploration and know enough about it to classify themselves as "very well informed" about it. This segment of fifteen million people, combined with an additional twenty-four million adults who are interested in, but inadequately informed about, space, adds up to a substantial number of voters. A poll released by NASA in late 1986 showed that 89% of the public want to resume shuttle flights, 85% support the decision to build a new shuttle, and 78% support the space station. More interesting is that 71% feel it is important for the United States to be ahead of the Soviet Union, and 60% favor the government's spending whatever is necessary in order to maintain our leadership.[8] Whether this level of public interest will result in a budget increase in the era of Gramm-Rudman remains debatable.

However, even a budget increase will not allow the same sort of preeminence that was possible in the *Apollo* era. The multifaceted nature of the space program generates many space races rather than just one, and even an increased budget and the capabilities of American technology will not allow one country to win all of them. What makes most sense is already becoming visible as an emerging trend: international cooperation, with different countries developing different technological strengths. Recently, the details of a rather comprehensive space cooperation agreement between the United States and the Soviet Union have been worked out.[9] The old notion of "pride and power" will be recast in a new mold, where a leading role in a cooperative venture will take the place of a solo, exclusively American expedition like *Apollo*.

Related to our international posture in space is the need to maintain a balanced space program. When planetary science was threatened in 1980, I joined my fellow members of the ruling Council of the

American Astronomical Society in deploring the arbitrary dismantling of any area of space science. We were specifically referring to planetary science at that point, but I think that a similar case can be made for any activity in NASA's space science and applications programs. It would seem obvious that NASA needs to support space science and applications research so that scientists and engineers will be ready to do something with the space station once it is in orbit. However, space scientists have felt the need to keep up a continuous chorus of well-publicized, appropriately timed protests and statements of concern in order to ensure that space science is not squeezed out.[10]

The last issue addressed in the current debate about space policy is the militarization of space. During the *Apollo* era, when many perceptions of the space program solidified, military expenditures on space were a minute fraction of the civilian space program. During the late 1970s and early 1980s, U.S. Defense Department expenditures on space accelerated, with the Pentagon overtaking NASA in 1981. In part thanks to Star Wars, the Defense Department now spends twice as much on space as NASA does, a total of $16 billion a year.

What does the rapid buildup in the military space program mean for NASA and the civilian space program? In my view, not as much as people think. I have not spent a great deal of time discussing space militarization in this book, in part because there are some excellent other books that describe our military space program.[11] However, except for the demand on launch services, the civilian and military space programs are largely separate. If there are major deployments connected with the Strategic Defense Initiative (SDI) in the 1992–1995 time period, these could compete with the deployment of the space station for slots on space shuttle flights.

However, I don't see this as a problem because I think that the pace of the SDI program will slow considerably after President Reagan leaves office, no matter who wins the 1988 election. I have deliberately waited until this chapter to give my own opinions on the SDI program, as I wanted the earlier description of SDI to be reasonably objective. It's clear to everyone knowledgeable about SDI that it is not an impenetrable shield that can protect us from ten thousand missiles. It also seems clear that some money needs to be spent on research on these technologies. The issues involve the pace, style, funding level, and technological direction of the SDI program.

My belief is that the funding levels that have been requested by the Reagan administration are clearly excessive and that the program has been oversold. Elaborate, expensive, and basically useless ''demonstrations'' of beam weapons technologies have dominated the publicity on SDI. I'm not impressed when a laser zaps a rocket sitting stationary on a firing range. What I want to know is whether, given that an X-ray laser can be built, a battle management scheme can be figured out that will allow it to shoot down more than one rocket when a bunch of half-asleep young second lieutenants are faced with an attack at 4 A.M. on a Sunday morning. Under such circumstances, how many incoming missiles can be destroyed? We know that an SDI system could stop one missile. It also seems clear that SDI could not stop ten thousand missiles, especially after the boost phase, when each missile deploys a hundred decoys, generating a million targets. The critical question is whether SDI can stop enough missiles to be a useful weapons system.

The complexity of the battle management scheme and computer programs increases rapidly as the number of targets increases. I have had considerable experience in writing and working with complicated computer programs. Many computer scientists have spent more time than I thinking about SDI, and I share their views that the battle management task is the most serious barrier to the successful deployment of an SDI system. The question that needs to be answered now is not whether an X-ray laser can be built, but what kind of computer system is needed to control SDI and how many incoming missiles it could cope with. A system that could protect against an attack by several hundred missiles might well be worth deploying in some future world where the nuclear arsenals of the superpowers have been reduced to that level. Such a system could protect people from the accidental launch of one or a few missiles and could provide some degree of protection against the possibility of an irresponsible country's acquiring a few bombs and using them for blackmail.

A second problem for the future is the possible development of antisatellite weapons systems. The Reagan administration chose not to pursue a moratorium on the development of these systems in the early 1980s, and some critics point to the possibility of yet another costly arms race in space.[12] However, there are a number of possible courses of action being discussed, including the development of some "rules of the road" for satellites and ASAT's comparable to the rules of the road that govern close encounters between American and Soviet warships at sea.

## AFTER THE SPACE STATION: THE TWENTY-FIRST CENTURY

By the year 2000, the space station will be in orbit and working. I believe that it could cause some profound changes in how we do business in space, that it could open the door to a much broader participation in the space program. Some hints of such broad participation began to appear in the space shuttle and *Salyut* programs, but in the American case at least these were cut short by *Challenger*.

The major barrier to a broader utilization of space has been the high initial cost. Up to now, anyone wanting to do business in space, whether a private company or a university research group, generally has needed to design, build, and pay for a machine that could do everything: get power from the sun, relay the information back to earth, maintain a reasonable operating temperature, point in particular directions in the sky if necessary, and at last make whatever measurements or do whatever experiments are desired. Many components of individual satellites—the systems that provide power, thermal control, and data relay—are much the same no matter what the mission of the satellite is. The hope is that the space station will provide the infrastructure so that these common elements of spacecraft need not be built individually for each space mission. The opportunities to tinker, to try things, will be there, and this increased access to space will open up many new frontiers in space applications and space science.

## COMMUNICATIONS

The communications industry is the only space industry that can be called mature. Growth of markets can be reasonably predicted, and much of the technology that we will be using for the next several decades is either available or immediately foreseeable. Depending on how the space station comes along, it may be possible that larger structures, such as the orbital antenna farms visualized by B. I. Edelson and W. L. Morgan in the late 1970s, will replace the current satellites, where each satellite has one owner and does one job.[13]

The technology will allow for just about any conceivable communication need, for a price. Two developments—direct broadcast television and videoconferencing—which are under way now may become considerably more widespread in the near future. Only modest improvements in technology will allow small, three-foot dishes on rooftops to receive TV signals broadcast directly from satellites, expanding the home entertainment menu. A number of companies have bought video equipment so that executives in different parts of the country can meet without spending half the day on airplanes; an estimated $187 million of such equipment was sold in 1984.[14]

A bit more visionary, but still only a modest extension of the cellular car phones which have proven quite popular, are pocket telephones also called "Dick Tracy radios." Forty years ago, comic-strip artist Chester Gould portrayed a wristwatch communicator that his hero Dick Tracy could use to talk to headquarters; toward the end of Gould's life, the radio was upgraded to a TV set. Given a big enough antenna in the sky, you might possibly be forever freed from the tyranny of the telephone lines. Poke a few buttons on your wrist radio, and a signal will be sent up to a satellite overhead, letting you call your friends anywhere in the world. No matter where your friend is, his or her Dick Tracy radio would pick up your signal. This could mean the end of "telephone tag" where two people end up leaving endless callback messages with secretaries or answering machines. It could also mean that there's nowhere to hide, no way to "be in a meeting" when you're actually on a golf course, no way for authors like me to hide at home and work on a book like this, unless a new kind of business etiquette were developed.

Global communications may expand the economic horizons of some Third World countries. In April 1986, 35 engineers started working in Bangalore, India, designing programs that Texas Instruments uses to design computer chips.[15] Communications technology allows people in remote areas to make a living in a technological occupation because their products, consisting of information, can be exported over satellite links rather than by means of roads, railroads, airports, or boat docks. Reliable transportation, especially for delicate electronic equipment, is hard to come by in many impoverished countries. A poor country with educated workers can leap over the Industrial Revolution and can join the Information Age, with no need to invest in smokestack products like steel mills.

No book like this would be complete without at least one outrageous speculation, so let me make one. Communications satellite technology may make a highly controlled, closed society impossible fifty years from now. A country like the Soviet Union exercises control by limiting the flow of people and goods through its territory. If people have to be next to each other to talk, control of the movement of people can mean the complete control of the society. However, let's consider what happens when every Russian can buy a Dick Tracy radio in the same way that they can now buy Western blue jeans. The Soviet government has been able to control the importation and distribution of expensive things like Xerox machines and computers, but when it comes to something relatively cheap like blue jeans it's been a lot less successful. The black market thrives on such imports. If the new communication technology becomes cheap enough to break through the Iron Curtain (and I admit that this is a big if), then it could produce fundamental changes in how the Soviet society works. Whether this would produce large-scale political change in countries like the Soviet Union is a bit more problematical, but those in power would have to change the tactics they use to suppress dissension, and they might not be able to suppress it at all.

The Soviet Union is already facing a similar problem as it seeks to adjust its educational system to the computer age. Two very visible individuals, Evgeny Velikhov (vice-president of the Soviet Academy of Sciences and an adviser to Gorbachev) and A. P. Ershov (head of the computer center at the Siberian branch of the academy) are leading a widely publicized program to improve computer literacy in the schools. Some observers think that the Soviet Union can benefit from contemporary computer technology and keep the computers locked up. I doubt it.[16]

Space communications is one of the best understood areas of applied, profit-making space technology. Other applications of space such as improved weather forecasting and navigation offer similar, if not as explicit, promises. One area has been discussed extensively as the next frontier for space commercialization, and it deserves a separate section of its own.

## MATERIALS PROCESSING IN SPACE

The microgravity business is one area where technological optimists have painted lucrative, rosy pictures of billion-dollar industries. The hype seems to be dying down a little in the mid-1980s. Charles Gould wrote in 1981 that one could fit all the stuff which has actually been produced in space (the crystals, the microspheres, the metals, the purified proteins, and so on) into a briefcase. Gould wryly contrasted this limited space output with the volume of books that have been written about it, which could fill up a telephone booth and more.[17] Some of these books, such as a recent report by the Business Higher Education Forum and several reports by the Congressional Office of Technology Assessment,[18] are quite sober assessments. Others are way up in the blue sky.

The list of potential space products extends far beyond those explored so far: the perfectly round plastic spheres, drugs, and large perfect crystals. The dreams include new glasses, high-temperature turbine blades, high-strength magnets, superlarge integrated circuits, foamed metals, "space wood" (a hypothetical material that is much stronger than natural wood but that is as easy to cut, drill, machine, and join as wood or plastic), and ultralarge membranes.[19] In contrast, the list of real space products that have been sold to real paying customers has one prosaic entry: perfectly round plastic balls that can calibrate the size of things in microphotographs.

In the short run, my expectations are guarded. The excessive hype has undoubtedly scared many bankers off. Few companies have sufficiently deep pockets to do anything without some financial backing, even if NASA is willing to give them free flight opportunities through a Joint Endeavor Agreement. Much of what has been done in this area has been done carelessly. The kind of applied or basic research that has been done by companies like John Deere, in which the space environment is used to study manufacturing processes rather than to actually make stuff, looks much more promising to me in the short run.

Is there anything visible in the long run which could possibly change the prospects for microgravity? Many of the blue-sky predictions were made in the mid 1970s, when there was a real expectation that the cost of a shuttle launch would drop to two hundred dollars per pound or even less. Lots of stuff we use on earth costs that much. Gold is a few hundred dollars an *ounce*. Crystalline, semiconductor-grade silicon costs around $200 per pound[20] (1985 dollars), and the electronics industry uses it in enormous quantities. If the cost of space manufacturing can be brought down to the level of hundreds of dollars per pound, a large number of markets may open up.

What might be the impact of such new materials? Forecasting such markets is very difficult. Consider some past examples. If you asked an electronics engineer fifty years ago, in 1936, what could be done with crystalline silicon, the answer would most likely have been "Nothing. It's a lousy conductor. It's not an insulator." If you asked a clothing designer in 1936 about synthetic fabrics, the response would be "Plastic pants? That stuff is heavy and inflexible. No way." If you mentioned something like Gore-tex®, which lets sweat out but keeps the rain off, you'd be met with a blank stare. Yet all these products are here, now, and have made a major impact on the market.

The key to the future of the manufacturing component of the microgravity program is low cost. It is remotely possible that the combination of a space station and a hypersonic transport, which would both reduce the cost of launching material to orbit and permit a manufacturer to leave the processing equipment in orbit for a while, would reduce the cost of space manufacturing to the point where large-scale space manufacturing could become a reality. Whether or not these possibilities can be realized will probably not be known until the twenty-first century. At present, even the proponents and practitioners of materials processing in space are presently taking a cautious approach. James Rose, Director of McDonnell Douglas's Electrophoresis Operations in Space, and Terence Fitzpatrick write, "No one yet knows when materials processing in space will come of age as a commercial activity.''[21] I think it's reasonably clear that using the microgravity environment as a lab for studying the behavior of materials, with a goal of better understanding what happens when you turn gravity on, will provide important insights to the materials industry. Space manufacturing on a large scale is an entirely different matter, no one knows now whether this potential area of activity in space is making pie in the sky or whether it is the next industrial revolution.

I'll end this section on space applications with a quick description of one of the more bizarre space industries: the cemetery in the skies. Two groups of morticians have signed contracts with a private launching firm, Space Services, Inc., to launch a space mausoleum to bury people's remains in space. Ashes produced in the usual way are further reduced in volume and are placed in capsules measuring one centimeter by three centimeters. When thirteen thousand of these are collected, they are put into a satellite and launched into orbit. The satellite is coated with reflective material so descendants can see it easily. The cost is $3,900 per funeral, not outrageously high. We'll see if anybody signs up.[22]

## SPACE SCIENCE

The future of space science is a lot harder to forecast than the future of space applications. My discussion of communications and materials processing stems from what we'll be able to do with currently available technology in an era when the space station makes it easier and cheaper to work in space. Science, in contrast, involves new discoveries, and the past exploration of our Solar System has taught us to expect the unexpected. It is this thrill of new discoveries that makes the exploration of space such an exhilarating human venture.

One way to sense the pace and type of future discoveries is to recount the number of discoveries made in the past. The first American spaceflight, *Explorer 1* in 1958, discovered that the earth was not orbiting in an empty sea of space but was surrounded by a magnificently complex magnetosphere, which protects astronauts in low earth orbit from the dangers of solar flares. The discovery pace continued in the 1960s, with the discovery of the hot surface of Venus, the intense exploration of the moon, the beginnings of space astronomy, and the global perspective provided by meteorology satellites. In the 1970s and 1980s, the pace continued. Let me list some highlights of the fifteen years between 1971 and 1986:

- 1971: *Mariner 9* discovered the biggest volcano in the Solar System and ancient riverbeds on the planet Mars.
- 1972: The *Copernicus* satellite discovered that most of the space in the Milky Way Galaxy is filled with hot gas.
- 1973: Photographs from *Skylab* revealed holes in the sun's corona or outer atmosphere.
- 1974: Analysis of X-ray and visible light measurements suggested that the double star system Cygnus X-1 contains a black hole.
- 1975: *Mariner 10* discovered globe-girdling cliffs on the planet Mercury, apparently produced when the whole planet shrunk.
- 1976: *Viking* returned pictures from the surface of Mars and determined the composition of the Martian soil and polar caps.
- 1977: The Kuiper Airborne Observatory discovered that the planet Uranus has rings.
- 1978: *Pioneer Venus* mapped the surface of this planet, showing huge, raised, continentlike plateaus.
- 1978: *Seasat* mapped the wind fields over the Pacific Ocean in detail.
- 1979: *Voyager* discovered Jupiter's ring and active volcanoes on Jupiter's large satellite Io.
- 1980: *Voyager* discovered innumerable ringlets within Saturn's rings and geological activity on Saturn's satellites.
- 1981: Satellite images of the sun, analyzed in this year, revealed that a comet hit the sun and disintegrated in 1979.
- 1982–1983: The severest global El Niño event in a hundred years became the best observed in history, being monitored closely by earth observation satellites in order to determine the cause of this global climatic catastrophe.
- 1983: LMC X-3 was identified as the second black hole candidate.
- 1984: The Infrared Astronomy Satellite (IRAS) revealed pro-

toplanetary disks around several nearby stars and galaxies that emit 99% of their energy in the infrared.

- 1986: *Voyager* revealed complex geological activity on the innermost satellite of Uranus and a magnetic field that is tilted sixty degrees from the planet's axis and is offset from the center.
- 1986: Several probes to Halley's comet returned the first pictures of a cometary nucleus ever taken and showed that the nucleus is much darker than had been thought.

This list of major discoveries, with at least one per year for fifteen years, is only a mere indication of how active the space science field has been. Although the news media tend to focus on major advances, many areas of science (such as the space biology program) progress by lots of small steps, no one of which attracts attention in quite the same way as the ones I've listed. No major area of space science is beginning to wind down, to reach the point where each new object or new measurement is simply a refinement of something previously known. The list is biased toward astronomy, partly because I'm most familiar with astronomy and partly because space astronomy, more than other fields, has generated "discoveries" of new objects that can be set off in lists like this.

Let me list some potential future newspaper headlines, drawn partly from a list published by the National Commission on Space:[23]

- *Lyman* satellite discovers fossil remnants of the Big Bang explosion that marked the beginning of the universe.
- *Quasat* radio telescopes in space produce an image of the surroundings of the black hole at the center of our Milky Way Galaxy.
- Amino acids, the building blocks of proteins, are discovered in the Uranian ocean.
- The first planet outside the Solar System is discovered.
- Pluto has methane volcanoes.
- Links between solar activity and weather are understood.
- Monthly hurricane predictions are accurate within twelve hours and 100 miles.

- Space medicine discovers a cure for bone thinning in astronauts, and the treatment is applicable to people on earth.
- A new state of matter is produced in a space laboratory.

The reason for the long-range optimism about space science is that routine access to space will make it a lot easier to do space science from orbit. When the space station is in place, we can start building repairable and refurbishable instruments and telescopes instead of the disposable satellites we have now. Looking somewhat further ahead to the twenty-first century, we can visualize instruments that will be partially constructed in space. Astronomers in particular are limited by the size of the space shuttle's cargo bay. Already, a high-priority astronomical mission is the Large Deployable Reflector project, a proposed ten-meter telescope in earth orbit. Such a telescope cannot be launched by the space shuttle in one piece; it has to be "deployed," meaning built or assembled, in space.

I should be cautious, however, about the planetary exploration program. The space station will help the planetary program do only a sample-return mission to Mars. I suspect that the ambition of some space enthusiasts to settle the inner Solar System will drive the planetary program toward Mars and away from a program balanced between the inner planets, the outer planets, and the primitive bodies—the kind of program that planetary astronomers want for scientific reasons.

Why do space science? There are a number of reasons, some of which range well beyond the simple need to satisfy human curiosity. First, there is the intellectual interest. Part of the essence of being human is a need to know where we are in the universe and how the universe works. Go out on a clear starry night, look up at the stars, and you cannot help but wonder just how humanity fits into the universe.

A second reason for doing space science, or science in general, is that science develops the general understanding of how the world works, and this understanding is then put to practical use. For example, the image-processing techniques developed in connection with the various planetary probes have been used in medicine. Our understanding, primitive as it is, of what freon does to the thickness of the ozone layer came from studies of the atmosphere of Venus. If we go even further back, the laws of motion that are the basis of modern engineering were discovered by Isaac Newton, who was trying to understand the orbits of the moon and the planets.

There is another, often less appreciated, connection between science and technology that has been given the perhaps unfortunate name of *spinoff*. Historically, investment in any technologically challenging project has led to the development of a number of new devices. NASA's contention has been that a number of useful products, like the material used in the Silverdome at Pontiac, Michigan, are "spun off" from the space program, an indirect but inevitable result of NASA's investment in high-technology enterprises. Critics argue that the benefits of spinoff are really an illusion, that direct investment in fabric technology or cardiac pacemakers or whatever would be a cheaper way of devising these gadgets. Thanks to a number of conversations with astronomer John McGraw, I think I understand how spinoff works. I am now convinced that the benefits from spinoff are very real, and that spinoff is a way in which science contributes to economic development, often in a very short time.

One of the primary barriers to technological innovation is the reluctance of many consumers to be the guinea pigs who try out brand-new products. A widely quoted piece of advice given to someone buying a personal computer is "Never buy Serial Number 1 of anything." In other words, the advice is to let someone else buy the first model of the device that emerges from the production line. Of course, if everybody followed this advice, innovation would be impossible. Spinoff works because NASA is the customer that will buy Serial Number 1, that will sometimes even force the development of new technological devices because of the demands that it places on the capacity of humans to design machines.

A story about the McDonald Observatory illustrates this quite clearly. In the early 1970s, Ed Nather, an astronomer who received his training in nuclear physics, was equipping the observatory's telescopes with computers. He had bought a computer from a new company called Data General, and one night, it developed problems. Nather unscrewed the cover from the power supply and, in the course of fixing the computer, discovered that he had bought Data General's very first minicomputer. The serial number was 00001. That computer is now on display at Data General's headquarters. The company's sales have grown thirty times since 1972 and were $1.2 billion in 1985. Basic science paid off, in real dollar terms, and in a very short time.

## MARS: THE NEXT LONG-TERM GOAL?

I have mentioned the possibility of sending a human crew to Mars several times in this book. Such an expedition, probably an international one, has been widely discussed as the next long-range goal for NASA beyond the space station. A presidential commission headed by Spiro Agnew reported to Nixon in 1969 that a trip to Mars should be the next long-range goal. The recent report of the National Commission on Space, headed by Thomas Paine, suggested a similar focus. Both presidential commissions visualized such a trip as being part of a larger effort to expand human frontiers into the inner Solar System, not as a single-minded crash program like the one-shot *Apollo* lunar-landing program.

Can we go to Mars? Mars, at its closest, comes within thirty-five million miles of the earth, over one hundred times the distance to the moon. This greater distance does not mean that a Mars expedition has to cost one hundred times the cost of *Apollo*. The difficulty of going somewhere in space is best measured in terms of the velocity change required, not in miles, and in these terms, Mars is not much more difficult to get to than the moon (see Chapter 3). The major challenge is the time it will take to get there. NASA has identified four different ways of getting to Mars, with different travel times. The most extended mission profile has astronauts taking nine months to get to Mars, reaching it when it is on the other side of the sun from Earth. They would stay a year and a half, waiting for Mars and Earth to get into position so that they could then take another nine-month journey to get back. Other mission profiles, which require more fuel or a gravity assist from Venus, shave a month or two from each travel leg and involve spending a much shorter time on Mars. A final possibility is to launch two vehicles to Mars. The astronauts would leave one vehicle, descend to Mars in a separate descent module, and then take off from the surface in time to meet the vehicle that they will return to Earth in. The National Commission on Space mentioned the possibility of setting up a number of "cycling spaceships" that would stay permanently in orbit between Earth and Mars. Someone who wanted to go from one planet to the other would match velocities with it, transfer to the cycling spaceship, and go.[24]

Extended spaceflight presents a number of problems that must be solved before a trip to Mars is undertaken. We have to find some medical or exercise treatment to prevent bone thinning, the gradual loss of calcium from the human skeleton that persists through long-duration spaceflight, or else pay the penalty of using a massive spinning wheel or cylinder as a Martian spaceship. The psychological hazards of long-duration spaceflight and the potential instability of a small isolated group need to be understood better so that we can select crews intelligently. We need to understand how to treat diseases in space, as it will no longer be possible to return a sick astronaut to earth in a few days. The U.S. space program has not even begun to develop a closed life-support system, where water can be reused instead of being dumped overboard. A Martian ship does not need to be a completely closed ecosystem, a "spaceship earth" where everything is recycled, but clearly, the "water loop" needs to be closed much better, so that wash water, cabin humidity, and possibly even urine can be redistilled and reused rather than being dumped overboard as was the case with *Skylab*.

The cost of the trip to Mars is hard to estimate at this point, but it's less than one might think. Most advocates of a manned mission to Mars see it as the first step toward the settlement of the red planet. Therefore, they suggest a mission profile that would develop some infrastructure that could be used later on as a piece of a Mars base. Typical estimates of the amount of mass required is several million kilograms. The Saturn V rocket that took the *Apollo* astronauts to the moon had a lift-off weight of 6,423,000 pounds, or about 3 million kilograms. The Mars mission, then, involves about as much mass as the *Apollo* project did, and the costs should not be very different. A detailed study in 1981 by NASA's Humboldt Mandell indicates that it may well be cheaper to go to Mars, in a range of $25–$50 billion as compared with the $75-billion cost of *Apollo*.[25] (I have converted Mandell's figures to 1986 dollars.) The National Commission on Space believes that its complete agenda—which includes sending people to Mars, among other ventures—could be accomplished with a modest increase in NASA's current budget, increasing to $20 billion dollars per year by 1995 (from the present $8 billion) and then growing slowly, less rapidly than the projected real growth in the U.S. gross national product (GNP) of 2.4% per year. The percentage of GNP invested in space would always be less than half the percentage invested during the *Apollo* years.

Why should we go to Mars? The technological optimists have a simple answer: "We should to go Mars because we can." However, there are many feats that are possible technologically but that don't make sense in economic or human terms. Advocates of a mission to Mars mention other reasons, some of which even contradict each other. On a Martian expedition, humans could:

- Explore Mars scientifically.
- Search for life.
- Reach beyond the niche in which we were born.
- Establish an independent, semi-isolated branch of humanity.
- "Terraform" the planet and make it readily habitable.
- Get to Mars before the Russians do.
- Cooperate with the Russians in getting there.[26]

These different magnets that reach out and pull various people toward Mars fall, basically, into three groups: scientific, human, and political. Let's explore them.

I cannot support a purely scientific justification for sending people to Mars because robot probes could do almost as good a job for a fraction of the cost. There is a great deal of scientific interest in Mars, as a planet that is similar to Earth in its initial mass and position within the Solar System, but different in its present condition. If we go, a great deal of planetary science will, in fact, be done by robot prospectors, which will pave the way for a human expedition. Although there are things that human visitors can do that automated landers can't do, our capabilities of teleoperation have advanced far beyond the days of the *Apollo* lunar landing, and the science by itself, which could not (and never did) justify the *Apollo* program, certainly can't justify sending people rather than robots to Mars. The justification for a trip to Mars has to go beyond science.

When I read what's written by advocates of a Mars expedition, it is what I'm calling the "human" reasons to go that play a large role. The Martian expedition is seen as part of a much broader push to colonize or settle the inner Solar System. It is seen as the logical descendant of previous moves of humanity into hostile environments, perhaps going all the way back ten million years to the time when humans first used animal skins as warm clothing and moved away from the tropical forest that our ancestors first lived in. There is an important difference, however. The costs of the first human colonies were thousands of times smaller than will be the costs of the first space colonies (see Chapter 13).

The most radical proposal connected with Martian settlement involves "terraforming" Mars, intentionally changing its environment to make it more habitable. This idea, first put forward twenty years ago by Carl Sagan, involves extracting volatile water vapor and carbon dioxide from the Martian surface in order to thicken its atmosphere. Mars is cold because its atmosphere is too thin to create a significant greenhouse effect. The initial idea was to spread carbon black on the Martian polar caps, causing them to evaporate and produce a thick atmosphere. Once the atmosphere is made thick, the hypothesis is that the "new" Mars will be quite stable.

I doubt that terraforming will be quite that easy. Let's look at some numbers. What we need to do is to dump a lot more water into the Martian atmosphere. The terrestrial greenhouse effect is produced largely by water and carbon dioxide, and Mars has the carbon dioxide already. If we could find on Mars a huge circular slab of ice, 100 meters thick and 230 kilometers in radius, and if we could figure out a way to transform this into water vapor, we could give Mars an atmosphere that would have as much water and carbon dioxide above each square centimeter of Mars as Earth has.[27] To make Mars truly Earth-like, we would then have to figure out a way of putting the necessary oxygen and, possibly, nitrogen into the atmosphere. One way of doing this is to use green plants, algae, or microbes to create the oxygen from carbon dioxide. Possibly, once we warmed Mars up, the carbon dioxide that is currently locked up, chemically absorbed on the rocks, would evaporate into the atmosphere. If we needed to find the carbon dioxide somewhere on Mars, some future astroengineer would need to process a 1-kilometer-thick, 320-kilometer-in-radius circular slab of dry ice. Mars probably contains the necessary volatiles, though we aren't absolutely sure; the layered deposits at the poles are up to 5 kilometers thick and are currently interpreted as being dust cemented by ice.[28] The permanent northern polar cap, almost certainly water ice, has a radius of 500 kilometers; the southern cap, which may be dry ice, has a radius of 175 kilometers.

This sort of astroengineering is visionary indeed, and our knowledge of the condition of Mars is far too limited for us to know whether it is possible in principle, not to mention practical. This sort of planetary engineering was the subject of a NASA summer study in the mid-1970s. Robert MacElroy was not encouraged to think small, and the study discussed the possibility of using genetic engineering techniques to create a supermicrobe that would thrive in the existing environment and do the necessary photosynthesis.[29]

The last set of reasons for going to Mars has to do with the terrestrial political situation. The Soviet Union has not announced any plans to send people to Mars, but rumors of such plans are widespread. One basis for such rumors is widely quoted offhand comments by individual Soviet scientists. The science magazine *Discover* quoted Georgiy Skryabin, scientific secretary of the Soviet Academy of Sciences, as saying "We're going to Mars." Former Senator Harrison Schmitt has said that Anatoly Alexandrov, former head of the Soviet Academy, "has said they are interested."[30] The National Commission on Space cited this theme, but in a more muted tone:

> Although there are conflicting accounts by high-ranking Soviet scientists as to when such a mission might be attempted (some have said in the near future, others have indicated it will not be until after the turn of the century), it seems clear that they will visit Mars within the next 50 years.[31]

Another source of these rumors is the Soviet preoccupation with long-duration spaceflight; there is no apparent reason to have left cosmonauts in *Salyut* for six months unless they are planning to send people to some distant place like Mars. Of course, the Soviets could simply be investigating the possibilities of sending people to Mars rather than definitely planning on an expedition. (Incidentally, that's what I think we should be doing.)

It's clear to me that the Soviets are interested in Mars. If it turns out to make sense to send people there in the next fifty years, they probably will. However, I think we need to be careful in assessing Soviet intentions. A statement like "We're going to Mars" could well mean "We're sending a mission to Mars" and may not necessarily imply a human expedition. The history of the *Apollo* program also calls for a guarded rather than a literal interpretation of Russian comments regarding their intentions.

There are two ways of reacting to the apparent Soviet interest in Mars that have been discussed by the advocates of American participation in a mission to Mars. One view has it that the Soviets are competition, and if anyone is to land on Mars, it should be people from the free world. In other words, the proposal is that the Soviet interest should stimulate an *Apollo*-style space race.

A more interesting approach has been taken by Senator Spark Matsunaga of Hawaii and others, who see the Soviet interest in Mars as a beautiful opportunity to demonstrate the possibility of American-Soviet cooperation. Matsunaga, in particular, has been active in introducing resolutions in Congress that support the general principle of international cooperation in space. The relative strengths of the two nations' space programs dovetail quite well at this time. The Soviet Union has had far more experience with long-duration spaceflight and is far more familiar with the difficulties that the crew will face. However, the Soviets have had a great deal of difficulty in actually building hardware that lasts a long time in space; in this respect, the United States is in the lead.

Realistically, what are the prospects for a cooperative mission? The two nations did fly the joint *Apollo–Soyuz* project in 1976. In 1977, the two nations signed an agreement calling for a joint shuttle–*Salyut* mission and an international space-platform program. In 1978, the United States pulled out. Senator Matsunaga contends that space cooperation was canceled because of Poland, because the Carter administration sought to punish the Soviets for its policies toward dissidents. In the past, the space program has become a political football, used to reward or punish the Soviets by administrations that see cooperative agreements as favorable exclusively to the Soviets. If real Soviet-American cooperation is to succeed, present and future administrations must recognize that America no longer has a preeminent position in space and that cooperation in a mission to Mars would benefit both sides.

I'm quite sure that, someday, humans will leave their footprints in the sands of Mars. How soon, and in what mode, is an open question. If we decide that the human destiny is to settle the inner Solar System, to establish self-sustaining bases on the moon, Mars, and possibly even the asteroids, then a manned mission to Mars will probably come sooner rather than later. If space settlement does not happen that soon, then a trip to Mars would most likely be a symbolic one. "Pride and power" could motivate an expedition to Mars. Another possibility is that a trip to Mars could be a symbol of international cooperation. In either case, the expedition would probably be limited to a single visit, or a few at most, like the *Apollo* program.

## SETTLEMENT OF THE INNER SOLAR SYSTEM

A mission to Mars is seen by many as one step in the settlement, or colonization, of the inner Solar System. The National Commission on Space has laid out a complete agenda for establishing low-cost access to the inner solar system. This bold plan, established before *Challenger,* does not include solar sailing, terraforming, or large-scale space colonies, but it does contain many other ideas that have been discussed in the community. The National Commission's vision is that of a human presence in the solar system by 2030, which is somewhat comparable in scale to the European presence in the United States in 1620, just after the Pilgrims landed. The timetable calls for a permanent lunar base by 2005, which will be used as a source for the raw materials needed for some of the bulky components of spaceships which will be used on the Mars run. An initial human outpost on Mars is visualized by 2016, with a full "Mars base" set for 2030. It remains to be seen whether it is a useful blueprint for the future.

Planetary scientists Bruce Murray, Michael Malin, and Ronald Greeley have posed two questions that are the key to a decision on whether humans should establish outposts on Mars, or anywhere else beyond Earth for that matter: (1) Can we carry on important activities there better than on earth? (2) Can much of the materials needed for life support be obtained locally?[32] The two questions are related because they are the two basic sides of a cost–benefit analysis. If the activities carried on in a space colony are really important, the need to transport all the materials for life support to that space colony can perhaps be justified.

We have only partial answers to both questions. Let's take the local availability of life support materials first. We believe that there is enough water and carbon dioxide on Mars for a human expedition to use. However, is the water readily accessible, or would we have to ship tons of diamond-tipped drills and drilling rigs to get it out of deep layers of permafrost? Are there places on the moon where the rocks contain a great deal of water in usable form? We don't know.

We do know enough about Mars, the moon, and other destinations in outer space to be sure that the Martian explorers will have a much more difficult time than Columbus did; remote sensing has given us a clear picture of a rather bleak planet. Bleak planets, however, are not necessarily useless. Millions of years ago, humans learned how to wear clothes and leave the tropical equatorial climate that we were born in. Building structures on Mars or on the moon that will allow us to survive in these hostile environments is more difficult than using animal skins to keep us warm. Earlier, I mentioned the 1975 summer studies that estimated that it would cost $40 million (1986 dollars) per person to build a colony at L-5, in the Earth–moon system. Establishing a self-sustaining colony on Mars will probably be at least that expensive.

This brings us to the other question: whether humans can do things on Mars, the moon, or wherever that can't be done equally well anywhere else. What we do in space, the main focus of this book, will determine in the long run where we go. The future, if any, of space industrialization holds the key to this question. Exploration of the near and distant universe could not, by itself, justify the tremendous costs of a human expedition to Mars, or even a lunar base. Advocates of a mission to Mars often mention as an analogy the Antarctic, where our bases in Antarctica were established for primarily scientific reasons. I think this analogy is a mistaken one. The costs of maintaining humans in the Antarctic are much less than the costs of going to Mars. The

National Science Foundation spends $110 million per year on the Antarctic research program, a cost of a little more than $1 million per person.[33] A Mars expedition is roughly a thousand times as expensive. In addition, there are geopolitical reasons for various nations to maintain an outpost in Antarctica, as there is a possibility of using that continent for economic or even military purposes.

Although the vision of humans landing on Mars may possibly be valuable enough to support a political decision to send one expedition there, the colonization or settlement of Mars or the moon will not occur unless the colonies can provide some sort of tangible return on the investment. Currently, there are no space industries that require permanent human presence on Mars or the moon. The equivalent of the North Atlantic codfish, South American gold, or East Asian spices simply isn't there yet, especially when we look beyond earth orbit. The communications satellite industry will at most require humans to be repairers in low earth orbit, not colonists on Mars. Of the other activities discussed in this book, virtually all require humans only to be repairers on the space station. A permanently inhabited station in geosynchronous orbit may possibly benefit the earth observation program in space sciences or may possibly be of use to the communications industry. My view, then, is that it is certainly premature to regard the settlement of the inner Solar System as inevitable.

There is one somewhat more speculative possibility. If large-scale space manufacturing happens (remember that this is a very big if), a lunar base might be justifiable on economic grounds. Such a base could be a low-cost source of raw materials for the space factory. The existence of such a base could make the Mars run quite inexpensive indeed and could make the Antarctic analogy much more apt than it currently seems.

Of course, future space industries may appear on the horizon. We don't know enough about Mars to rule out the possibility that it could be the source of some exotic material that would justify an astronomical freight charge. It may turn out that it will be easier to establish a base on Mars than we currently think, and that the more modest benefits of establishing a research station would justify an expedition.

What should be done, in my view? The National Commission on Space has set forth a bold vision that contains a large number of worthwhile missions, not just a document that says we should go to Mars and do nothing else. There are enough reasons to entertain the possibility of at least a one-shot, symbolic mission to Mars to suggest that Mars be adopted as a long-range goal for the American space program. At the same time, we need to recognize that the uncertainties of a Martian expedition are such that the balance and breadth of NASA's programs need to be maintained. No one has suggested that the Mars project dominate the space program in the way that *Apollo* did; in fact, the report of the National Commission specifically rejects such a narrow focus.

The adoption of such a long-range goal would then guide the shorter range decisions that are made as specific projects are selected in the remainder of this century. Many of the projects that planetary scientists have proposed as the next logical steps are necessary as the first steps toward a Martian expedition anyway. The space shuttle manifest will guide us through the mid-1990s, and the remainder of this century can be spent exploring Mars by means of automatic probes, orbiters, and penetrators. Parallel to this effort, we need to gain additional understanding of long-duration spaceflight. Then, as the twenty-first century dawns, we can determine whether it's worthwhile to go to Mars.

*   *   *

For the next ten years, NASA will have its hands full rebuilding the space shuttle, building the space station, and maintaining a space science and applications program so that we can do something with all of that infrastructure once it's there. In the longer run, the capabilities of the space station will permit major advances in space applications. Television signals broadcast from satellites and Dick Tracy radios are possibilities, or more than possibilities. In space science, we can't forecast the discovery of something that is now unknown, but the pace of the last fifteen years shows no sign of slackening.

An issue for the twenty-first century is the question of whether we want to send humans to Mars. Some people associated with the space program think we must go; others, including myself, are more guarded with respect to an immediate commitment. The major unanswered question regarding the settlement of the inner Solar System is what the colonists will do to justify the tremendous investment in establishing their presence in a remote, hostile environment. Whether or not we decide right now that we're going to Mars, the initial steps are the same: exploration of the red planet by remotely operated orbiters, penetrators, and possibly even landers, along with the maintenance of a balanced space program. We can determine just what Mars is like and what resources a human expedition will need. Then, and only then, can we make the decisions about our future in space.

# Summary of Part IV

The way we do business in space is changing, and it will change even more once the space station is up there. Several years ago, the choice was simple: we could send an expensive astronaut or a dumb machine, and there was no middle ground. However, the roles that humans can play in space are becoming much more varied, and the capacities of teleoperation can expand the ways in which humans and machines can interact to get the job done better.

The next ten years will be hard times for people in the space business. Shuttle launch slots will be scarce while the shuttle is fixed and the fleet rebuilt; it will take ten years to launch most of the missions that are currently in progress and waiting for a ride into orbit. After 1995, the capabilities of the space station will, we hope, make it much easier to gain access to space.

Once the space station is constructed, humanity will face a major decision about whether human settlement in space is to extend beyond the station itself and its associated platforms in low earth orbit. The present status of space industrialization does not suggest a compelling need, or even a reasonable justification, for settling the inner Solar Systems, but new space industries may well appear on the horizon. In addition, there may well be other justifications for at least a symbolic human mission to Mars at some time early in the twenty-first century.

Even if we don't go to Mars, the space station will make it possible for humanity to move beyond the cradle of Earth, to begin living and working in space, exploring the near and distant universe, and using the space environment for a wide variety of applications.

# Reference Notes

## CHAPTER 1

1. The survey was done by Market Opinion Research and was presented by NASA administrator James Fletcher in an address to the National Issues Forum on the Space Program, The Brookings Institution, Washington, DC, on 1 October 1986.
2. Largely because I like to check such assertions, I picked eight major figures, guided by the table of contents of Samuel Eliot Morison's *The European Discovery of America* (New York: Oxford University Press, 1971). These pioneer explorers of America are Ericsson, John Cabot, Columbus, Cartier, Verazzano, Hudson, Magellan, and Drake. Four of the eight were killed in action: Cabot and Hudson disappeared at sea, and Verazzano and Magellan were massacred by natives. The other four died of natural causes. Another indication of the hazards of exploration is the fate of the crew that set forth with Magellan in 1517 in the first successful attempt to sail around the world, described by Morison as the greatest journey in the history of humanity. Most of this crew, including Magellan, didn't survive the trip.
3. The sequence of events is largely drawn from the *Report to the President by the Presidential Commission on the Space Shuttle Challenger Accident* (Washington, DC: Government Printing Office, 1986), Chapter 3.
4. David K. Sanger, "NASA Breaks Some More Bad News," *The New York Times,* 3 August 1986, sec. 4, p. 7; Paul Recer, "Doomed Crew May Have Known Fate," *Wilmington (Del.) Evening Journal,* (29 July 1986):A1, A4; Philip M. Boffey, "NASA Discusses Seeming Conflicts in Disclosures on Astronauts' Deaths," *The New York Times,* (2 August 1986):5.
5. *Presidential Challenger Commission,* 184.
6. *Presidential Challenger Commission,* 120–121.
7. Philip M. Boffey, "NASA Set to Alter Booster Joint, Rejecting a Radical New

Design," *The New York Times,* (3 July 1986):A1, A24; "NASA Picks New Rocket Design for Space Shuttle," *New York Times,* (13 August 1986):A17.
8. Quoted in the *Presidential Challenger Commission,* 125.
9. Ibid., 249–252.
10. Ibid., 139.
11. Ibid., 93.
12. Ibid., 148.
13. M. Mitchell Waldrop, "NASA Responds to the Rogers Commission," *Science* 233, (1 August 1986):512–513.

# CHAPTER 2

1. Walter A. McDougall, . . . *the Heavens and the Earth: A Political History of the Space Age* (New York: Basic Books, 1985), 74–140.
2. McDougall, 112–131; James E. Oberg, *Red Star in Orbit* (New York: Random House, 1981), 14–38.
3. Joseph D. Atkinson, Jr., and Jay M. Shafritz, *The Real Stuff: A History of NASA's Astronaut Recruitment Program* (New York: Praeger, 1985), pp. 149–152.
4. John F. Kennedy, Address to Congress on Urgent National Needs, 25 May 1961; quoted in Theodore R. Simpson, ed., *The Space Station: An Idea Whose Time Has Come* (New York: IEEE Press, 1985), 255.
5. Quoted in John Logsdon, "Opportunities for Policy Historians: The Evolution of the U.S. Civilian Space Program," in *A Spacefaring People,* ed. Alex Roland (Washington, D.C.: Government Printing Office, 1984), 84. See also John Logsdon, *The Decision to Go to the Moon* (Cambridge: MIT Press, 1970).
6. Ronald Reagan, National Space Policy Statement, 4 July 1982; quoted in Simpson, 273–279.
7. The source for the budget figures is Paul B. Stares, *The Militarization of Outer Space: U.S. Policy 1945–1984* (Ithaca, NY: Cornell University Press, 1984), 255–258, updated by John Logsdon and others in presentations to the National Issues Forum at the Brookings Institution, October 1986.
8. Spark M. Matsunaga, *The Mars Project* (New York: Hill & Wang, 1986), 49.
9. Samuel Florman, *The Existential Pleasures of Engineering* (New York: St. Martin's Press, 1976), especially chapters 8–11.
10. The phrase "pride and power," taken from the title of Vernon Van Dyke, *Pride and Power: the Rationale of the Space Program* (Urbana: University of Illinois Press, 1964), has been echoed by many, most recently by John Logsdon in his presentation to the National Issues Forum, Brookings Institution, 1 October 1986.
11. Matsunaga, Chapter 18.
12. Office of Technology Assessment, *International Cooperation and Competition in*

*Space Activities* (Washington, DC: U.S. Congress, Office of Technology Assessment, OTA-ISC-239, July 1985), 33.

13. Sessions at meetings of the American Association for the Advancement of Science (AAAS) are taped. This quotation is from a session on "Moving Industry into Space: I" at the 1984 annual meeting. A tape of the session (#84AAAS-58) can be obtained from Mobiltape, 1741 Gardena Ave., Glendale, CA 91204.

14. James E. Oberg and Alcestis R. Oberg, *Pioneering Space: Living on the Next Frontier* (New York: McGraw-Hill, 1986), 3.

15. See Ray Allen Billington, *The Genesis of the Frontier Thesis: A Study in Historical Creativity* (San Marino, CA: Huntington Library, 1971), 1–8; Ray Allen Billington, *Westward Expansion: A History of the American Frontier*, 2nd ed. (New York: Macmillan, 1960), 1–14.

16. James Van Allen, "Space Science, Space Technology, and the Space Station," *Scientific American* 254, no. 1 (January 1986):32–39; "Myths and Realities of Space Flight," *Science* 232 (30 May 1986):1075–1076. For responses to the *Science* article see *Science 233* (8 August 1986):610–611.

17. Karl Henize, in an interview with Leif Robinson of *Sky and Telescope* magazine, elaborates on his perspective in *Sky and Telescope* 72, no. 5 (November 1986):446–449.

18. John F. Kennedy, speech at Rice University on September 12, 1962, quoted in Theodore R. Simpson, ed., *The Space Station: An Idea Whose Time Has Come* (New York: IEEE Press, 1984), 260–261.

19. Quoted in Francis X. Clines, *The New York Times,* (17 June 1986):C5.

20. From the Aeronautics and Space Report of the President (1980), quoted in John A. Simpson, "Space Science and Exploration: A Historical Perspective," in Alex Roland, ed., *A Spacefaring People* (Washington, DC: Government Printing Office, 1985), NASA SP-4405, 18.

21. Marcia Smith, talk presented to the 1983 AAAS meeting.

22. See D. Dickson, "Redesign of Ariane is Under Way," *Science* 233 (25 July 1986):411–412; *Aviation Week and Space Technology* (9 June 1986):46 ff.; Ray Williamson *et al., International Cooperation and Competition in Civilian Space Activities,* (Washington, DC: U.S. Congress, Office of Technology Assessment, OTA-ISC-239, July 1985), 110–116.

23. For descriptions of Japanese technology, see Williamson *et al.,* 77–83; Donald E. Fink, "Japan's Space Initiative," *Aviation Week and Space Technology,* (14 July 1986):13, 18ff.

24. Masao Uamanouchi and Yukihiko Takenaka, "Space Development Activities in Japan," in M. Nagatomo, ed., *Proceedings of the Fourteenth International Symposium on Space Technology and Science* (Tokyo: Agne Publishing, 1984), 62–63 (in English).

25. John F. Burns, *The New York Times,* (19 May 1986):D10. American restrictions on technology transfer may prevent this launch from actually happening.

26. Donald E. Fink, "China's Technology Leap," *Aviation Week and Space Technology* (26 May 1986):11.
27. D. Dickson, "Australia Bids to Reenter the Space Race," *Science* 231 (3 January 1986):16–17.

## CHAPTER 3

1. See *NASA Space Systems Technology Model,* 6th ed., NASA TM 88174 (Washington, DC: NASA, 1985), 1–31 to 1–67.
2. William J. Broad, "Blow to Security Seen in Loss of Titan Missile," *The New York Times,* (20 April 1986):1, 32; ibid., (22 April 1986):C1, C3; R. Jeffrey Smith, "Titan Accident Disrupts Military Space Program," *Science* 232, (9 May 1986):702–704.
3. M. Mitchell Waldrop, "After Challenger: Painful Choices," *Science* 232, (13 June 1986):1335–1337.
4. Theresa M. Foley, "Reagan Bars Shuttle from Competing for New Satellite Launch Contracts," *Aviation Week and Space Technology,* (25 August 1986):22–23.
5. A. Roland, "The Shuttle: Triumph or Turkey?" *Discover,* 6 (November 1985):29.
6. Alex Roland, "The Space Shuttle Program: A Policy Failure?" *Science* 233, (10 May 1986):1099–1106.
7. Roland, "Triumph or Turkey," 79–80.
8. *Report to the President by the Presidential Commission on the Space Shuttle Challenger Accident* (Washington, DC: Government Printing Office, 1986), 164–177.
9. See, for example, G. Harry Stine, *The Third Industrial Revolution* (New York: Putnam, 1975), 37, for a prediction of $38 per pound in 1974 dollars (this amount corresponds to $82 per pound in 1985 dollars).
10. R. Michael Hord, *Handbook of Space Technology: Status and Projections* (Boca Raton, FL: CRC Press, 1985), 5, shows a figure of about $6,000/kg (1980 dollars = $3,500/lb in 1985 dollars) for a Delta launch in 1970; A. Roland, "The Shuttle: Triumph or Turkey," *Discover* 6, no. 11 (November 1985), quotes a lower range of $500–$1,000/lb (1970 dollars), or $1,400–$2,800/lb (1985 dollars). Various speakers at a meeting of the American Association for the Advancement of Science in 1984 quoted numbers in the $1,000–$2,000-per-pound range (talks by T. Rogers, J. Townsend, and M. Schneiderman, "Moving Industry into Space," recorded on audiotapes 84AAAS 58–63, available from Mobiltape Co., 1741 Gardena Ave., Glendale, CA 91204), consistent with the shuttle costs cited in Roland's article.
11. Roland, "Triumph or Turkey?" 45–49.
12. Allen provided a fascinating account of this episode in his book (written with

Russell Martin), *Entering Space: An Astronaut's Odyssey*, rev. ed. (New York: Stewart, Tabori, & Chang, 1985).

13. National Commission on Space, *Pioneering the Space Frontier*, (New York: Bantam Books, 1986), 110.

14. "United Technologies Could Ready Unmanned Carrier in Three Years," *Aviation Week* (24 November 1986):57; "United Technologies Designs Heavy-Lift Launch Vehicle," *Aviation Week* (17 November 1986):19.

15. David Dickson, "France Pushes Europe towards Manned Space Flight," *Science* 231, (17 January 1986):209–210; see also *Aviation Week and Space Technology*, (3 February 1986):27; J.-C. Cretenet, "Ariane 5—Hermes," in L. J. Carter and P. M. Bainum, eds., *Space: A Developing Role for Europe* (San Diego: Univelt, 1984), 165–186.

16. C. Covault, *Aviation Week and Space Technology*, (14 April 1986):6–18; *Aviation Week*, (1 September 1986):42; John Noble Wilford, *The New York Times*, (12 August 1986):C1, C9.

17. See National Commission on Space, 112–115; Jerry Grey, "The New Orient Express," *Discover*, (January 1986):73–81.

18. G. L. Grodzovskii, Yu. N. Ivanov, and V. V. Tokarev, *Mechanics of Spaceflight Low-Thrust* (Jerusalem: Israel Program for Scientific Translations, 1969), 40–44.

19. R. Michael Hord, *CRC Handbook of Space Technology: Status and Projections*, (Boca Raton, FL: CRC Press, 1985), 35–44. The figure which is the only mention of solar sailing in this book is on page 36.

20. *NASA Space Systems Technology Model*, 6th ed., NASA TM 88174 (Washington, DC: NASA, 1985).

# CHAPTER 4

1. S. Astrain, "Global Overview of Satellite Communications," in J. Alper and J. N. Pelton, eds., *The Intelsat Global Satellite System*, Vol. 93 of *Progress in Astronautics and Aeronautics* (New York: American Institute of Aeronautics and Astronautics, 1984), 3.

2. See, for example, "Voices from the Sky," originally published in *Spaceflight*, March 1968, and reprinted in Alper and Pelton, 7–20; see also A. C. Clarke, *Ascent into Orbit: A Scientific Autobiography* (New York: Wiley, 1984).

3. Clarke, 11.

4. E. W. Ploman, *Space, Earth, and Communication* (Westport, CT: Quorum Books), 125.

5. John McPhee, *Coming into the Country* (New York: Farrar, Straus, & Giroux, 1977).

6. From a tape recording of "Space Industrialization: the Growing Community," 1983 meeting of the American Association for the Advancement of Science. Tape recording available from Mobiltape, 1741 Gardena Ave., Glendale, CA 91204.

7. Christopher S. Wren, "Computer Age Comes to Eskimos in Canada," *The New York Times,* (27 March 1986):A15.

8. There are now at least two magazines for home satellite TV enthusiasts. Many of these data come from *TV Satellite Videoworld* 1, no. 1 (September 1985), and *STV* 3, no. 7 (July 1985). *OnSat* is a weekly programming guide.

9. Mark Long, *A Guide to Satellite TV* (Mendocino, CA: Quantum Publishing, 1985), 54.

10. *TV Satellite Videoworld* 1, no. 1 (September 1985).

11. See the magazines listed earlier, and also David Owen, "Satellite Television," *The Atlantic* 255, no. 6 (June 1985):45–62.

12. *TV Satellite Videoworld* (September 1985):10–12.

13. M. Gustafson, *STV* (July 1985):99.

14. J. Goldhirsh, *Johns Hopkins APL Technical Digest* 5, no. 4 (October 1984):360–363.

15. Reginald Stuart, "Rules on Satellite Orbits Irk Developing Nations," *The New York Times,* (8 August 1985):D1–D2.

16. M. Wallick, "Leading the Way," *Wilmington (Del.) News Journal,* (19 May 1986):A1, A8.

17. Jay C. Lowndes, "Joint Venture Installs Earth Stations for Shopping Mall Tele-conferencing," *Aviation Week and Space Technology,* (21 April 1986):57.

18. M. J. Richter, "Satellite Signals Let Meetings Stay Home," *Hartford Courant,* (20 July 1986):C1, C2.

19. J. Ott, "Videoconference Use Expands to Meet Business Needs," *Aviation Week and Space Technology,* (22 July 1985):157–163; *Forbes,* (21 October 1985):8.

# CHAPTER 5

1. Yes, it checks. NOAA's weather data for 1980 give twenty-nine days with more than a tenth of an inch of rain for the 330 days in all months of the year excluding August at the University of Arizona station in Tucson.

2. R. A. Kerr, "Pity the Poor Weatherman," *Science* 228, (10 May 1985):704–706; see also R. A. Kerr, "Forecasting the Weather a Bit Better," *Science* 228, (5 April 1985):40–41; R. A. Kerr, "Weather Satellites Coming of Age," *Science* 229, (19 July 1985):255–257.

3. U.S. Congress, Office of Technology Assessment, *Civilian Space Policy and Applications* (Washington, DC: Government Printing Office, 1982), 341–342.

4. For the purposes of this example, I'm assuming that it costs the utility 5 cents to generate each kilowatt hour. The actual cost of electricity varies from one part of the country to another but is really somewhat higher because you have to add the cost of transmission and billing. A 10% saving is a saving of 0.5 cents per kilowatt hour, which (when multiplied by America's use of 600 kilowatt hours per year) adds up to $3 billion.

5. W. L. Smith, W. P. Bishop, V. F. Dvorak, C. M. Hayden, J. H. McElroy, F. R. Mosher, V. J. Oliver, J. F. Purdom, and D. Q. Wark, "The Meteorological Satellite: Overview of 25 Years of Operation," *Science* 231, (31 January 1986):455–462.

6. M. Mitchell Waldrop, "A Silver Lining for the Weather Satellites," *Science,* 226, (14 December 1984):1289–1291.

7. Perry's testimony and the suggestion that the Iranian rescue mission could have been saved by satellite navigation are contained in Thomas Karas, *The New High Ground: Systems and Weapons of Space Age War* (New York: Harper & Row, 1983), 124–126.

8. National Oceanic and Atmospheric Administration, *NOAA Satellite Programs Briefing (August 1983)* (Washington, DC: Department of Commerce, 1983), 26–27, 70–71.

9. The Doppler effect is a technical name for an effect that causes the frequency of a signal to change, depending on whether the source of the signal is moving toward or away from you. Listen to the siren on a police car, or the whistle on a train, and you will notice the musical pitch drop when the vehicle passes you (if you have a musical ear), because the frequency is boosted by the Doppler effect when the siren moves toward you and is lowered by this effect when the siren moves away from you. Don't worry: you don't need to understand the details (I hope) in order to get my main point either here or elsewhere.

10. Robert Nichols, "Rescue from Space: Search and Rescue Satellites," *STV* 3, no. 9 (September 1985):54–59.

11. General Advisory Committee [to the President] on Arms Control and Disarmament, "A Quarter Century of Soviet Compliance Practices under Arms Control Commitments," in W. C. Potter, ed., *Verification and Arms Control* (Lexington, MA: D. C. Heath, Lexington Books, 1985), 239–250.

12. B. D. Blair and G. D. Brewer, "Verifying SALT Agreements," in W. C. Potter, ed., *Verification and SALT: The Challenge of Strategic Deception* (Boulder, CO: Westview Press, 1980), 9.

13. D. A. Wilkening, "Monitoring Bombers and Cruise Missiles," in Potter, *Verification and Arms Control,* 110–111.

14. B. G. Blair and G. D. Brewer, "Verifying SALT Agreements," in Potter, *Verification and SALT,* 19.

15. J. Richelson, "Technical Collection and Arms Control," in Potter, *Verification and Arms Control,* 169–216. This is a good summary of satellite capabilities.

16. Blair and Brewer, 16.

17. *Department of Defense Appropriations for 1983, Part 5,* 16; *Department of Defense Appropriations for 1984, Part 8,* 337; House Committee on Armed Services, *Department of Energy National Security and Military Applications of Nuclear Energy Authorization Act of 1984* (Washington, DC: U.S. Government Printing Office, 1984), 383–384.

18. J. Richelson, 190–197.

19. Ronald Reagan, quoted in U.S. Congress, Office of Technology Assessment, *Ballistic Missile Defense Technologies,* OTA-ISC-254 (Washington, DC: U.S. Government Printing Office, September 1985), 297–299.

20. U.S. Congress, Office of Technology Assessment, *Ballistic Missile Defense Technologies;* Jeff Hecht, *Beam Weapons* (New York: Plenum Press, 1984).

21. G. Yonas, "Strategic Defense Initiative: The Politics and Science of Weapons in Space," *Physics Today* (June 1985):24–32; Wolfgang K. H. Panofsky, "The Strategic Defense Initiative: Perception vs. Reality," 34–45.

22. Hecht, 353, based on U.S. Defense Department data.

23. Most of these numbers are from D. Walker, J. Bruce, and D. Cook, "SDI: Progress and Challenges," unclassified staff report submitted to Senator William Proxmire, Senator J. Bennett Johnston, and Senator Lawton Chiles, 17 March 1986. I obtained a copy from Senator Proxmire's office. The FY 1987 appropriation is taken from the *New York Times,* (31 October 1986).

24. Reagan's speech is quoted in "Excerpts from Reagan's Speech at Jersey School," *The New York Times,* (20 June 1986):A8.

25. J. E. Cushman, Jr., "Weinberger Terms Missile Defense Test a Success," *The New York Times,* (2 July 1986):A8.

26. U.S. Congress, Office of Technology Assessment, *Ballistic Missile Defense,* 286–289.

27. Quoted by Robert Scheer in *The Los Angeles Times,* (22 September 1985):1, 14.

28. Walker, Bruce, and Cook, 12 (emphasis in the original).

29. Quoted in Mark Crawford, "In Defense of 'Star Wars,'" *Science* 228, (3 May 1985):563.

30. Freeman Dyson, *Weapons and Hope* (New York: Harper & Row, 1984), especially Chap. 22.

31. U.S. Congress, Office of Technology Assessment, *Ballistic Missile Defense,* 223.

32. The list is modified slightly from Hecht, 86.

33. M. Mitchell Waldrop, "Resolving the Star Wars Software Dilemma," *Science* 232, (9 May 1986):710–713.

34. This list is largely taken from Jeremy J. Stone, "The Four Faces of Star Wars: Anatomy of a Debate," *FAS Public Interest Report* 38, no. 3 (March 1985):1, 3, 6–9, and U.S. Congress, Office of Technology Assessment, *Ballistic Missile Defense,* 5.

35. U.S. Congress, Office of Technology Assessment, *Ballistic Missile Defense,* 59–61.

36. See Paul B. Stares, *The Militarization of Space: U.S. Policy, 1945–1984* (Ithaca, NY: Cornell University Press, 1985); U.S. Congress, Office of Technology Assessment, *Anti-Satellite Weapons, Countermeasures, and Arms Control,* OTA-ISC-281 (Washington, DC: U.S. Government Printing Office) 52, 91–102.

# CHAPTER 6

1. J. D. Rosendhal, private communication, June 1985.
2. D. McLain, National Marine Fisheries Service, quoted in D. James Baker *et al.*, Satellite Planning Committee of Joint Oceanographic Institutions, Inc., *Oceanography from Space: A Research Strategy for the Decade 1985–1995* (Executive Summary), (Washington, DC: Joint Oceanographic Institutions, 1984), 6–7. [If you need the address it's 2100 Pennsylvania Ave NW, Washington, DC 20037.]
3. For a review of El Niño, see Colin S. Ramage, "El Niño," *Scientific American* 254, no. 6 (June 1986):76–83.
4. For a series of articles on the 1982–1983 El Niño, see M. A. Cane, "Oceanographic Events during El Niño," *Science* 222 (1983):1189–1194; E. M. Rasmussen and J. M. Wallace, "Meteorological Aspects of the El Niño/Southern Oscillation," *Science* 222 (1983):1195–1202; R. T. Barber and F. P. Chavez, "Biological Consequences of El Niño," *Science* 222 (1983):1203–1208.
5. Robert C. Beal, "Ocean Research with Synthetic Aperture Radar," *Johns Hopkins APL Technical Digest* 6, no. 4 (October–December 1985):293–299.
6. John R. Apel, "Satellite Sensing of Ocean Surface Dynamics," *Annual Review of Earth and Planetary Science* 8 (1984):303.
7. Science and Mission Requirements Working Group for the Earth Observing System, *Earth Observing System*, NASA Technical Memorandum 86129 (Greenbelt, MD: NASA-Goddard Space Flight Center, for sale by the National Technical Information Service, Springfield, VA), 16–19.
8. Earth System Sciences Commiteee, NASA Advisory Council, *Earth System Science: A Program for Global Change (Overview)* (Washington, DC: NASA, 1986).
9. Unfortunately I've had a great deal of difficulty finding a popular or even a semipopular exposition of these magnetospheric phenomena. Some reviews for specialists are D. J. Williams, "The *OPEN* [Origin of Plasma in the Earth's Neighborhood] Program: An Example of the Scientific Rationale for Future Solar-Terrestrial Research Programs," in M. Dryer and E. Tandberg-Hanssen, eds., *Solar and Interplanetary Dynamics* (Dordrecht, Holland: Reidel, 1980), 507–522; M. A. Pomerantz, "Cosmic Rays," in L. J. Lanzerotti and C. G. Park, eds., *Upper Atmosphere Research in Antarctica*, Antarctic Research Series Vol. 29 (1979), pp. 12–35. A voluminous set of committee reports is Stirling Colgate *et al., Space Plasma Physics: The Study of Solar System Plasmas*, report of the Space Science Board of the National Research Council, Vols. 1 and 2 (Washington, DC: National Academy of Science Press, 1978). A more recent committee report is S. M. Krimigis *et al., An Implementation Plan for Priorities in Solar-System Space Physics* (Washington, DC: National Academy Press, 1985).
10. European Space Agency, *European Space Science: Horizon 2000*, ESA SP-1070 (Paris: European Space Agency, 1984), 3–18.

11. Quotations not otherwise credited are from an interview with Dr. Klemas conducted in the fall of 1985.
12. V. Klemas, "Remote Sensing of Coastal Fronts and Their Effects on Oil Dispersion," *International Journal of Remote Sensing* 1, no. 1 (1980):11–28.
13. V. Klemas and W. D. Philpot, "Drift and Dispersion Studies of Ocean-Dumped Waste Using Landsat Imagery and Current Drogues," *Photogrammetric Engineering and Remote Sensing* 47, no. 4 (1981):533–542.
14. T. Mauro, "Skycam Launches Disputes," *Wilmington (Del.) News-Journal Papers,* (19 February 1986):A12.

# CHAPTER 7

1. For a reasonably sober description of the early experiments, see R. J. Naumann and H. W. Herring, *Materials Processing in Space: Early Experiments* (Washington, DC: NASA, 1980), NASA SP-443. G. Harry Stine's book *The Third Industrial Revolution* (New York: Putnam, 1975) is a fascinating description of the possibilities in this area, but he fails to warn the reader that many of the possibilities he talks about will not work out for one reason or another.
2. Paul Ritt, in "Space Industrialization: the Growing Community," presented at the 1983 meeting of the American Association for the Advancement of Science. A tape of the meeting is available as tape 83AAAS-49 and -50 from Mobiltape Co., 1741 Gardena Ave., Glendale, CA 91204.
3. James J. Haggerty, *Spinoff 1984* (Washington, DC: NASA, 1984), 52–55; available from the U.S. Government Printing Office.
4. Alcestis Oberg, *Spacefarers of the '80s and '90s* (New York: Columbia University Press, 1985), 148.
5. James J. Haggerty, *Spinoff 1985* (Washington, DC: NASA, 1985), 53.
6. L. G. Napolitano, "Marangoni Convection in Space Microgravity Environments," *Science* 225, (13 July 1984):197–198; also see David Shapland and Michael Rycroft, *Spacelab: Research in Earth Orbit* (Cambridge: Cambridge University Press, 1984), 146–147.
7. J. E. Hart, J. Toomre, A. E. Deane, N. E. Hurlburt, G. A. Glatzmaier, G. H. Fichtl, W. W. Fowlis, and P. A. Gilman, "Laboratory Experiments on Planetary and Stellar Convection Performed on Spacelab 3," *Science* 234, (3 October 1986):61–64.
8. Stine, 114–116.
9. This diagram and discussion are based on D. Richman, talk presented to the 1983 AAAS meeting, in their session "Space Industrialization: The Growing Community," as well as on Z. Prusik, "Continuous Flow-through Electrophoresis," in Z. Deyl, ed., and F. M. Everaerts, Z. Prusik, and P. J. Svendsen, co-eds., *Electrophoresis: A Survey of Techniques and Applications,* Vol. 18 of *Journal of*

*Chromatography Library* (Amsterdam: Elsevier, 1979), 229–251. An excellent summary of the electrophoresis experiment is given by J. Rose and T. Fitzpatrick, "The Potential of Materials Processing Using the Space Environment," in *Space Stations and Space Platforms: Concepts, Design, Infrastructure, and Uses,* Vol. 99 of *Progress in Astronautics and Aeronautics* (New York: American Institute of Aeronautics and Astronautics, 1985), 167–200.

10. See D. Richman's talk, ibid., and A. R. Thomson, "Free-Flow Electrophoresis," in Colin F. Simpson and Mary Whittaker, eds., *Electrophoretic Techniques* (London: Academic Press), 253–273.

11. Oberg, 143.

12. David Osborne, "Business in Space," *The Atlantic Monthly* 255, no. 5 (May 1985):47.

13. "Drug Made on Space Shuttle Now Being Tested in Animals," *The New York Times,* (10 June 1985):B7.

14. Gail Bronson, "Mission Irrelevant," *Forbes,* (24 March 1986):176–177.

15. Rose and Fitzpatrick, 180. The Get Away Special program is described in more detail in Chapter 14.

16. James J. Haggerty, *Spinoff '84,* 37.

17. "Space-Processed Latex Spheres Sold," *Aviation Week and Space Technology,* 123, (22 July 1985):22.

18. Haggerty, *Spinoff 1984,* 37.

19. James R. Chiles, "Small Wonder: The Magnificent Ball Bearing," *American Heritage of Invention and Technology* 1, no. 1 (Summer 1985):59–63.

20. Osborne, 48–49.

21. Bronson, 176.

22. A. R. Thompson, "Free Flow Electrophoresis," in Colin F. Simpson and Mary Whittaker, eds., *Electrophoretic Techniques* (London: Academic Press, 1983), 253–274.

23. John Townsend (of Fairchild Space Co.) in a talk to the 1984 AAAS meeting, in the session "Moving Industry Into Space," tape no. 84AAAS 58–63.

24. G. Kolata, "The Great Crystal Caper," *Science* 229, (26 July 1985):370–371; see also additional criticism by R. Leberman and a rebuttal by Robert J. Naumann, Robert S. Snyder, Charles E. Bugg, Lawrence J. DeLucas, and F. L. Suddath, *Science* 230, (25 October 1985):373–376.

25. Kolata, 371.

# CHAPTER 8

1. Daniel Boorstin, *The Discoverers* (New York: Random House, 1983), 235–236.

2. Samuel E. Morison, *The European Discovery of America: The Southern Voyages, A.D. 1492–1616* (New York: Oxford University Press, 1974), 467.

3. Some good book-length coverage of the era of planetary exploration can be found in J. Kelly Beatty, B. O'Leary, and A. Chaikin, eds., *The New Solar System*, 2nd ed. (Cambridge, MA: Sky Publishing, 1982); Clark R. Chapman, *The Inner Planets* (New York: Scribners', 1982); Bruce Murray, Michael Malin, and Ronald Greeley, *Earthlike Planets: Surfaces of Mercury, Venus, Earth, Moon, Mars* (San Francisco: Freeman, 1981); David Morrison and Jane Samz, *Voyage to Jupiter*, NASA SP-439 (Washington, DC: Government Printing Office, 1980); David Morrison, *Voyages to Saturn*, NASA SP-451 (Washington, DC: Government Printing Office, 1982).

4. N. F. Ness, "The Magnetic Field of Mercury," in E. A. Muller, ed., *Highlights of Astronomy*, Vol. 4, Part 1 (Dordrecht, Holland: Reidel, 1977), 179–190; N. F. Ness, "The Magnetic Fields of Mercury, Mars, and Moon," *Annual Review of Earth and Planetary Sciences*, Vol. 7 (Palo Alto, CA: Annual Reviews, Inc., 1979), pp. 249–288.

5. David Morrison and Jane Samz, *Voyage To Jupiter*, NASA SP-439 (Washington, DC: U.S. Government Printing Office, 1980), 23–27.

6. The name Pioneer refers to a particular class of spacecraft. A complete list of planetary missions is provided as a table in Chapter 10.

7. B. A. Smith *et al.* (40 authors), "Voyager 2 in the Uranian System: Imaging Science Results," *Science* 233, (4 July 1986):43–64.

8. David Morrison, Torrence V. Johnson, Eugene M. Shoemaker, Laurence A. Soderblom, Peter Thomas, Joseph Veverka, and Bradford A. Smith, "Satellites of Saturn: Geological Perspectives," in Tom Gehrels and Mildred Shapley Mathews, eds., *Saturn* (Tucson: University of Arizona Press, 1984), 640–670.

9. Clark R. Chapman, *Planets of Rock and Ice* (New York: Scribners', 1982), 136.

10. See, e.g., Tycho T. von Rosenvinge, John C. Brandt, and Robert W. Farquhar, "The International Cometary Explorer Mission to Comet Giacobini-Zinner," *Science* 232, (18 April 1986):353–356.

11. R. J. Hynds, S. W. H. Cowley, T. R. Sanderson, K.-P. Wenzwel, and J. J. Van Rooijen, "Observations of Energetic Ions from Comet Giacobini-Zinner," *Science* 232, (18 April 1986):361–365.

12. See Note 7, above, and also F. M. Ipavich, A. B. Galvin, G. Gloeckler, D. Hovestadt, B. Klecker, and M. Scholer, "Comet Giacobini-Zinner: In Situ Observations of Energetic Heavy Ions," *Science* 232, (18 April 1986):366–369. Other articles in this issue of *Science* contain additional results from the Giacobini-Zinner mission; see also J. Kelly Beatty, "Comet G-Z: The Inside Story," *Sky and Telescope* 70 (November 1985):426–427.

13. See, for example, "Steve Edberg: IHW's 'Mr. Comet,'" in *Astronomy* 14, no. 7 (July 1986):24–31; D. DiCicco, "A User's Guide to Halley's Comet," *Sky and Telescope* (September 1983):211.

14. R. Z. Sagdeev, J. Blamont, A. A. Galeev, V. I. Moroz, V. D. Shapiro, V. I. Shevchenko, and K. Szego, "Vega Spacecraft Encounters with Comet Halley,"

*Nature* 321, (15 May 1986):259–261; K. Hirao and T. Itoh, "The Planet A Encounters," *Nature* 321, (15 May 1986):294–296; R. Reinhard, "The Giotto Encounter with Comet Halley," *Nature* 321, (15 May 1986):313–317. This issue of *Nature* contains a number of other articles on the scientific results from the comet encounters. For a shorter, less specialized account of these missions, see Richard A. Kerr, "Giotto Finds a Big Black Snowball at Halley," *Science* 231, (28 March 1986):1502–1503, or J. Kelly Beatty, "An Inside Look at Halley's Comet," *Sky and Telescope* 71 (May 1986):438–443.

15. R. Jastrow, "Exploring the Moon," in Paul A. Hanle and Von Del Chamberlain, eds., *Space Science Comes of Age* (Washington, DC: Smithsonian Institution Press, 1981), 47.

16. Solar System Exploration Committee, *Planetary Exploration Through the Year 2000* (Washington, DC: Government Printing Office, 1983).

17. Planetary scientists had hoped that the Comet Rendezvous/Asteroid Flyby mission would be included in the budget request for the 1987 fiscal year, submitted to Congress in January 1986. Although commitments to TOPEX and ISTP pushed it out of that budget request, before *Challenger* its status as a "new start" in the fiscal 1988 budget (submitted to Congress in January 1987) seemed assured. See M. Mitchell Waldrop, "Budget Decision Threatens Planetary Plan," *Science* 230, 1 (November 1985):526; Maria Neugebauer, "Space Missions to Comets," *Physics Today,* November 1985:38–44.

18. John Noble Wilford, "NASA Drops Plan to Launch Rocket from the Shuttle," *New York Times,* (20 June 1986):A1, A13.

19. See Eliot Marshall, "Shooting Plutonium into Space," *Science* 231, 21 March 1986:1358–1359.

20. European Space Agency, *European Space Science: Horizon 2000,* ESA SP-1070 (Paris: European Space Agency, 1985), 55–57; "Matra Studies Scenario for Comet Nucleus Sample Return Mission," *Aviation Week and Space Technology,* (20 October 1986):99.

21. M. Mitchell Waldrop, "A Soviet Plan for Exploring the Planets," *Science* 228, (10 May 1985):698–699.

22. "Vesta Mission Will Explore Mars, Asteroids," *Aviation Week and Space Technology,* (27 October 1986):54.

23. European Space Agency, 10.

24. *New York Times,* (24 June 1986):C3.

25. T. A. Heppenheimer, *Colonies in Space* (Harrisburg, PA: Stackpole Books, 1977); Gerard K. O'Neill, *2081* (New York: Simon & Schuster, 1981). See also Chapter 13.

26. John S. Lewis, quoted in W. K. Hartmann, "Solar System Resources," in Ben R. Finney and Eric M. Jones, eds., *Interstellar Migration and the Human Experience,* (Berkeley: University of California Press, 1985), 26–42.

27. Solar System Exploration Committee, 67.
28. Ibid., 65.

# CHAPTER 9

1. For a detailed list of missions, see Table 8.
2. See particularly the *Venera 15* and *16* map of Maxwell Montes in Yu. N. Alexandrov, A. A. Crymov, V. A. Kotelnikov, G. M. Petrov, O. N. Rzhga, A. I. Sidorenko, V. P. Sinilo, A. I. Aakharov, E. A. Akim, A. T. Basilveski, S. A. Kadnichanski, and Yu. S. Tjuflin, "Venus: Detailed Mapping of Maxwell Montes Region," *Science* 231, (14 March 1986):1271–1273.
3. See M. H. Carr, *The Surface of Mars* (New Haven, CT: Yale University Press, 1981); M. H. Carr and R. Greeley, *Volcanic Features of Hawaii: A Basis for Comparison with Mars,* NASA SP-403 (Washington, DC: Government Printing Office, 1980).
4. James C. G. Walker, "Atmospheric Evolution of the Terrestrial Planets," in C. Ponnampzruma, ed., *Comparative Planetology* (New York: Academic Press, 1978), 144.
5. See Richard A. Kerr, "Venus Is Looking More like Earth Than Mars," *Science* 232, (9 May 1986):709–710. This is a brief report on the seventeenth lunar and planetary science conference, 17–21 March 1986, in Houston; abstracts of the conference are available from the Lunar and Planetary Institute, 3303 NASA Road One, Houston, TX 77058-4399.
6. See Alexandrov *et al.*, notes 1 and 2; Solar System Exploration Committee, *Planetary Exploration through the Year 2000* (Washington, DC: Government Printing Office, 1983), 140.
7. If you're interested in specifics, the terrestrial air pressure at an altitude of 8 kilometers, about the height of Mount Everest, is 356 millibars, about a third of its value at sea level. The Martian air pressure varies from 7 to 10 millibars.
8. Data from James B. Pollack, "Atmospheres of the Terrestrial Planets," and "Titan," both in J. Kelly Beatty, Brian O'Leary, and Andrew Chaikin, eds., *The New Solar System* (Cambridge, MA: Sky Publishing, 1982), and from D. M. Hunten, M. G. Tomasko, F. M. Flasar, R. E. Samuelson, D. F. Strobel, and D. J. Stevenson, "Titan," in Tom Gehrels and Mildred Shapley Matthews, eds., *Saturn* (Tucson: University of Arizona Press, 1984), p. 675. The data for Titan are quite uncertain; the presence of argon in the atmosphere is only inferred, as the radio data indicate that something heavier than nitrogen is there and argon is the only sensible possibility.
9. C. Covault, "White House, Kremlin May Revive Cooperative Space Programs," *Aviation Week and Space Technology,* (6 October 1986):23–24.
10. You can see such pictures in J. B. Pollack, "Titan," in J. K. Beatty, B. O'Leary, and A. Chaikin, eds., *The New Solar System,* 2nd ed. (Cambridge, UK, and

Cambridge, MA: Cambridge University Press and Sky Publishing, 1982), 161–167.

11. D. M. Hunten, M. G. Tomasko, F. M. Flasar, R. E. Samuelson, D. F. Strobel, and D. J. Stevenson, "Titan," in Gehrels and Matthews, 671–759; see especially Section VI in that chapter (written by Stevenson).

12. Solar System Exploration Committee, 152–153.

13. Samuel Taylor Coleridge, "The Rime of the Ancient Mariner" (Peter Pauper Press, n.d.), 17.

14. For a general history of the *Viking* program, see Edward C. Ezell and Linda Naumann Ezell, *On Mars: Exploration of the Red Planet 1958–1978*, NASA SP-4212 (Washington, DC: U.S. Government Printing Office, 1980).

15. "The Evolving Role of Man in Space," tape transcript available as tape 83AAAS 17 and 18 from Mobiltape Co., 1741 Gardena Ave., Glendale, CA 91204.

16. Harold P. Klein, "The Search for Life on Mars," reprinted in Michael H. Carr, *The Surface of Mars* (New Haven, CT: Yale University Press, 1981), 195, reprinted from *Reviews of Geophysics and Space Physics* 17 (1979):1655–1662.

17. David Morrison, "Planetary Exploration in the Space Station Era," in I. Bekey and D. Herman, eds., *Space Stations and Space Platforms: Concepts, Design, Infrastructure, and Uses,* Vol. 99 in *Progress in Astronautics and Aeronautics* (New York: American Institute of Aeronautics and Astronautics, 1985), 131–149.

18. These phases of planetary exploration were first defined by the Committee on Planetary and Lunar Exploration (COMPLEX) of the National Academy of Sciences. See COMPLEX, *Strategy for Exploration of the Inner Planets: 1977–1987* (Washington, DC: National Academy of Sciences, 1978). Another excellent discussion of these phases is given in Noel Hinner's, "The Golden Age of Planetary Exploration," in J. Kelly Beatty, B. O'Leary, and A. Chaikin, eds., *The New Solar System,* 2nd ed. (Cambridge, MA: Sky Publishing, 1982), 3–10.

19. Solar System Exploration Committee, *Planetary Exploration through the Year 2000* (Washington, DC: Government Printing Office, 1983), 16.

20. Penelope J. Boston, ed., *The Case for Mars* (San Diego: Univelt, for the American Astronautical Society), xv–xvi.

21. W. Sullivan, "Russian Spacecraft Scheduled to Hover Above Martian Moon," *The New York Times,* (25 June 1985):C6; M. M. Waldrop, "A Soviet Plan for Exploring the Planets," *Science* 228, (10 May 1985):698–702. C. Covault, "Soviets in Houston Reveal New Lunar, Mars, Asteroid Flights," *Aviation Week and Space Technology,* (1 April 1985):18–19.

22. See the SSEC report, pp. 143–147, and N. Longdon, ed., *Space Science: Horizon 2000* (Noordwijk: European Space Agency, 1984), 53–54.

23. T. Mutch, "The Viking Lander Imaging Investigation: An Anecdotal Account," in *The Martian Landscape,* NASA SP-425 (Washington, DC: Government Printing Office, 1978), 31; National Commission on Space, *Pioneering the Space Frontier* (New York: Bantam Books, 1986), 67.

24. This mission is described in Morrison, 141–146.

25. Craig Covault, "White House, Kremlin May Revive Cooperative Space Programs," *Aviation Week and Space Technology* (6 October 1986):23–24.

# CHAPTER 10

1. The title of this chapter is a phrase used by Alan Meltzer of Rensselaer Polytechnic Institute.

2. The transcript is quoted in James Elliot and Richard Kerr, *Rings: Discoveries from Galileo to Voyager* (Cambridge: MIT Press, 1984), 10. The book indicates that a tape of the conversation is available from the MIT Press.

3. Ibid., 20.

4. Ibid., 92–105.

5. William I. McLaughlin, "Prediscovery Evidence of Planetary Rings," *Journal of the British Interplanetary Society* 33 (1980):287–294.

6. M. H. Acuna and N. F. Ness, "The Main Magnetic Field of Jupiter," *Journal of Geophysical Research* 81 (1976):2917.

7. David Morrison and Jane Samz, *Voyage to Jupiter,* NASA SP-439, (Washington, DC: Government Printing Office, 1980), 85.

8. For an explanation of how such arcs can exist, and for a set of references to discovery papers, see P. Goldreich, S. Tremaine, and N. Borderies, *Astronomical Journal* 92 (1976):490–494.

9. Quoted in R. Greenberg and A. Brahic, "Introduction," to R. Greenberg and A. Brahic, eds., *Planetary Rings* (Tucson: University of Arizona Press, 1984), 5.

10. S. J. Weidenschilling, C. R. Chapman, D. R. Davis, and R. Greenberg, "Ring Particles: Collisional Interactions and Physical Nature" in Greenberg and Brahic, 367 ff.; J. A. Burns, M. K. Showalter, and G. E. Marfill, "The Ethereal Rings of Jupiter and Saturn," in Greenberg and Brahic, 200–272; B. A. Smith *et al.,* "Voyager 2 in the Uranian System: Imaging Science Results," *Science* 233, (4 July 1986):43–64.

11. Joseph A. Burns, "Planetary Rings," in J. Kelly Beatty, B. O'Leary, and A. Chaikin, eds., *The New Solar System,* 2nd ed. (Cambridge, MA: Sky Publishing, 1982), 129–142.

12. Andrew P. Ingersoll, "Jupiter and Saturn," in Beatty *et al.,* 117–128; see also C. M. Yeates, T. V. Johnson, L. Colin, F. P. Fanale, L. Frank, and D. M. Hunten, *Galileo: Exploration of Jupiter's System,* NASA SP-479 (Washington, DC: Government Printing Office, 1985), 17–31.

13. D. Morrison and J. Samz, 117–138.

14. A. P. Ingersoll, R. F. Beebe, B. J. Conrath, and G. E. Hunt, "Structure and Dynamics of Saturn's Atmosphere," in Gehrels and Matthews, 217.

15. See Garry Hunt and Patrick Moore, *Saturn* (New York: Rand McNally, 1982), 22.

16. R. Hanel, B. Conrath, F. M. Flasar, V. Kunde, W. Maguire, J. Pearl, J. Pirraglia, R. Samuelson, D. Cruickshank, D. Gautier, P. Gierasch, L. Horn, and P. Schulte, "Infrared Observations of the Uranian System," *Science* 233, (4 July 1986):70–74.

17. N. F. Ness, M. H. Acuna, K. W. Behannon, L. F. Burlaga, J. E. P. Connerney, R. P. Lepping, and F. M. Neubauer, "Magnetic Fields at Uranus," *Science* 233, (4 July 1986):85–88.

18. This table and much of the comparative discussion of planetary magnetism are drawn from L. J. Lanzerotti and S. M. Krimigis, "Comparative Magnetospheres," *Physics Today* (November 1985):24–32, updated from Ness *et al.*, 86. For discussions of the origin of planetary magnetic fields, see N. F. Ness, "The Magnetic Fields of Mercury, Mars, and the Moon," *Annual Review of Earth and Planetary Sciences* 7 (1979):249–288.

19. David Morrison, "Planetary Exploration in the Space Station Era," in Bekey and Herman, eds., 131–149.

# CHAPTER 11

1. For surveys of space astronomy, or surveys of astronomy that include results from space missions, see Bevan M. French and Stephen P. Maran (eds.), *A Meeting with the Universe* (Washington, DC: Government Printing Office, n.d., but written around 1980). For accounts of particular subfields of astronomy that include results from space see H. Shipman, *Black Holes, Quasars, and the Universe*, 2nd ed. (Boston: Houghton Mifflin, 1980).

2. The numbers refer to the position of this object in the sky, and "H" refers to the *HEAO-1* sky survey. The results of this research project are described in detail in J. A. Nousek, H. L. Shipman, J. B. Holberg, J. Liebert, S. H. Pravdo, N. E. White, and P. Giommi, "H1504+65: An Extraordinarily Hot Compact Star Devoid of Hydrogen and Helium," *Astrophysical Journal* 309 (1986), 230–240.

3. For details on Cygnus X-1 (and other information about black holes), see Harry L. Shipman, *Black Holes, Quasars, and the Universe*; George Greenstein, *Frozen Star* (New York: Freundlich Books, 1983).

4. See, e.g., J. B. Hutchings, D. Crampton, and A. P. Cowley, "LMC X-3: A Black Hole in a Neighbor Galaxy," *Mercury* (July–August 1984):106–107.

5. For some brief discussions of the hazards of radiation, see Arnauld E. Nicogossian and James F. Parker, *Space Physiology and Medicine*, NASA SP-447 (Washington, DC: Government Printing Office, 1982), 43–47, 293–302; or Shields Warren and Douglas Grahn, "Ionizing Radiation," in James F. Parker, Jr., and Vita R. West, eds., *Bioastronautics Data Book*, NASA SP-3006 (Washington, DC: Government Printing Office, 1973), 417–435.

6. For a discussion of radiation hazards, see, for example, Bernard L. Cohen,

*Nuclear Science and Society* (Garden City, NY: Doubleday Anchor, 1974), Chs. 4 and 5. Incidentally, the life expectancy of the *Skylab 4* astronauts is diminished by about ten days as a result of their exposure to radiation and the resulting increased cancer risk.

7. An extensive discussion of these effects can be found in L. J. Lanzerotti, ed., "Impacts of Ionospheric/Magnetospheric Physics on Terrestrial Science and Technology," in S. A. Colgate *et al.*, eds., *Space Plasma Physics: The Study of Solar System Plasmas,* vol. 2 (Washington, DC: National Academy of Sciences, 1978), 1177–1268; see also pp. 85–87 in vol. 1 of this report.

8. See "Solar Physics," Ch. 1 of *Challenges to Astronomy and Astrophysics: Working Documents of the Astronomy Survey Committee* (Washington, DC: National Academy of Science, 1983), 18, 25, 42, and especially 48–50.

9. A contrasting view to that in Note 8, also by a "blue-ribbon" panel of the National Academy of Sciences, is expressed in Committee on Solar and Space Physics, Space Science Board, *An Implementation Plan for Priorities in Solar-System Space Physics* (Washington, DC: National Academy Press, 1985), 10.

10. For a comprehensive account of the discoveries of the *Copernicus* satellite telescope, see Lyman Spitzer, *Searching Between the Stars* (New Haven, CT: Yale University Press, 1982).

11. W. Cash, P. Charles, S. Bowyer, F. Walter, G. Garmire, and G. Riegler, "The X-Ray Superbubble in Cygnus," *Astrophysical Journal Letters* 238 (1980), L71–L74. See also Wallace H. Tucker, *The Star Splitters: The High Energy Astronomy Observatories,* NASA SP-466 (Washington, DC: Government Printing Office, 1984), 99–102.

12. B. Katz *et al.,* "Optical Flashes in Perseus," *Astrophysical Journal Letters* 307 (1986), L33–L37.

13. There are many good popular accounts of modern cosmology, modern ideas regarding the origin of the universe. See, for example (in alphabetical order), Eric Chaisson, *Cosmic Dawn: The Origins of Matter and Life* (Boston: Little, Brown, 1981); E. Harrison, *Cosmology: The Science of the Universe* (Cambridge, UK: Cambridge University Press, 1981); H. Pagels, *Perfect Symmetry: The Search for the Beginning of Time* (New York: Simon & Schuster, 1985); H. Reeves, *Atoms of Silence: An Exploration of Cosmic Evolution,* trans. Ruth Lewis and John S. Lewis (Cambridge: MIT Press, 1983); Shipman, *Black Holes, Quasars, and the Universe;* J. S. Trefil, *The Moment of Creation* (New York: Scribners', 1983); S. Weinberg, *The First Three Minutes* (New York: Basic Books, 1976).

14. Kim McDonald, "Delay of Space-Shuttle Flights till 1988 Deals 'Devastating' Blow to Astronomers," *The Chronicle of Higher Education,* (23 July 1986):7.

15. C. Covault, "U.S. Scientists Cite Reagan for Losing Space Preeminence," *Aviation Week and Space Technology,* (3 November 1986):40–42.

16. *European Space Science—Horizon 2000,* ESA SP-1070 (Paris: European Space Agency, 1984), 3.

## CHAPTER 12

1. See, for example, Carl Sagan, *The Cosmic Connection* (New York: Dell, 1973); Carl Sagan, *Cosmos* (New York: Random House, 1980).

2. R. T. Rood and J. S. Trefil, *Are We Alone?* (New York: Scribners', 1982); M. H. Hart and B. Zuckerman, eds., *Extraterrestrials: Where Are They?* (New York: Pergamon, 1982).

3. For a very brief account of these early writings, see Lewis White Beck, "Extraterrestrial Intelligent Life," in Edward Regis, Jr., ed., *Extraterrestrials: Science and Alien Intelligence* (Cambridge, UK: Cambridge University Press, 1985), 3–18; a longer history is given in Steven Dick, *Plurality of Worlds: The Origins of the Extraterrestrial Life Debate from Democritus to Kant* (Cambridge, UK: Cambridge University Press, 1982).

4. These considerations are based on star densities quoted in C. Allen, *Astrophysical Quantities*, 3rd ed. (London: Athlone Press, 1973); W. Heintz, *Double Stars* (Dordrecht: Reidel, 1978).

5. Astronomer Bob Harrington agrees with me that multiple stars should be considered possible good suns. See R. S. Harrington, "Planetary Orbits in Multiple Star Systems," in J. Billingham, eds., *Life in the Universe*, NASA Conference Publication 2156 (Washington, DC: Government Printing Office, 1981), 119–124.

6. See, for instance, I. S. Shklovskii and C. Sagan, *Intelligent Life in the Universe* (San Francisco: Holden Day, 1966); H. Shipman, *The Restless Universe: An Introduction to Astronomy* (Boston: Houghton Mifflin, 1977), Ch. 16.

7. See P. Cloud, *Cosmos, Earth, and Man* (New Haven, CT: Yale University Press, 1978), 117–140. Where there's enough oxygen around, iron tends to form $Fe_2O_3$, but in an oxygen-poor environment, it makes FeO. Fossil rocks contain layers that show that there was FeO around.

8. See, e.g., Duwayne M. Anderson, "Role of Interfacial Water and Water in Thin Films in the Origin of Life," in J. Billingham, ed., *Life in the Universe*, NASA Conference Publication 2156 (Washington, DC: Government Printing Office, 1981).

9. For an excellent, but somewhat technical, discussion of the importance of self-reproducing chemical communities, see E. Argyle, "Chance and the Origin of Life," in *Extraterrestrials: Where Are They?*" (New York: Pergamon Press, 1982), 100–112; reprinted from *Origins of Life* 8, no. 4 (1977):287–298.

10. George Gaylord Simpson, "The Nonprevalence of Humanoids," *Science* 143, (21 February 1964):769–775; E. Mayr, "The Probability of Extraterrestrial Intelligent Life," in Regis, 23–30. For a counterargument by an evolutionary biologist, see David M. Raup, "ETI without Intelligence," in Regis, 31–42.

11. See, for instance, Billingham, ed., *Life in the Universe;* Don Goldsmith and Tobias Owen, *The Search for Life in the Universe* (Menlo Park, CA: Ben-

jamin/Cummings, 1980); Regis, *Extraterrestrials;* The classic in this field is Carl Sagan and Iosif S. Shklovskii, *Intelligent Life in the Universe.*

12. Many of the earlier references are listed in Frank J. Tipler, "Extraterrestrial Intelligent Beings Do Not Exist," *Quarterly Journal of the Royal Astronomical Society* 21 (1980):267–281.

13. For more elaborate counterarguments, see F. Drake, "N Is Neither Very Small nor Very Large," in M. D. Papagiannis, ed., *Strategies for the Search for Life in the Universe* (Dordrecht, Holland: Reidel, 1980):27–34; Carl Sagan and William I. Newman, "The Solipsist Approach to Extraterrestrial Intelligence," *Quarterly Journal of the Royal Astronomical Society* 24, no. 2 (1983):113–121.

14. Carlton was released by the Phillies early in the year, but under the terms of his contract, another team that signed him had to pay only a prorated portion of the major league minimum salary, and the Phillies had to pick up the rest of his contract. Carlton had given the Phillies many good years through the 1970s and early 1980s, so the million that they paid him in 1986 could be considered deferred compensation for his previous service with the club, not just money down the drain.

15. G. Cocconi and P. Morrison, "Searching for Interstellar Communication," *Nature* 184 (1959):184–186.

16. There are several books that demonstrate that current UFO reports have explanations that do not involve extraterrestrial spacecraft. See Edward Condon *et al.*, *Scientific Study of Unidentified Flying Objects* (New York: Bantam, 1969); Philip Klass, *UFO's: The Public Deceived* (Buffalo, NY: Prometheus Books, 1983); Philip J. Klass, *UFO's Explained* (New York: Vintage, 1974); D. Menzel and E. Taves, *The U.F.O. Enigma* (New York: Doubleday, 1977); Robert Sheaffer, *The UFO Verdict* (Buffalo, NY: Prometheus Books, 1980); and William H. Stiebing, Jr., *Ancient Astronauts, Cosmic Collisions, and Other Popular Theories about Man's Past* (Buffalo, NY: Prometheus Books, 1984). If you must read something that argues that UFOs are extraterrestrial spacecraft, J. Allen Hynek, *The UFO Experience: A Scientific Inquiry* (Chicago: Regnery, 1972) is the best of the lot, better than looking at the supermarket checkout counter for the *National Enquirer* or buying one of the many UFO books that share the *Enquirer*'s disdain for the facts.

17. J. Tarter, "Searching for Extraterrestrials," in Regis, 176. Much of the description of past and present searches is based on this article.

18. B. M. Oliver and J. Billingham, *Project Cyclops Report,* NASA CR 114445, 1973. Available from J. Billingham, Code LT, NASA-Ames Research Center, Moffett Field, CA: 94035.

19. William Sweet, "Spielberg Funds Search for ETs by Harvard Smithsonian," *Physics Today* (January 1986):80–81; Werner Von Braun and Frederick I. Ordway III, *History of Rocketry and Space Travel* (New York: T. Y. Crowell, 1969), 64.

20. J. Tarter and B. Zuckerman, "Microwave Searches in the U.S.A. and Canada,"

in M. Papagiannis, ed., *Strategies for the Search for Life in the Universe* (Boston: Reidel, 1980), 81–92.

# CHAPTER 13

1. Stephen B. Hall, ed., *The Human Role in Space* (Park Ridge, NJ: Noyes Publications, 1985); published by arrangement with NASA-Marshall Space Flight Center.
2. M. Smith, "The Evolving Role of Man in Space," in Theodore R. Simpson, eds., *The Space Station: An Idea Whose Time Has Come* (New York: IEEE Press, 1985), 3–31; talk to the 1983 meeting of the AAAS, available on tapes AAAS83-17 and -18. Many of the roles discussed in this chapter are roles Smith identified, though I have added a few in considering the future.
3. James E. Oberg and Alcestis R. Oberg, *Pioneering Space: Living on the Next Frontier* (New York: McGraw-Hill, 1986), 41–62.
4. Joseph P. Allen with Russell Martin, *Entering Space: An Astronaut's Odyssey*, 2nd ed. (New York: Stewart, Tabori, & Chang, 1985).
5. James E. Oberg, *Red Star in Orbit* (New York: Random House, 1981), 183–201.
6. S. Matsunaga, *The Mars Project: Journeys beyond the Cold War* (New York: Hill & Wang, 1986), 164.
7. Kerry Mark Joels and Gregory P. Kennedy, *The Space Shuttle Operator's Manual* (New York: Ballantine Books, 1982), 6.2–6.6); Oberg, 99–102.
8. J. Flinn, *Wilmington Morning News*, 29 March 1986; S. Nichols, *Wilmington Sunday News-Journal*, (31 August 1986):G2.
9. G. K. O'Neill, "The Colonization of Space," *Physics Today* (September 1974): 32–40.
10. See Gerard K. O'Neill, *The High Frontier: Human Colonies in Space* (New York: Morrow, 1976); O'Neill, *2081: A Hopeful View of the Future* (New York: Simon & Schuster, 1981); T. A. Heppenheimer, *Colonies in Space* (Harrisburg, PA: Stakpole Books, 1977); Richard D. Johnson and Charles Holbrow, eds., *Space Settlements: A Design Study*, NASA SP-413 (Washington, DC: Government Printing Office, 1977); Gerard O'Neill, John Billingham, Brian O'Leary, and Beulah Gossett, eds., *Space Resources and Space Settlements*, NASA SP-428 (Washington, DC: Government Printing Office, 1979).
11. A discussion of the merits of various proposed colony locations is contained in Heppenheimer, 109–113.
12. For costs of the colony, see NASA SP-413, pp. 139–153.
13. For a guarded, but generally skeptical, view of the solar-power satellite system, see U.S. Congress, Office of Technology Assessment, *Solar Power Satellites* (Washington, DC: Government Printing Office, 1981). The comparison of the efficiencies of terrestrial and space solar generators is on p. 128; the environmental hazards are reviewed in Chapter 8.

14. James A. Van Allen, "Myths and Realities of Space Flight," *Science* 232, (30 May 1986):1075–1076; for letters, see *Science* 233, (8 August 1986):610–611. An earlier reference to Van Allen's views in contrast with my own is given in Chapter 2 of this book.

15. S. E. Morison, *Admiral of the Ocean Sea* (Boston: Little, Brown, 1942).

16. F. Dyson, *Disturbing the Universe* (New York: Harper & Row, 1979), 118–126.

17. Recall Chapter 2.

18. Tom Wolfe, *The Right Stuff* (New York: Bantam, 1983), Ch. 4.

19. A summary of physiological responses to spaceflight is given in Arnauld E. Nicogossian and James F. Parker, *Space Physiology and Medicine*, NASA SP-447 (Washington, DC: Government Printing Office, 1982), 127–139.

20. Nicogossian and Parker, 206.

21. Michael Modell and Jack M. Spurlock, "Rationale for Evaluating a Closed Food Chain for Space Habitats," in T. Stephen Cheston and David L. Winter, eds., *Human Factors of Outer Space Production* (Boulder, CO: Westview Press, 1980), AAAS Selected Symposium Series, Vol. 50, pp. 134–142.

22. Harlan F. Brose, "Environmental Control and Life Support (ECLS) Design Optimization Approach," in Mireille Gerard and Pamela W. Edwards, eds., *Space Station: Policy, Planning, and Utilization* (New York: American Institute of Aeronautics and Astronautics, 1983), 191.

23. Quoted in Oberg and Oberg, 129.

24. Oberg and Oberg, Ch. 12.

25. See Yvonne Clearwater, "A Human Place in Outer Space," *Psychology Today* 19, no. 7 (July 1985):34–43; Mary M. Connors, Albert A. Harrison, and Faren R. Akins, *Living Aloft: Human Requirements for Extended Spaceflight*, NASA SP-483 (Washington, DC: Government Printing Office, 1985).

26. D. Kent, personal interview, 14 May 1986.

27. Connors *et al.*, 152.

28. Clearwater, 43.

# CHAPTER 14

1. Many of the technical data on the capabilities of *Salyut* and *Skylab* comes from Philip P. Chandler, Leonard David, and Courtland S. Lewis, "MOL, Skylab, and Salyut," in Theodore R. Simpson, ed., *The Space Station: An Idea Whose Time Has Come* (New York: IEEE Press, 1984), 31–49.

2. William J. Broad, *The New York Times*, (17 July 1986):A19; J. Kelly Beatty, "The First 100 Days of Mir," *Sky and Telescope* (August 1986):134; A. Chaikin, "Life in Orbit," *Omni* 8, no. 12 (September 1986):66–71. Andrew Chaikin is an excellent, trustworthy science writer, and his article should be reliable despite its

appearance in *Omni,* a magazine that has occasionally published some far-out speculations and has not made their nature clear.

3.  C. Covault and T. M. Foley, "NASA Station Design Focuses on Assembly, Early Activation," *Aviation Week and Space Technology,* (22 September 1986):16–21.

4.  See, e.g., R. Kline, R. McCaffrey, and D. B. Stein, "A Summary of Potential Designs of Space Stations and Platforms," in I. Bekey and D. Herman, eds., *Space Stations and Space Platforms—Concepts, Design, Infrastructure, and Uses,* Vol. 99 in *Progress in Astronautics and Aeronautics* (New York: American Institute of Aeronautics and Astronautics, 1985), 267–351; NASA Space Station Task Force, *Space Station Program: Description, Applications, and Opportunities* (Park Ridge, NJ: Noyes Publications, 1985), 259–270; John D. Hodge and Claude C. Priest, "The U.S. Space Station," in Simpson, 119–158.

5.  M. Mitchell Waldrop, "NASA Unveils Space Station Concept," *Science* 232, (30 May 1986):1089.

6.  U.S. Congress, Office of Technology Assessment, *Civilian Space Stations and the U.S. Future in Space,* OTA-STI-241 (Washington, DC: Office of Technology Assessment, November 1984), lists a number of possible uses of the external tank on pp. 77–82.

7.  See Chapter 3 for a discussion of various transfer vehicles.

8.  Much of the information on SPAS and other space platforms comes from U.S. Congress, Office of Technology Assessment, 66ff.

9.  Roger Soderman, "Satellites on a Shoe String," presented in "Moving Industry into Space," Session 2, at the 1984 meeting of the American Association for the Advancement of Science. Quotations from tape 84AAAS-60, available from Mobiltape, 1741 Gardena Ave., Glendale, CA 91204.

10. NASA-Goddard Space Flight Center, *Attached Shuttle Payload Carriers,* brochure available from Attached Shuttle Payload Program, Code 420, NASA-Goddard Space Flight Center, Greenbelt, MD 20771.

11. Theresa Foley, *Aviation Week and Space Technology,* (1 September 1986):40–41.

12. M. Mitchell Waldrop, "Asking for the Moon," *Science* 226 (23 November 1984):948–949; Paul W. Keaton, "An International Research Laboratory on the Moon," Report LA-9143-MS; Keaton, "A Moon Base/Mars Base Transportation Depot," LA-10552-MS; Keaton, "Low-Thrust Rocket Trajectories," LA-10625-MS; National Commission on Space, *Pioneering the Space Frontier* (New York, Bantam, 1986) 62–65. The reports, made for the Los Alamos Scientific Laboratory, are available from the National Technical Information Service, Springfield, VA.

13. Ray A. Billington, *Westward Expansion: A History of the American Frontier,* 2nd ed. (New York: Macmillan, 1960), 334–336.

# CHAPTER 15

1. Arthur C. Clarke, *Profiles of the Future* (New York: Bantam, 1963), Ch. 1. Newcomb is quoted on pp. 2–3, and Woolley is quoted on p. 8.

2. See, for instance, T. R. Simpson, *The Space Station: An Idea Whose Time Has Come* (New York: IEEE Press, 1985), xi.

3. Craig Covault, "New Manifest for Space Shuttle Generates Payload Sponsor Debate," *Aviation Week and Space Technology*, (13 October 1986):22–23; M. Mitchell Waldrop, "NASA Announces a New Schedule for the Shuttle," *Science* 234, (17 October 1986):279–280.

4. National Commission on Space, *Pioneering the Space Frontier* (New York: Bantam, 1986), 2.

5. See Samuel E. Morison, *Admiral of the Ocean Sea* (Boston: Little, Brown, 1942), 44–64.

6. Robert Anderson and Marvin Goldberger (Study Co-Chairmen), *Space: America's New Competitive Frontier* (Washington, DC: Business Higher Education Forum, 1986), available from the Forum at Suite 800, One Dupont Circle, Washington, DC 20036.

7. J. Logsdon, presentation to the National Issues Forum, The Brookings Institution, 1 October 1986. See also Vernon Van Dyke, *Pride and Power: The Rationale of the Space Program* (Urbana: University of Illinois Press, 1964).

8. Jon D. Miller, "The Information Needs of the Public Concerning Space Exploration," a special report for NASA, July 20, 1982; see also Miller's talk to the 1983 AAAS Meeting, as part of the symposium "The Evolving Role of Man in Space," recorded on tape as tape 83AAAS-17 and -18, available from Mobiltape, 1741 Gardena Ave., Glendale, CA 91204; Jon D. Miller, R. W. Suchner, and A. M. Voelker, *Citizenship in an Age of Science* (New York: Pergamon, 1980).

9. Craig Covault, "U.S., Soviet Negotiators Agree to New Space Cooperation Pact," *Aviation Week* (10 November 1986):27–28.

10. Theresa Foley, "Scientists Warn NASA of Threats to Space Station Usefulness," *Aviation Week* (24 November 1986):18–19.

11. See Paul B. Stares, *The Militarization of Space: U.S. Policy 1945–1984* (Ithaca, NY: Cornell University Press, 1985); Thomas Karas, *The New High Ground* (New York: Simon & Schuster, 1983).

12. Stares, 236–254.

13. B. I. Edelson and W. L. Morgan, "Orbital Antenna Farms," *Astronautics and Aeronautics* 15 (September 1977):20–23.

14. D. Neustadt, "Action! Camera! It's Time for a Video Meeting," *The New York Times*, (10 November 1985):6F–7F.

15. Marc Beauchamp, "Planet Computer," *Forbes*, (24 February 1986):61–63.

16. Constance Holden, "Soviets Launch Computer Literacy Drive," *Science* 231, (10 January 1986):109–110.

17. Charles Gould, "Large Scale Human Benefits of Space Industrialization," in B. J. Bluth and S. R. McNeal, *Update on Space,* Vol. 1 (Granada Hills, CA: National Behavior Systems, 1981), 43.

18. Business Higher Education Forum, *Space: America's New Competitive Frontier* (Washington, DC: Business Higher Education Forum, 1986); for a summary see J. Eberhart, "Report on a Report: Wisdom between the Lines" *Science News* 129, (19 April 1986):250.

19. Gould, 44; G. Harry Stine, *The Third Industrial Revolution* (New York: Putnam's, 1975), 100–128.

20. Dennis Costello and Paul Rappaport, "The Technological and Economic Development of Photovoltaics," *Annual Review of Energy* 5 (1980):339 (with the cost converted to 1985 dollars).

21. James T. Rose and Terrence D. Fitzpatrick, "The Potential of Materials Processing Using the Space Environment," in I. Bekey and D. Herman, eds., *Space Stations and Space Platforms: Concepts, Design, Infrastructure, and Uses,* Vol. 99 in *Progress in Astronautics and Aeronautics* (New York: American Institute of Aeronautics and Astronautics, 1985), 169.

22. M. Mitchell Waldrop, "Ashes to Ashes to Orbit," *Science* 227, (8 February 1985):615.

23. National Commission on Space, 29.

24. A number of mission profiles are discussed in Michael Lemonick, "Mission to Mars," *Science Digest* (March 1986):26–35, 84. See also R. L. Staehle, "An Expedition to Mars Employing Shuttle-Era Systems, Solar Sails, and Aerocapture," in Penelope J. Boston, ed., *The Case for Mars* (San Diego: Univelt, 1981), 91–108.

25. Humboldt C. Mandell, Jr., "The Cost of Landing Man on Mars," in Boston, 281–292, 1981.

26. These reasons are extracted from "Special Report: Mars," a series of articles in *Discover* (September 1984):12–27, and Boston, x–xii.

27. The numbers work out, approximately, as follows: The surface area of Mars, which has a radius of 3,396 kilometers, is approximately $10^8$ square kilometers. The surface pressure on the earth of 15 pounds per square inch corresponds to an atmospheric mass of $10^{10}$ kilograms above each square kilometer of Earth's surface. So if we were to supply Mars with a terrestrial atmosphere, we would need $10^{10}$ kilograms per square kilometer $\times$ $10^8$ square kilometers or $10^{18}$ kilograms of material. But the gases that produce the terrestrial greenhouse effect are primarily carbon dioxide and water (M. Neiburger, J. G. Edinger, and W. D. Bonner, *Understanding Our Atmospheric Environment,* San Francisco: W. H. Freeman, 1971, p. 61). It turns out that Earth and Mars have a very similar global inventory of carbon dioxide in the atmosphere; the Martian atmosphere is thinner but is pure $CO_2$. So what we need to do is to supply the water, which is 1.6% of Earth's atmosphere, on the average, or 1.6 by $10^{16}$ kilograms of water. A circular slab of ice about 230 kilometers in radius and 100 meters thick has a volume of

1.6 by $10^{19}$ cubic centimeters and, as ice has a density of 1 gram per cubic centimeter, contains the necessary amount of water. If we needed to get the dry ice to provide the oxygen, we would need thirteen times more oxygen than water (because oxygen is 21% of Earth's atmosphere), and as dry ice is only two-thirds oxygen, we would need twenty times as much dry ice as water ice. A 1-kilometer-thick circular slab with a radius of 320 kilometers would contain twenty times as much material as the water slab. For a further discussion of terraforming, see Carl Sagan, *Cosmos* (New York: Ballantine, 1980), 108–110.

28. M. H. Carr, *The Surface of Mars* (New Haven, CT: Yale University Press, 1981), Chs. 12 and 13.

29. Arthur L. Robinson, "Colonizing Mars: The Age of Planetary Engineering Begins," *Science* 205, (18 February 1977):208, summarizing M. M. Averner and R. D. MacElroy, eds., *On the Habitability of Mars: An Approach to Planetary Ecosynthesis,* NASA SP-414 (Springfield, VA: National Technical Information Service), 1976.

30. Skryabin's statement is quoted in A. Chaikin, "Mars or Bust," *Discover* (September 1984):17; Schmitt's is quoted by Michael Lemonick, "Mission to Mars," *Science Digest* (March 1986):31.

31. National Commission on Space, *Pioneering the Space Frontier* (New York: Bantam, 1986), 161.

32. Bruce Murray, Michael C. Malin, and Ronald Greeley, *Earthlike Planets: Surfaces of Mercury, Venus, Earth, Moon, Mars* (San Francisco: Freeman, 1981), 350.

33. This number refers to actual expenses in fiscal 1985 and is taken from *Physics Today* 39, no. 5 (May 1986):58.

# List of Acronyms (LOA)

AAAS—American Association for the Advancement of Science
ABM—Antiballistic missile
*ANS*—Astronomical Netherlands Satellite
ASAT—antisatellite weapons system
*AXAF*—Advanced X-ray Astrophysics Facility
BMD—ballistic missile defense
CNES—Centre National d'Études Spatiales (French space agency)
*COBE*—Cosmic Background Explorer
*COS*—Celestial Observation Satellite
*CRAF*—Comet Rendezvous/Asteroid Flyby
DOD—Department of Defense
ELT—emergency location transmitter
EOS—Electrophoresis Operations in Space
EOS—Earth Observation System
EOSAT—Earth Observation Satellite Company
EPIRB—emergency position-indicating radio beacon
ERTS—Earth Resources Technology Satellite
ESA—European Space Agency
ESSA—Environmental Science Service Administration
ET—external tank (for the space shuttle)
ETI—extraterrestrial intelligence
EURECA—European Retrievable Carrier
*EUVE*—Extreme Ultraviolet Explorer
EVA—extravehicular activity
*EXOSAT—European X-Ray Observation Satellite*
FDA—Food and Drug Administration
FROD—functionally related observable differences
GAS—Get Away Special

GEO—geostationary (or geosynchronous) earth orbit
GEODSS—Ground Based Electro-Optical Deep Space Surveillance system
GNP—gross national product
*GOES*—Geostationary Operational Environmental Satellite (for meteorology)
GPS—global positioning system
GRM—Geopotential Research Mission
*GRO*—Gamma Ray Observatory
GTE—General Telephone and Electronics
*HEAO*—High Energy Astronomical Observatory
HOTOL—Horizontal Takeoff and Landing (refers to the British program that hopes to
    develop a spaceplane)
*HST*—Hubble Space Telescope
ICBM—intercontinental ballistic missile
*ICE—International Cometary Explorer*
IEEE—Institute for Electrical and Electronics Engineers
IGY—International Geophysical Year
IMP—Interplanetary Monitoring Platform
INTELSAT—International Telecommunications Satellite Corporation
*IRAS*—Infrared Astronomy Satellite
ISAS—Institute for Space and Astronautical Sciences (science part of the Japanese
    space agency)
*ISEE—International Sun-Earth Explorer*
*ISO*—Infrared Scientific Observatory
ISTP—International Solar-Terrestrial Physics program
*ITOS*—Improved TIROS Operational Satellite (meteorology)
*IUE*—International Ultraviolet Explorer
JEA—Joint Endeavor Agreement
KH—Keyhole (reconnaissance satellite)
LDEF—Long Duration Exposure Facility
LDR—Large Deployable Reflector
LEO—low earth orbit
LMC—large magellanic cloud
LOA—list of acronyms
MESA—Modular Experimental Platform for Science and Applications
MLR—Monodisperse Latex Reactor
MMU—Manned Maneuvering Unit
MPS—materials processing in space
NASA—National Aeronautics and Space Administration
NAVSTAR—DOD satellite navigation system (see GPS)
NOAA—National Oceanic and Atmospheric Administration
NROSS—Navy Remote Ocean Sensing System
*OAO*—Orbiting Astronomical Observatory

*OGO*—Orbiting Geophysical Observatory
OMV—orbital maneuvering vehicle (low orbit to low orbit)
*OSCAR—Orbiting Satellite Carrying Amateur Radio*
OSO—Orbiting Solar Observatory
OTA—Office of Technology Assessment of the U.S. Congress
*Ph-D—Phobos/Deimos* mission
PPS—precise positioning system (more accurate version of GPS)
RAE—Radio Astronomy Explorer
*ROSAT*—Roentgen satellit (German X-ray satellite observatory; NASA is a junior partner)
ROTV—reusable orbital transfer vehicle (low orbit to geosynchronous orbit)
RTG—radioisotope thermoelectric generator
SAGE—Stratospheric Aerosol and Gas Experiment
SAR—synthetic aperture radar
SARSAT—Search and Rescue Satellite
SAS—Small Astronomy Satellite
SAS—space adaptation syndrome
SDI—Strategic Defense Initiative
SDLV—Shuttle Derived Launch Vehicle
SETI—search for extraterrestrial intelligence
SIRTF—Satellite Infrared Telescope Facility
SPAS—Space Pallet Satellite
*SPOT*—Système Probatoire pour l'Observation de la Terre (French satellite that takes high-resolution pictures of the earth)
SPS—standard positioning system
SSEC—Solar System Exploration Committee
TDRS—Tracking and Data Relay Satellite (NASA communications satellite)
THURIS—The Human Role in Space (NASA study)
TIROS—Television and Infrared Observation Satellite (meteorological imaging)
*TOPEX*—Ocean Topography Experiment
*UARS*—Upper Atmospheric Research Satellite
UFO—unidentified flying object
UHF—ultrahigh frequency (TV channels 14–82)
VHF—very high frequency (TV channels 2–13)
WARC—World Administrative Radio Conference
*XTE*—X-Ray Timing Explorer

# Index